Marine Metagenomics

Takashi Gojobori • Tokio Wada
Takanori Kobayashi • Katsuhiko Mineta
Editors

Marine Metagenomics

Technological Aspects and Applications

Springer

Editors
Takashi Gojobori
Computational Bioscience Research Center
King Abdullah University of Science
and Technology
Thuwal, Saudi Arabia

Takanori Kobayashi
Kitasato University School of Marine
Biosciences
Sagamihara, Japan

Tokio Wada
Japan Fisheries Research and Education
Agency (Retired)
Yokohama, Japan

Katsuhiko Mineta
Computational Bioscience Research Center
King Abdullah University of Science
and Technology
Thuwal, Saudi Arabia

ISBN 978-981-13-8136-2 ISBN 978-981-13-8134-8 (eBook)
https://doi.org/10.1007/978-981-13-8134-8

This Springer imprint is published by the registered company Springer Nature Singapore Pte Ltd.
The registered company address is: 152 Beach Road, #21-01/04 Gateway East, Singapore 189721, Singapore

Preface

Metagenomic approaches have been so popular that not only biome in human guts but also microbes in soil, seawater, and air samples have become significant targets of the studies of microbial heterogeneity in understanding ecological background, evolutionary adaptation, and environmental monitoring of microbial communities.

In particular, marine metagenomics, the metagenomic studies of seawaters, have been actively developed with an acutely increasing speed, mostly because of a wide spectrum of significant applications. For example, a pioneering study of Sargasso Sea by Dr. Craig Venter disclosed a vast amount of novel species and genes of marine microorganisms that have been not identified in any means before. Recent studies of exploratory examination of microbes over a number of different locations in the globe, such as Tara Ocean and Malaspina Ocean Exploration Projects, have shown various new discoveries.

Moreover, changes in the marine environment due to the global warming effect and the pollution have become a major global problem. To maintain the healthy marine ecosystem, the advanced environmental monitoring and assessing systems are required. As a new monitoring technology, this metagenomic method has been expected for comprehensive analyses of the DNA of microorganisms living in seawater. This method drew attention as a means of understanding the dynamics of the marine environment.

As for the technical advancements, the amplicon-oriented methodologies of using 16S and 18S rRNAs as probes have been well established. Because of tremendous advancements of next-generation sequencer, more amplified coverage of sequencing genomic fragments has been possible in a very cost-efficient way. For this reason, the random shotgun methods of marine metagenomics have been getting more popularity, particularly because functional examination of obtained sequences of genomic fragments is possible when they are discussed as novel genes.

According to our knowledge, on the contrary, there are no textbooks or reference books on advanced methodologies of marine metagenomics so far, and it was an obstacle for new researchers to step into this research area. Here we publish textbook on marine metagenomics, whose readers will be a wide range from undergraduate students to advanced researchers. But, the main targets will be from the graduate students to marine scientists including environmental scientists. The purposes of this book are to introduce and to illustrate the state-of-the-art marine metagenome researches. We explain the methods of marine metagenomic analyses in an easy-to-understand manner and introduce the latest knowledge of marine metagenomics based on the studies of the authors' projects. We hope that this book can be used as a primer for new researchers and as a hands-on manual for experimental methods.

I strongly hope that this textbook should become a seminal milestone in this research fields and its relevant research domains.

Thuwal, Saudi Arabia — Takashi Gojobori
Yokohama, Japan — Tokio Wada
Sagamihara, Japan — Takanori Kobayashi
Thuwal, Saudi Arabia — Katsuhiko Mineta
Winter, 2019

Contents

Part I

Technological Aspects of Marine Metagenomics: Sample Collection and Preparation Methods

Metagenomic Methods: From Seawater to the Database

1

Md. Shaheed Reza, Atsushi Kobiyama, Jonaira Rashid, Yuichiro Yamada, Yuri Ikeda, Daisuke Ikeda, Nanami Mizusawa, Saki Yanagisawa, Kazuho Ikeo, Shigeru Sato, Takehiko Ogata, Toshiaki Kudo, Shinnosuke Kaga, Shiho Watanabe, Kimiaki Naiki, Yoshimasa Kaga, Satoshi Segawa, Katsuhiko Mineta, Vladimir Bajic, Takashi Gojobori, and Shugo Watabe

Abstract

In this article, methods or techniques of metagenomics including targeted 16S/18S rRNA analyses and shotgun sequencing will be discussed. It is sometimes difficult, especially for beginners, to follow the manufacturer's recommendation as mentioned in the protocol and to go through different steps from the preparation of starting material (e.g., DNA), library preparation, and so on. We

Md. S. Reza
Kitasato University School of Marine Biosciences, Sagamihara, Japan

Department of Fisheries Technology, Bangladesh Agricultural University, Mymensingh, Bangladesh

A. Kobiyama · J. Rashid · Y. Yamada · Y. Ikeda · D. Ikeda · N. Mizusawa · S. Yanagisawa
S. Sato · T. Ogata · T. Kudo
Kitasato University School of Marine Biosciences, Sagamihara, Japan

K. Ikeo
Kitasato University School of Marine Biosciences, Sagamihara, Japan

National Institute of Genetics, Shizuoka, Japan

S. Kaga
Iwate Fisheries Technology Center, Iwate, Japan

Ofunato Fisheries Promotion Center, Iwate Prefectural Government, Iwate, Japan

S. Watanabe · K. Naiki · Y. Kaga · S. Segawa
Iwate Fisheries Technology Center, Iwate, Japan

K. Mineta · V. Bajic · T. Gojobori
Computational Bioscience Research Center, King Abdullah University of Science and Technology, Thuwal, Saudi Arabia

S. Watabe (✉)
Kitasato University School of Marine Biosciences, Sagamihara, Japan
e-mail: swatabe@kitasato-u.ac.jp

T. Gojobori et al. (eds.), *Marine Metagenomics*,
https://doi.org/10.1007/978-981-13-8134-8_1

will try to explain all the steps in detail and share our experience here. It all starts with collection of samples and collection of ecological/environmental metadata followed by sample fractionation (optional), extraction of DNA, sequencing, and finally data analyses to interpret results. Sample collection has always been the most important part of a study as it requires proper planning, a good workforce to execute, permission(s) of sampling from appropriate authority, and precaution(s) about endangered species during sampling. Here, we first describe methodology for a shallow river and in the later section methodology for a deep marine bay. In either case, slight modifications can be made to succeed in sampling. Determination of physicochemical parameters as metadata simultaneously is also an important task. These samples are then processed to extract DNA which needs to be representative of all cells present in the sample. Finally, sequencing is done by a next-generation sequencer, and data analyses are completed. Through these methods, scientists are now able to overcome the unculturability problem of more than 99% of environmental microorganisms and uncovered functional gene diversity of environmental microorganisms.

Keywords
Free-living microorganisms · Particle-associated microorganisms · Shotgun metagenomics · Size fractionation · Targeted metagenomics

1.1 Methods

Two approaches are used to study environmental samples including soil, water, and air:

1. Targeted 16S or 18S rRNA analysis
2. Shotgun sequencing

For either of these approaches, methodology of metagenomic workflow remains more or less similar, and they can be applied for any environment from shallow river to deep lake, sea or bay, air, and sediment.

1.1.1 Methodology for 16S rRNA Environmental Sequence Analyses

1.1.1.1 Sample Collection

The foremost part of environmental metagenomic study is sampling. Before planning for sampling, proper permission should be taken from the city government office, mayor's office, etc. Consent from fishery cooperatives may sometimes be necessary. In the case of sampling from deep waters, safety of the researchers is an important issue. Appropriate lifesaving gears should be taken. Care should be given to preserve endangered species or wildlife. Finally, proper cautions should be taken to ensure collection of representative samples as it may directly affect downstream analyses.

Fig. 1.1 Metagenomic sampling from a shallow river. (**a**) Monitoring environmental parameters, (**b**) recording environmental data, (**c**) sampling from surface layer of the river, (**d**) prefiltration using 100 μm filter, (**e**) pouring water sample into a sterile bottle, and (**f**) collection of sediments from riverbed of the Tama River, Tokyo, Japan

During collection of samples, proper care should be taken to minimize contamination (Fig. 1.1). It may be suggested to use sterilized glassware for collecting water samples from a river or sea. Alternatively, water collection vessels can be rinsed with water three times before being used for sample collection. In case of collection of sediments from the river or seabed, it is always better to use "Marine Biology Dredge," e.g., Ekman Dredge, when the water is deep. Otherwise, sediments can be directly scrapped from the riverbed when the water depth is low.

1.1.1.2 Transport and Preservation of Sample

It is always important to quickly transport the samples to the laboratory for processing. Care should be taken not to allow rising of the temperature inside the samplers. It is suggested to use ice packs and coolers with an icebox for transportation, thereby minimizing multiplication of microorganisms inside the bottle. Sometimes, it may become necessary to preserve it as samples may be collected from places far from the laboratory. In case of lack of on-site sample preparation facilities, the samples can be preserved at 4 °C for 8–10 h.

1.1.1.3 Sample Treatment and Filtration

Size fractionation of water is an important part of metagenomic analyses as it separates planktonic microorganisms according to their size. It is a routine to use size fractionation, often equipped with a prefilter of larger pore size (typically 1.0–100 μm) and smaller pore sizes (0.1–0.2 μm) over a sample volume ranging from 0.05 to 20 l (Ganesh et al. 2014; Padilla et al. 2015; Reza et al. 2018a) (Fig. 1.2). It is generally accepted that free-living bacterioplankton that consist of more than 90% of the total bacterial population in the marine and freshwater ecosystems are trapped on the small-size pore filters, while their particle-associated counterparts get trapped on the larger-size pore filters as they remain attached to suspended particles particularly in the 5-μm filter fraction during filtration (Reza et al. 2018a). While the wide range of sample volumes used for filtration in different samplings is typical for coastal and open water environments (Padilla et al. 2015), the use of prefilter by larger pore size is generally practiced to remove debris that may be present in the sample (Reza et al. 2018b).

Different types of water filtration systems are available in the market such as:

(i) Nalgene™ polysulfone reusable bottle top filter with receiver (Thermo Fisher Scientific, Waltham, MA, USA)
(ii) Advantec purification system (Advantec Co. Ltd., Tokyo, Japan)
(iii) Sterivex™ cartridge filters (Millipore, Merck KGaA, Darmstadt, Germany)

In the case of using Nalgene™ polysulfone reusable bottle top filter with a receiver, ~350 ml of water is poured onto the filter with receiver (Thermo Fisher Scientific) and filtered sequentially through 45-mm diameter each of 5-, 0.8-, and 0.2-μm pore-sized Isopore™ membrane filters (Merck Millipore Ltd., Tullagreen, Ireland) using a vacuum pump. The pressure of the vacuum pump should be monitored carefully as it may disrupt cells of many soft-bodied protists and eukaryotes.

1.1.1.4 DNA Extraction

Generally microorganisms adhere to the filters, and different methods are used to extract DNA from these filters. Here we will discuss Qiagen's bead-beating

Fig. 1.2 Size fractionation of water samples. (**a**) Filtration of water, (**b**) trapping microorganisms on a filter, (**c**) rolling a filter paper, and (**d**) insertion of a filter paper into a 5-ml PowerWater Bead Tube

technique (Qiagen GmbH, Hilden, Germany). After removing the filters from the filtration unit using sterile forceps, the following steps can be used:

(i) First the filter membrane should be inserted into the 5-ml PowerWater® Bead Tube and subsequently be rolled in such a way that its top side faces inward.
(ii) Then 1 ml of Solution PW1 preheated at 55 °C is added to the PowerWater® Bead Tube.
(iii) The PowerWater® Bead Tube is secured horizontally to a Vortex adapter and vortexed at the maximum speed for 5 min.
(iv) Then the tubes are centrifuged at ≤4000 × *g* for 1 min at room temperature. The speed will depend on the capability of the centrifuge in use.
(v) The supernatant is transferred to a clean 2-ml Collection Tube by drawing up the supernatant using a 1-ml pipette tip by placing it down into the beads. Here, placing the pipette tip down into the beads and pipetting more than once are required to ensure the removal of the supernatant. Any carryover of beads will not affect subsequent steps. In this way, recovery between 600 and 650 μl of the supernatant is possible depending on the type of filter membranes used.
(vi) Then the tubes are centrifuged at 13,000 × *g* for 1 min.
(vii) To avoid the pellet, the supernatant is transferred to a clean 2-ml Collection Tube, and 200 μl of Solution PW2 are added, and the mixture is vortexed briefly to mix. They are subsequently incubated at 4 °C for 5 min and centrifuged at 13,000 × *g* for 1 min.
(viii) Then the supernatant is transferred to a clean 2-ml Collection Tube by carefully avoiding the pellet.
(ix) To the 2-ml Collection Tube, 650 μl of Solution PW3 is added, and the mixture is vortexed briefly to mix. Then 650 μl of supernatant is loaded onto a Spin Filter and centrifuged at 13,000 × *g* for 1 min.
(x) The flow is discarded through, and the process is repeated until all the supernatant has been loaded onto the Spin Filter. In this way, a total of two loads for each sample processed are required.
(xi) Then the Spin Filter basket is placed into a clean 2-ml Collection Tube, and 650 μl of Solution PW4 is added and centrifuged at 13,000 × *g* for 1 min.
(xii) The flow through is discarded, and 650μl of Solution PW5 is again added, and the mixture is centrifuged at 13,000 × *g* for 1 min. The flow through is again discarded and centrifuged again at 13,000 × *g* for 2 min to remove residual wash.
(xiii) Finally the Spin Filter basket is placed into a clean 2-ml Collection Tube and centrifuged at 13,000 × *g* for 1 min after adding 100 μl of Solution PW6 to the center of the white filter membrane.
(xiv) The Spin Filter basket is discarded, and the obtained DNA is now ready for any downstream application.

After extraction, DNA concentration needs to be measured as most of the library preparation kits require a specified amount of DNA for downstream analyses. For example, Nextera XT DNA Sample Preparation Kit (Illumina Inc., San Diego, CA, USA) requires only 1 ng of input DNA. In such cases, DNA quantitation can be done with a Qubit 2.0 Fluorometer (Invitrogen, Carlsbad, CA, USA) using Qubit dsDNA HS Assay Kit (Invitrogen) that uses highly sensitive and accurate fluorescence-based Qubit quantitation assays for quantitation of DNA, RNA, and protein. For each assay, it may be necessary to run new standards for calibrating the Qubit 2.0 Fluorometer. For later uses, the values from the previously done calibration can be used once calibration is completed for the first time. The process is briefly described below:

(i) Calibration of the Qubit 2.0 Fluorometer requires the preparation of the appropriate standard solutions. We need to set up two assay tubes: one for standards and the other for our sample. To prepare the Qubit Working Solution by diluting the Qubit reagent 1:200 in Qubit buffer, we need 200 μl of working solution each for standard and sample. We need to add 190 μl of working solution and 10 μl of standard solution (supplied with the kit) to make total volume in each assay tube to 200 μl. As for assay of our sample, we need to add 180–199 μl working solution and 1–20 μl of our sample to make the total volume to 200 μl. Generally, we use 1–2 μl of our sample and 198–199 μl of working solution for each sample assay.
(ii) Then all tubes are to be vortexed for 2–3 s and incubated at room temperature for 2 min.
(iii) After completion, they should be inserted into the Qubit 2.0 Fluorometer to take readings.

The extracted DNA can be stored at −20 °C, and aliquots of the extracted DNA can be used for metagenomic 16S rRNA sequencing.

1.1.1.5 PCR Amplification of 16S rRNA Gene and Next-Generation Sequencing

To determine the diversity and composition of the bacterial communities in the size-fractionated microorganism samples, different variable regions such as V1, V2, or V3 of the bacterial 16S rRNA gene can be targeted in metagenomic studies. The selection of appropriate primer is very important, as success of such study depends on the selection and design of primers. In this section, we discuss about sequencing methodology using Ion PGM (Thermo Fisher Scientific) and MiSeq (Illumina Inc., San Diego, CA) platforms.

(i) For Ion PGM:

Polymerase chain reaction (PCR) is performed as follows: 20 μl each of PCR mixture comprises 0.1 μl Ex *Taq* HS DNA polymerase (Takara, Otsu, Japan), 2 μl of 10× Ex *Taq* DNA polymerase buffer (Takara), 1.6 μl of 2.5 mM dNTP, 12.3 μl of nuclease-free water, 1 μl of 2 mM adaptor-labeled forward primer 1 μl of 2 mM adaptor-labeled and barcoded reverse primer, and 2 μl of undiluted template DNA. PCR conditions can consist of 5 min incubation at 95 °C followed by 30 cycles of 95 °C for 30 s, 55 °C for 30 s, and 72 °C for 30 s with the last cycle employing an extension for 5 min at 72 °C. Required number of independent PCRs can be carried out for the same amount of sample with a combination of forward primers and the common reverse primer.

Following amplification, all PCR products are examined for their size and specificity by electrophoresis on 2.5% w/v agarose and gel purified using Fast-Gene Gel/PCR Extraction Kit (Nippon Genetics Co. Ltd., Tokyo, Japan). Prior to sequencing, all amplicon types are assessed for fragment size distribution and DNA concentration using a Bioanalyzer 2100 (Agilent Technologies, Santa Clara, CA, USA). Ion Torrent sequencing is then performed according to the Ion Torrent standard workflow (Thermo Fisher Scientific). All barcoded amplicons are diluted to 13 pM and equimolarly pooled. Then 25 μl of amplicon library is used for downstream application using the Ion PGM Template OT2 400 Kit for use with the Ion OneTouch 2 System. This procedure results in the generation of template-positive Ion PGM Template OT2 400 Ion Sphere Particles. After enrichment of the particles, a sequencing run is performed with the Ion 318 Chip Kit v2 based on the Ion PGM Sequencing 400 Kit User Guide, and FASTQ files are generated with Torrent Suite software version 4.2.1.

(ii) For MiSeq:

For targeted 16S rRNA analyses using MiSeq, PCR libraries need to be constructed using forward (F) and reverse (R) indexed primes targeting variable regions between V1 and V3. For making barcoded libraries using Illumina's TruSeq or Nextera kits, 12 barcodes come with the library preparation kit that Illumina provides. Therefore, indexing primers included in the sequencing reagent kits are used for running MiSeq (Reza et al. 2018c). After PCR amplification, the amplicons are quantified using a Bioanalyzer and pooled in equimolar amounts as described in Illumina's "16S Metagenomic Sequencing Library Preparation" Guide (Part # 15044223 Rev. B). Finally, sequencing is performed as paired-end reads using an Illumina MiSeq platform and a MiSeq Reagent Kit v3 (600 cycles).

1.1.1.6 Sequence Analyses

For Ion PGM datasets, individual sequence reads are first filtered within the PGM software to remove low-quality and polyclonal sequences. To avoid possible bias and to minimize taxonomic ambiguity, a stringent criterion is generally set to use only longer reads where sequences are again quality trimmed to remove <270 bp using Genomics Workbench 8.0 (CLC bio, Cambridge, MA, USA). They are then processed by the NGS analysis pipeline of the SILVA rRNA gene database project (SILVAngs 1.2) (Quast et al. 2013). Downstream analyses are performed as described by Klindworth et al. (2013), where different tools in the pipeline including SINA for the alignment of sequences (Pruesse et al. 2012), CD-HIT for the clustering of sequences (Li and Godzik 2006), and BLAST for the classification of sequences (Camacho et al. 2009) are included.

As for 16S targeted libraries generated by MiSeq, all reads can be uploaded to BaseSpace server, and sequences of at least 100 bp are analyzed using the Illumina 16S Metagenomic App (v1.0.1) for taxonomic classification using an Illumina-curated version of Greengenes May 2013 reference taxonomy database.

1.1.1.7 Results Obtained for 16S rRNA Analyses for the Tama River, Tokyo, Japan

In this section, we present results of one of our studies conducted on Tama River, Tokyo, Japan, that employed next-generation sequencing technology targeting a 16S ribosomal RNA (rRNA) gene amplicon. Both the particle-associated and free-living portions collected from the river were studied separately after size fractionation of water samples trapped by 5.0-, 0.8-, and 0.2-μml filters. It was revealed that this urban river was most dominated by Proteobacteria, followed by Bacteroidetes, Actinobacteria, and Cyanobacteria (Fig. 1.3). Through this approach, we did not detect any archaeal and eukaryotic sequences as we used bacteria-specific primers (Reza et al. 2018c).

Our study was the first river metagenomic report from Japan. Through the brief analysis, we were able to identify genus-level taxonomic profile of the Tama River that can work as a baseline for the river system concerned and can be used to

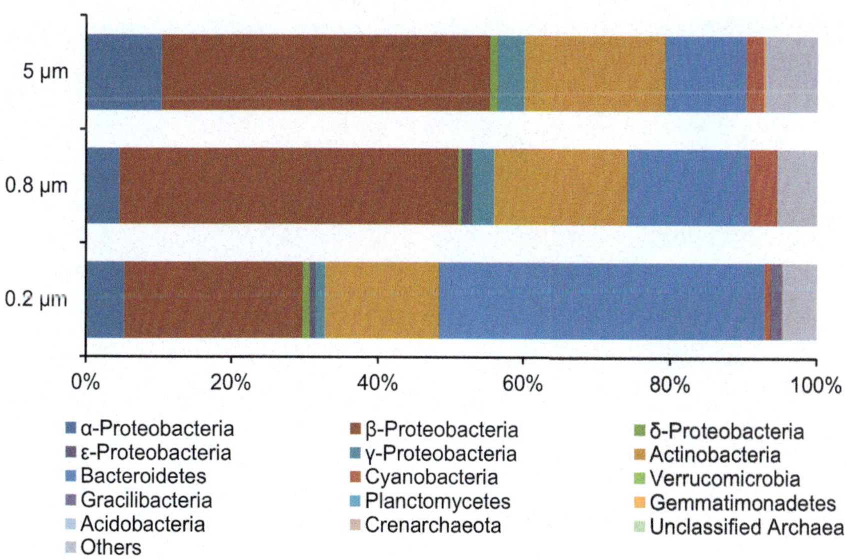

Fig. 1.3 Relative abundances of the major bacterial and archaeal phyla and classes in 5.0-, 0.8-, and 0.2-μm filter fractions from the Tama River based on the taxonomic identification of 16S rRNA gene fragments. The artificial group "Others" contains unidentified sequences, as well as sequences with a relative abundance lower than 1%. (From Reza et al. 2018c)

enhance the existing water quality monitoring efforts to improve well-being of the residents in the area.

1.1.2 Methodology for Shotgun Metagenomic Sequence Analyses

1.1.2.1 Sample Collection

Generally for deep water expedition in the marine water, no specific permits are required for sampling. This gives freedom for researchers who conduct sampling. However, proper permissions should be obtained for protecting endangered or protected species at sea. Usually vertical water sampler is used to collect water sample from a specific water depth. The volume of water sample collected varies depending on biomass present in the water. Generally 5–10 l of seawater is collected and prefiltered through a 100-μm filter to remove debris (Fig. 1.4). Corresponding seawater quality parameters, including temperature, salinity, and dissolved oxygen, are also measured on site using a water quality profiler. Simultaneously representative water samples are collected in separate sunlight-protected bottles for measurement of other water quality parameters including the abundance of chlorophyll *a* (chl-*a*) and heterotrophic bacteria (Yamada et al. 2017).

Fig. 1.4 Sampling of water at sea. (**a**) Travelling to sampling spot, (**b**) prefiltration using a 100-μm filter, and (**c**) and (**d**) sample collection in the Ofunato Bay, Iwate Prefecture, Japan

1.1.2.2 Sample Treatment and Filtration

In this section, we will discuss about filtration using Advantec purification system (Advantec Co. Ltd.). Generally the prefiltered water is filtered sequentially through 20-μm pore-sized Nylon Net Filter NY20 with a 47-mm diameter (Merck Millipore Ltd.) and through 5-, 0.8-, and 0.2-μm pore-sized Isopore Membrane Filters, 142 mm in diameter (Merck Millipore Ltd.), using a peristaltic pump (Masterflex L/S, Cole-Parmer International, Vernon Hills, IL, USA) (Fig. 1.5).

To minimize disruption of soft-bodied protists and other picoplankton, we maintained a low pump speed of 30–40 rpm. After filtration, all filters were stored at −80 °C until DNA extraction.

1.1.2.3 DNA Extraction

Genomic DNA is extracted from 0.2-μm filters using the PowerWater® DNA Isolation Kit (Mo Bio Laboratories, Inc., Carlsbad, CA, USA), following the manufacturer's recommendations as described earlier.

DNA concentrations also should be measured with Qubit dsDNA HS Assay Kit (Invitrogen), using a Qubit Fluorometer 2.0 (Invitrogen). The extracted DNA samples can be stored at −20 °C until use.

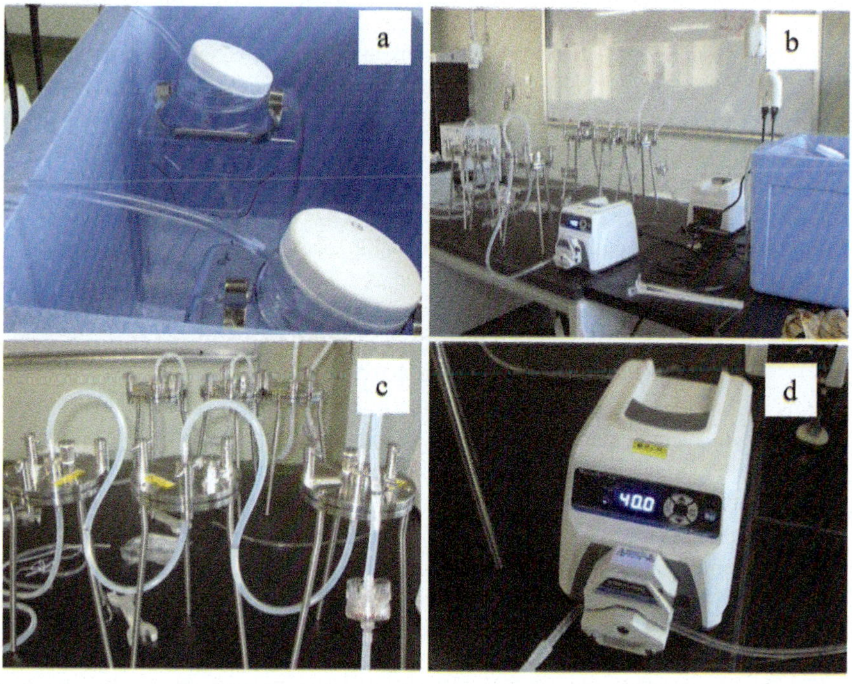

Fig. 1.5 Filtration of water. (**a**) Suction of water from tank, (**b**) and (**c**) sequential filtration using filtration system, and (**d**) peristaltic pump in use

1.1.2.4 Sequence processing and analyses

Extracted DNA aliquots are used for metagenomic shotgun sequencing. For library preparation by Nextera XT DNA Sample Preparation Kit (Illumina Inc.), 1 ng of gDNA is required. So sufficient amount of DNA should be diluted with sterile distilled water to adjust the concentration to 0.2 ng/μl.

Five microliters of gDNA at a concentration of 0.2 ng/μl are then taken, whereas DNA samples for metagenomics are prepared employing the Nextera XT DNA Sample Preparation Kit (Illumina Inc.) and sequenced as paired-end reads using an Illumina MiSeq platform and a MiSeq Reagent Kit v3 (600 cycles) (Illumina Inc.).

Illumina paired-end reads for each library are first joined by overlapping forward and reverse reads of the same DNA fragment (paired-end sequences) using the software FLASH (Magoc and Salzberg 2011) with default parameters (overlap minimum, 10 nt; maximum allowed ratio between the number of mismatched base pairs and the overlap length, 0.25). Quality filtering of these shotgun sequencing reads is then performed by removing reads <50 bp and quality trimmed to Phred 20 using the Genomics Workbench (CLC bio). BLASTn is performed using the quality- and size-filtered sequences against the NCBI-nt reference database. Taxonomic analysis at the genus level is then performed using MEGAN v5.10.3 (Huson et al. 2007), after parsing the BLAST output. Comparative analysis in MEGAN is also performed after normalizing counts per the recommendations of the authors. The least common ancestor (LCA) assignment algorithm is set at the parameters: minimum support = 1; minimum score = 50; top percent = 10; and maximum expected 1.0E – 10.

The obtained sequence matrices are generally used for correlating with various environmental factors including chl-*a* concentrations, heterotrophic bacterial abundances, and other metadata using statistical packages, e.g., R software in the Ofunato Bay (R Development Core Team 2008), based on experiment objectives (Rashid et al. 2018).

1.1.2.5 Results Obtained for a Shotgun Sequence Analyses for the Ofunato Bay, Iwate Prefecture, Japan

In this section, we are going to present the results of one of our shotgun metagenomic studies conducted on the Ofunato Bay, a deep enclosed bay in Iwate Prefecture, Japan (Reza et al. 2018b). Unlike our Tama River study, only free-living portions of the bay were reported as this portion has been considered as the most dominant among the total bacterial population. Phylogenetic differences among spring, summer, autumn, and winter bacterioplankton of the bay was studied that showed members of some group decreased at high water temperatures but increased at low temperatures, indicating seasonal change of the free-living bacterioplankton in the Ofunato Bay (Fig. 1.6).

As the Ofunato Bay is housed by Japan's famous buoy-and-rope type oyster (*Crassostrea gigas*) culture facilities, we were able to find strong locality signal for bacterial communities based on bacterial functional analyses (Kobiyama et al.

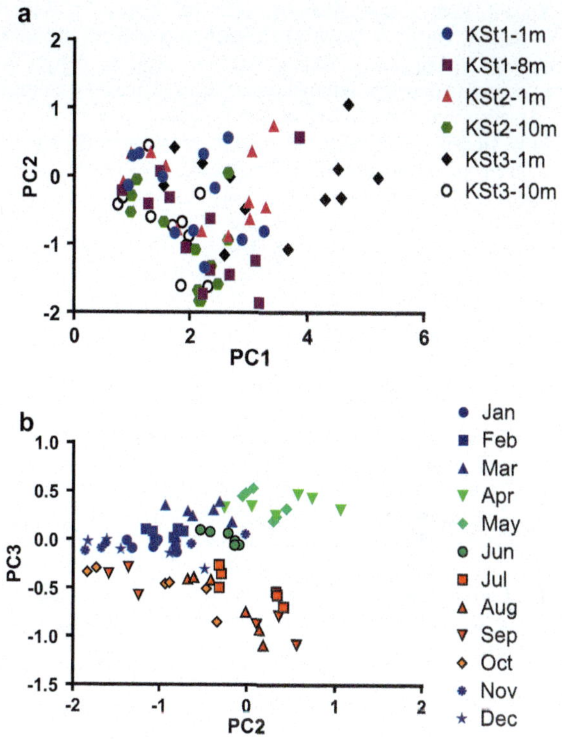

Fig. 1.6 Principal component analyses (PCAs) based on relative abundances of bacterial genera obtained from shotgun metagenomic data. Panel "a" shows that several datasets from KSt. 3 derived from 1-m water depth were clustered (dotted parabola), while Panel "b" shows that datasets derived from spring, summer, and winter were clustered together (parabola filled with green, red, and blue, respectively). Data employed was only from 0.2-μm-size filters targeting the free-living fractions from samples collected during January to December in the Ofunato Bay. (From Reza et al. 2018b)

2018). The dominant bacterial species *Candidatus* Pelagibacter ubique and *Planktomarina temperata* were reportedly found to show change seasonally, playing important roles in producing dimethyl sulfide (DMS) and methanethiol (MeSH) from dimethylsulfoniopropionate (DMSP) as signaling molecules for possible formation of scent of the tidewater or as fish attractants (Kudo et al. 2018).

References

Camacho C, Coulouris G, Avagyan V, Ma N, Papadopoulos J, Bealer K, Madden TL (2009) BLAST +: architecture and applications. BMC Bioinform 10:421–429

Ganesh S, Parris DJ, DeLong EF, Stewart FJ (2014) Metagenomic analysis of size-fractionated picoplankton in a marine oxygen minimum zone. ISME J 8:187–211

Huson DH, Auch AF, Qi J, Schuster SC (2007) MEGAN analysis of metagenomic data. Genome Res 17:377–386

Klindworth A, Pruesse E, Schweer T, Peplies J, Quast C, Horn M, Glöckner FO (2013) Evaluation of general 16S ribosomal RNA gene PCR primers for classical and next-generation sequencing based diversity studies. Nucleic Acids Res 41:e1

Kobiyama A, Ikeo K, Reza MS, Rashid J, Yamada Y, Ikeda Y, Ikeda D, Mizusawa N, Sato S, Ogata T, Kudo T, Kaga S, Watanabe S, Naiki K, Kaga Y, Mineta K, Bajic V, Gojobori T, Watabe S (2018) Metagenome-based diversity analyses suggest a strong locality signal for bacterial communities associated with oyster aquaculture farms in Ofunato Bay. Gene 665:149–154

Kudo T, Kobiyama A, Rashid J, Reza MS, Yamada Y, Ikeda Y, Ikeda D, Mizusawa N, Ikeo K, Sato S, Ogata T, Jimbo M, Kaga S, Watanabe S, Naiki K, Kaga Y, Segawa S, Mineta K, Bajic V, Gojobori T, Watabe S (2018) Seasonal changes in the abundance of bacterial genes related to dimethylsulfoniopropionate catabolism in seawater from Ofunato Bay as revealed by metagenomic analysis. Gene 665:174–184

Li W, Godzik A (2006) Cd-hit: a fast program for clustering and comparing large sets of protein or nucleotide sequences. Bioinformatics 22:1658–1659

Magoc T, Salzberg SL (2011) FLASH: fast length adjustment of short reads to improve genome assemblies. Bioinformatics 27:2957–2963

Padilla CC, Ganesh S, Gantt S, Huhman A, Parris DJ, Sarode N, Stewart FJ (2015) Standard filtration practices may significantly distort planktonic microbial diversity estimates. Front Microbiol 6:547

Pruesse E, Peplies J, Glöckner FO (2012) SINA: accurate high-throughput multiple sequence alignment of ribosomal RNA genes. Bioinformatics 28:1823–1829

Quast C, Pruesse E, Yilmaz P, Gerken J, Schweer T, Yarza P, Peplies J, Glöckner FO (2013) The SILVA ribosomal RNA gene database project: improved data processing and web-based tools. Nucleic Acids Res 41:D590–D596. (database issue)

Rashid J, Kobiyama A, Reza MS, Yamada Y, Ikeda Y, Ikeda D, Mizusawa N, Ikeo K, Sato S, Ogata T, Kudo T, Kaga S, Watanabe S, Naiki K, Kaga Y, Mineta K, Bajic V, Gojobori T, Watabe S (2018) Seasonal changes in the communities of photosynthetic picoeukaryotes in Ofunato Bay as revealed by shotgun metagenomic sequencing. Gene 665:127–132

Reza MS, Kobiyama A, Yamada Y, Ikeda Y, Ikeda D, Mizusawa N, Ikeo K, Sato S, Ogata T, Jimbo M, Kudo T, Kaga S, Watanabe S, Naiki K, Kaga Y, Mineta K, Bajic V, Gojobori T, Watabe S (2018a) Taxonomic profiles in metagenomic analyses of free-living microbial communities in the Ofunato Bay. Gene 665:192–200

Reza MS, Kobiyama A, Yamada Y, Ikeda Y, Ikeda D, Mizusawa N, Ikeo K, Sato S, Ogata T, Jimbo M, Kudo T, Kaga S, Watanabe S, Naiki K, Kaga Y, Mineta K, Bajic V, Gojobori T, Watabe S (2018b) Basin-scale seasonal changes in marine free-living bacterioplankton community in the Ofunato Bay. Gene 665:185–191

Reza MS, Mizusawa N, Kumano A, Oikawa C, Ouchi D, Kobiyama A, Yamada Y, Ikeda Y, Ikeda D, Ikeo K, Sato S, Ogata T, Kudo T, Jimbo M, Yasumoto K, Yoshitake K, Watabe S (2018c) Metagenomic analysis using 16S ribosomal RNA genes of a bacterial community in an urban stream, the Tama River, Tokyo. Fish Sci 84:563–577

R Development Core Team (2008) R: a language and environment for statistical computing. R Foundation for Statistical Computing, Vienna. ISBN 3-900051-07-0

Yamada Y, Kaga S, Kaga Y, Naiki K, Watanabe S (2017) Changes of seawater quality in Ofunto Bay, Iwate, after the 2011 off the Pacific coast of Tohoku Earthquake. J Oceanogr 73:11–24

Collection of Microbial DNA from Marine Sediments

2

Tomoko Sakami

Abstract

When collecting sediment samples, caution is needed to prevent contamination from outside microorganisms and altered quality of the sediment during deposition. Extraction from marine sediments involves the disruption of the microbial cells to dissolve the genomic DNA into extraction buffer and the purification of the DNA from the extraction buffer. Bacterial disruption can be achieved using cycles of freezing and thawing, pulverizing cells using beads, and degrading the cell membrane using detergents and/or enzymes, such as protease and/or lysozyme. DNA are purified from the extraction buffer by organic solvent extraction and/or adsorption to silica materials. Marine sediments often contain large amounts of humic substances. Polysaccharides are also contaminated in the DNA extract sometimes. These materials inhibit PCR, so removal treatments are required. DNA yield, extraction efficiency, and grade of purification differ in each method and depending to sediment sample qualities. Various commercial kits for DNA extraction from environmental samples are available. A method appropriate to the purpose of the study should be selected.

Keywords

Polysaccharide · Humic substance · PCR inhibition

When collecting sediment samples, caution is needed to prevent contamination from outside microorganisms and altered quality of the sediment during deposition. The tools and equipment used in sampling should be sterilized whenever possible.

T. Sakami (✉)
National Research Institute of Aquaculture, Japan Fisheries Research and Education Agency, Minami-ise, Mie, Japan
e-mail: sakami@affrc.go.jp

T. Gojobori et al. (eds.), *Marine Metagenomics*,
https://doi.org/10.1007/978-981-13-8134-8_2

Keeping the tools and equipment clean in an ordinal manner can be sufficient, since the numbers of contaminating microorganisms will be very low compared to the bacterial abundance in sediments, which can reach 10^9 cells/g. Conversely, marine sediments contain much water and organic matter, which can promote the rapid growth of some microorganisms at a normal temperature. This overgrowth could change the community composition. Therefore, collected samples should be stored in a cool and dark environment as soon as possible. Freezing using dry ice is preferable if a further process, such as fractionation of microorganisms, is not planned.

Dissolved oxygen concentration and organic matter abundance/quality change vertically in marine sediments. In muddy sediments, the change can occur suddenly in the immediate vicinity of the surface. Microbial communities will change corresponding to the environmental conditions. So, the depth of sample collection should be determined to fit to the purpose of the study. The collection depth can be selected by using a core-type sediment sampler or by inserting a small core on the surface of the sediment that was collected by a grab-type sediment sampler.

DNA can be extracted directly or indirectly. Direct extraction from marine sediments involves the disruption of the microbial cells and dissolution of the released genomic DNA into extraction buffer, followed by purification of the DNA from the buffer. To most accurately analyze microbial community diversity, DNA should be extracted equally from all the microbial cells irrespective of species. This can be a challenge in reality, since the extraction efficiency can vary between bacterial species. For example, the cell wall of gram-positive bacteria features a thick multilayered peptidoglycan component that confers greater resistance to physical disruption than other types of bacteria (Piscard et al. 1992). The disruption technique used should fully disrupt gram-positive bacteria at least to yield the full complement of genomic DNA from the microbial community. Bacterial disruption can be achieved using cycles of freezing and thawing, pulverizing cells using beads, and degrading the cell membrane using detergents like sodium dodecyl sulfate (SDS) and/or enzymes, such as protease and/or lysozyme. These approaches are often used in combination. SDS and freezing/thawing treatments can extract DNA effectively from sediment samples (Zhou et al. 1996). The ballistic use of beads often increases the yield of extracted DNA (Salonena et al. 2010; Alain et al. 2011; Lever et al. 2015).

Purifying DNA from the extraction buffer first involves the removal of water-insoluble materials including cell protein by organic solvent extraction. When shaken with a phenol-chloroform mixture, these water-insoluble components will partition into the hydrophobic phenol-chloroform layer. After purification, the DNA dissolved in the aqueous fraction is often concentrated by adding isopropanol or ethanol together with salts to precipitate DNA. Silica membrane or magnetic beads coated with silica materials are also used to concentrate DNA because DNA will absorb to silica and can be eluted after washing (Vandeventer et al. 2012; Miller et al. 1999). Marine sediments, especially muddy sediments, often contain large amounts of humic substances, which mingle with the extracted DNA and subsequently inhibit PCR (Watson and Blackwell 2000). These inhibitory materials can be removed by agarose gel electrophoresis and/or resin column treatments (Harry et al. 1999). When sediment samples contain plant debris or macrobenthos, such as polychaete, polysaccharides are often contaminated in the DNA extract because its chemical character is similar with that of nucleic acids (Jaufeerally-Fakim and Dookun 2000). To remove polysaccharides, cetrimonium bromide or hydroxyapatite can be used to differentiate the absorption efficiency between nucleic acid and polysaccharide according to salt concentration (Zhou et al. 1996; Jaufeerally-Fakim and Dookun 2000).

Extraction can be indirect. In this approach, microbial cells are isolated from a sediment sample prior to cell destruction. Sediment samples are shaken in sterilized seawater or buffer containing detergents to suspend the microbial cells and centrifuged moderately to pellet the minerals. In soil samples, gradient density centrifugation can be used for the efficient collection of microbial cells (Pillai et al. 1991). DNA yields are generally very low in the indirect method, typically two orders of magnitude lower than the direct method (Steffan et al. 1988; Gabor et al. 2003).

Differences in connection strengths between microbial cells and minerals can bias the community composition in the extracted DNA when using the indirect methods, although Gabor et al. (2003) reported no obvious differences between the direct and indirect methods in prokaryotic community diversity. On the other hand, the indirect method can obtain highly purified and non-fragmented DNA. Moreover, DNA debris that are not from living microbial cells but which have absorbed to minerals like silica can be eliminated by the indirect method. The method is recently used to make a metagenomic library, such as fosmid clones (Solomon et al. 2016).

To examine microbial community diversity using PCR amplicons like bacterial 16S rRNA gene analysis, a small amount of DNA is sufficient as long as the extracted DNA is fully representative of the microbial community. Dilution of a DNA extract can yield a good PCR result (Piscard et al. 1992). However, DNA should be conditioned to get a PCR product by normal reaction cycles (~30) because an over-dilution of template DNA can bias the composition of the amplified genes. Addition of bovine serum albumin (BSA) to PCR reaction cocktails can suppress inhibition caused by contaminants in DNA extracts. However, excessive BSA can inhibit the PCR reaction (Lekang et al. 2015). When DNA extracts are applied to quantitative PCR studies, an internal standard DNA, such as a pGEM vector, can be added to correct DNA extraction efficiency, since the efficiency of extraction varies markedly depending on sample characters and/or extraction methods (Hariganeya et al. 2013).

Various commercial kits for DNA extraction from environmental samples are available. DNA yield, extraction efficiency, and grade of purification differ between the kits. Lekang et al. (2015) reported that when using a commercial kit, small bacterial community composition can bias among sediment samples having different characteristics. However, the cost of the commercial kit was several times greater than that of manual extraction methods. As the composition of the microbial community that is obtained can be different depending on the method of DNA extraction (Luna et al. 2006), an appropriate one to the purpose of the study should be selected.

References

Alain K, Callac N, Ciobanu M, Reynaud Y, Duthoit F, Jebbar M (2011) DNA extractions from deep subseafloor sediments: novel cryogenic-millbased procedure and comparison to existing protocols. J Microbiol Methods 87(3):355–362

Gabor E, de Vries E, Janssen DB (2003) Efficient recovery of environmental DNA for expression cloning by indirect extraction methods. FEMS Microb Ecol 44(2):153–163

Hariganeya N, Tanimoto Y, Yamaguchi H, Nishimura T, Tawong W, Sakanari H, Yoshimatsu T, Sato S, Preston CM, Adachi M (2013) Quantitative PCR method for enumeration of cells of cryptic species of the toxic marine dinoflagellate *Ostreopsis* spp. in coastal waters of Japan. PLoS One 8(3):e57627

Harry M, Gambier B, Bourezgui Y, Garnier-Sillam E (1999) Evaluation of purification procedures for DNA extracted from organic rich samples: interference with humic substances. Analusis 27(5):439

Jaufeerally-Fakim J, Dookun A (2000) Extraction of high quality DNA from polysaccharides-secreting *Xanthomonads*. Sci Technol Res J 6:33–40

Lekang K, Thompson EM, Troedsson C (2015) A comparison of DNA extraction methods for biodiversity studies of eukaryotes in marine sediments. Aquat Microb Ecol 75:15–25

Lever MA, Torti A, Eickenbusch P, Michaud AB, Šantl-Temkiv T, Jørgensen BB (2015) A modular method for the extraction of DNA and RNA, and the separation of DNA pools from diverse environmental sample types. Front Microbiol 6:476

Luna GM, Dell'Anno A, Danovaro R (2006) DNA extraction procedure: a critical issue for bacterial diversity assessment in marine sediments. Environ Microbiol 8(2):308–320

Miller DN, Bryant JE, Madsen EL, Ghiorse WC (1999) Evaluation and optimization of DNA extraction and purification procedures for soil and sediment samples. Appl Environ Microbiol 65(11):4715–4724

Pillai SD, Josephson KL, Bailey RL, Gerba CP, Pepper IL (1991) Rapid method for processing soil samples for polymerase chain reaction amplification of specific gene sequences. Appl Environ Microbiol 57:2283–2286

Piscard C, Ponsonnet C, Paget E, Nesme X, Simonet P (1992) Detection and enumeration of bacteria in soil by direct DNA extraction and polymerase chain reaction. Appl Environ Microbiol 58(9):2717–2722

Salonena A, Nikkiläa J, Jalanka-Tuovinena J, Immonena O, Rajilić-Stojanovićb M, Kekkonenc RA, Palvaa A, de Vosa WM (2010) Comparative analysis of fecal DNA extraction methods with phylogenetic microarray: effective recovery of bacterial and archaeal DNA using mechanical cell lysis. J Microbiol Methods 81(2):127–134

Solomon S, Kachiprath B, Jayanath G, Sajeevan TP, Bright Singh IS, Philip R (2016) High-quality metagenomic DNA from marine sediment samples for genomic studies through a preprocessing approach. 3 Biotech 6:160

Steffan RJ, Goksoyr J, Bej AK, Atlas RM (1988) Recovery of DNA from soils and sediments. Appl Environ Microbiol 54:2908–2915

Vandeventer PE, Lin JS, Zwang TJ, Nadim A, Johal MS, Niemz A (2012) Multiphasic DNA adsorption to silica surfaces under varying buffer, pH, and ionic strength conditions. J Phys Chem B 116(19):5661–5670

Watson RJ, Blackwell B (2000) Purification and characterization of a common soil component which inhibits the polymerase chain reaction. Can J Microbiol 46(7):633–642

Zhou J, Bruns MA, Tiedje JM (1996) DNA recovery from soils of diverse composition. Appl Environ Microbiol 62(2):316–322

Primer Design, Evaluation of Primer Universality, and Estimation of Identification Power of Amplicon Sequences In Silico

3

Akifumi S. Tanabe, Satoshi Nagai, Yuki Hongo, Motoshige Yasuike, Yoji Nakamura, Atushi Fujiwara, and Seiji Katakura

Abstract

Primer design for polymerase chain reaction is an essential step for DNA barcoding, metabarcoding, molecular phylogenetics, and population genetic studies. Both forward and reverse primers need to match to annealing positions of template sequences of all subtaxa of target taxa in order to amplify target locus sequences. Although there are many existing such "universal" primer sets, the existing primer sets are not able to amplify target locus sequences of target taxa in some cases. Moreover, recent high-throughput sequencers cannot read long contiguous sequences and require new universal primer sets which can amplify relatively short barcode DNA sequences because most of the existing universal primer sets amplify longer sequences than high-throughput sequencers can read contiguously. Taking accumulation of nucleotide sequences of public DNA databases into consideration, retrieving target locus sequences of target taxa from public DNA databases, and designing primer set based on consensus sequence of retrieved sequences can be the solution for this problem. In this text, we provide methods and procedures to design universal primer pairs and to evaluate primer universality and identification power of amplicons.

Keywords

DNA barcoding · Metabarcoding · Environmental DNA (eDNA) · Metagenome · Marine eukaryote community

---

A. S. Tanabe (✉)
Center for Ecological Research, Kyoto University, Otsu, Shiga, Japan

S. Nagai · Y. Hongo · M. Yasuike · Y. Nakamura · A. Fujiwara
National Research Institute of Fisheries Science, Japan Fisheries Research and Education Agency, Yokohama, Kanagawa, Japan

S. Katakura
City of Mombetsu, Kaiyo-koryukan, Kaiyo-koen, Mombetsu, Hokkaido, Japan

T. Gojobori et al. (eds.), *Marine Metagenomics*,
https://doi.org/10.1007/978-981-13-8134-8_3

3.1 Introduction

Polymerase chain reaction using oligonucleotide primer set which is applicable across all or almost all target taxa is an essential step in ecological community surveys based on metabarcoding technique and still becoming increasingly more important. In this step, appropriate primer set is required, but designing method of such "universal" primer set and evaluation procedures of their suitability are not broadly shared. Therefore, we provide detailed methods and procedures of primer design, evaluation of primer universality, and estimation of identification power of amplicon sequences using international sequence database (hereafter, INSD) in this text. Some of these methods have been already used in Tanabe et al. (2016).

3.2 Prerequisites

In this text, Debian/GNU Linux or compatible environment is assumed as operating system. Additionally, several programs or packages listed in Table 3.1 and their dependent packages must be installed correctly. Moreover, several NCBI databases must be downloaded, extracted, and placed to correct path shown in Table 3.2. All commands shown below are terminal or console commands with bash or compatible shell environment. To download, extract, and install these required programs and

Table 3.1 Required programs

Package	Version	Public web site
Claident	0.2.2017.05.22	https://github.com/astanabe/Claident
Phylogears2	2.0.2016.09.06	https://github.com/astanabe/Phylogears
MAFFT	7.310	http://mafft.cbrc.jp/alignment/software/
Primer3	2.3.7	http://primer3.sourceforge.net/
VSEARCH	2.4.3	https://github.com/torognes/vsearch
ecoPCR	0.8.0	https://git.metabarcoding.org/obitools/ecopcr/wikis/home
EMBOSS	6.6.0	http://emboss.sourceforge.net/
BLAST+	2.6.0	ftp://ftp.ncbi.nlm.nih.gov/blast/executables/blast+/
R	3.4.1	https://cran.r-project.org/

Note that program version must not be the same as above

Table 3.2 Required databases

Database	Download date	Distributed site	Install path
NCBI Taxonomy	2017/08/11	ftp://ftp.ncbi.nih.gov/pub/taxonomy/	~/taxonomy
NCBI nt	2017/08/11	ftp://ftp.ncbi.nih.gov/blast/db/	~/blastdb

Note that database download date must not be the same as above data to the correct path, type the following commands:

```
# Download installer
wget \
https://github.com/astanabe/PrimerDesign/archive/master.tar.gz \
-O PrimerDesign-master.tar.gz
tar -xzf PrimerDesign-master.tar.gz
cd PrimerDesign-master
# Install dependent packages
# sudo password required
sh installDEPS.sh
# Install required programs
sh installPROGRAMS.sh
# Install required data
sh installDB.sh
# Delete temporary directory
cd ..
rm -rf PrimerDesign-master
```

These commands also install several utility scripts into ~/work. In the following sections, ~/work is assumed as working directory, and the number of CPU cores is assumed as 4. Because the above installer scripts install commands into ~/bin, the following commands need to be executed to add this path to PATH environment variable at the beginning of each analysis:

```
# Add ~/bin to PATH
export PATH=~/bin:$PATH
```

Additional information and update will be provided at https://github.com/astanabe/PrimerDesign.

3.3 Primer Design

In this section, we introduce methods and procedures to retrieve nucleotide sequences of target locus of target taxon and to pick universal primer pairs based on a degenerate consensus sequence. If the universal primer pair for nuclear ribosomal DNA or bacterial ribosomal DNA is needed, nucleotide sequences of target locus of target taxon should also be retrieved from SILVA Project (Quast et al. 2013) or Ribosomal Database Project (Cole et al. 2014).

3.3.1 Retrieving INSD Nucleotide Sequences of Target Locus of Target Taxon

In order to design appropriate primer set, nucleotide sequences of target locus of target taxon are required and can be obtained from INSD.

The first method retrieving nucleotide sequences of target locus of target taxon from INSD is taxon restricted nucleotide database search. For example, the following command searches nucleotide database by keywords at NCBI and retrieves GenBank ID (hereafter, GI) list:

```
# Search nucleotide database and retrieve GI
clretrievegi \
 --keywords='"ddbj embl genbank"[Filter] AND txid7088
[Organism:exp] AND mitochondrion AND "complete genome"' \
 Lepidoptera_mtgenome.txt
```

The file Lepidoptera_mtgenome.txt is the name of output file, and the --keywords argument value is used as search keywords for NCBI. "ddbj embl genbank"[Filter] is a keyphrase to exclude non-INSD sequence data. txid7088[Organism:exp], mitochondrion, and "complete genome" are keywords or keyphrases to limit search results to lepidopteran data entries, mitochondrial sequence entries, and complete genome sequences, respectively. AND is a logical AND operator. Therefore, the output file contains GI list of lepidopteran complete mitochondrial genome sequences. Note that 7088 is the NCBI Taxonomy ID of Lepidoptera. To know the NCBI Taxonomy ID of any taxa, search NCBI Taxonomy https://www.ncbi.nlm.nih.gov/taxonomy.

Then, the following command retrieves GenBank entries contained in input GI list:

```
# Retrieve GenBank entries
pgretrieveseq \
 Lepidoptera_mtgenome.txt \
 Lepidoptera_mtgenome.gb
```

Finally, the following command extracts 12S rRNA region from the retrieved GenBank data:

```
# Extract 12S rRNA region
extractfeat \
 -type rRNA \
 -tag product \
 -value "12S*|small*RNA|s-rRNA" \
 Lepidoptera_mtgenome.gb \
 Lepidoptera_12S_fulllength.fasta
```

GenBank format sequence file contains many "feature" information such as gene region (see also https://www.ncbi.nlm.nih.gov/Sitemap/samplerecord.html). The -type rRNA argument excludes non-rRNA sequences from the output, and the -tag product -value "12S*|small*RNA|s-rRNA" argument limits extracted sequences to the regions having 12S*, small*RNA, or s-rRNA as a value of product tag in feature information. In the value (inside of " ") of -value argument, | is a logical OR operator and space cannot be used, but the wild card character * can be used instead of space or any characters. To know more about extractfeat command, type man extractfeat, or see http://emboss.sourceforge.net/apps/cvs/emboss/apps/extractfeat.html.

Alternatively (or additionally), we can use taxon restricted BLAST search using one or more already available nucleotide sequences of target locus of target taxon as queries to obtain such sequences. Before executing BLAST, taxon restricted BLAST database must be constructed from NCBI nt which is partially nonredundant nucleotide sequence database excluding GSS, STS, PAT, EST, HTG, and WGS division sequences. For example, the following commands generate lepidopteran insect sequence database whose sequences have genus or lower taxonomic information:

```
# Construct taxonomy database of Lepidoptera
clmaketaxdb \
 --includetaxid=7088 \
 ~/taxonomy \
 Lepidoptera.taxdb
# Extract GIs of Lepidopteran sequences which have genus or
# lower taxonomic information
clretrievegi \
 --taxdb=Lepidoptera.taxdb \
 --includetaxa=genus,.+ \
 Lepidoptera_genus.txt
# Construct BLAST database of Lepidopteran sequences
cd ~/blastdb
blastdb_aliastool \
 -dbtype nucl \
 -db ./nt \
 -gilist ~/work/Lepidoptera_genus.txt \
 -out Lepidoptera_genus \
 -title Lepidoptera_genus
cd ~/work
```

Because NCBI Taxonomy ID of Lepidoptera is 7088, this value is given for --includetaxid argument of clmaketaxdb to restrict included taxa of taxonomy database within Lepidoptera. Then, the GIs of sequences which have genus-level taxonomic information are retrieved using clretrievegi from taxonomy database of Lepidoptera. Finally, lepidopteran sequence BLAST database which contains the sequences listed in GI list file is constructed using blastdb_aliastool as alias of NCBI nt.

The following command conducts nucleotide BLAST using reference 12S rRNA sequences contained in a FASTA file (Lepidoptera_12S_fulllength.fasta) as queries and saves 1,000 bp or longer aligned part of BLAST-hit sequences covering at least 95% of query to output file (Lepidoptera_12S_fulllength_blastn.fasta) which contains full-length or nearly full-length 12S rRNA sequences of Lepidoptera:

```
# Reduce the number of sequences using VSEARCH
vsearch \
 --notrunclabels \
 --id 0.95 \
 --strand both \
 --threads 4 \
 --cluster_fast Lepidoptera_12S_fulllength.fasta \
 --centroids Lepidoptera_12S_fulllength_reduced95.fasta
# Perform BLAST search
clblastseq \
 blastn \
  -task blastn \
  -word_size 11 \
  -evalue 1e-5 \
```

```
  -max_target_seqs 1000000 \
 end \
 --blastdb ~/blastdb/Lepidoptera_genus \
 --minalnlen=1000 \
 --minalnpcov=0.95 \
 --output=FASTA \
 --numthreads=4 \
 Lepidoptera_12S_fulllength_reduced95.fasta \
 Lepidoptera_12S_fulllength_blastn.fasta
```

In the above, the input sequences are clustered, and centroid sequences of clusters are extracted and used as query to reduce the number of times of BLAST search. Note that the final output file Lepidoptera_12S_fulllength_blastn.fasta may contain duplicate sequences because of multiple queries.

3.3.2 Picking Universal Primer Pairs Based on a Degenerate Consensus Sequence

Because the sequences retrieved by BLAST may contain duplicate sequences and the total number of retrieved sequences is often extremely large, sequence clustering based on sequence similarity/identity and extraction of centroid (representative) sequences of clusters are strongly recommended. This process reduces the total number of sequences. For example, the following command clusters input sequences (Lepidoptera_12S_fulllength_blastn.fasta) based on percent identity threshold of 95%, extracts centroid sequences, and saves them to output file (Lepidoptera_12S_fulllength_blastn_reduced95.fasta):

```
# Cluster sequences and extract centroid sequences of clusters
vsearch \
 --notrunclabels \
 --id 0.95 \
 --strand both \
 --threads 4 \
 --cluster_fast Lepidoptera_12S_fulllength_blastn.fasta \
 --centroids Lepidoptera_12S_fulllength_blastn_reduced95.fasta
```

The sequence file is often contaminated reverse strand sequences. Therefore, strand standardization is strongly recommended for multiple sequence alignment. The following command standardizes strand of input sequences to the strand of the first sequence of input file:

```
# Standardize strand
pgstanstrand \
 Lepidoptera_12S_fulllength_blastn_reduced95.fasta \
 Lepidoptera_12S_fulllength_blastn_reduced95_stanstrand.fasta
```

Then, multiple sequence alignment can be conducted like below:

```
# Multiple sequence alignment
mafft \
  --auto \
  --thread 4 \
  Lepidoptera_12S_fulllength_blastn_reduced95_stanstrand.fasta \
  > \
  Lepidoptera_12S_fulllength_blastn_reduced95_
    stanstrand_aligned.fasta
```

Finally, degenerate consensus sequence is generated, and up to 50 universal primer pairs are picked based on the consensus sequence by the following command:

```
# Pick primer set
pgpickprimer \
  --maxpick=50 \
  --consensus=95 \
  --sizerange=300-500 \
  --tmrange=45-60 \
  Lepidoptera_12S_fulllength_blastn_reduced95_stanstrand_
aligned.fasta \
  Lepidoptera_12S_fulllength_blastn_reduced95_stanstrand_aligned_
primers.fasta
```

The --maxpick and --consensus are arguments for specifying the maximum number of picked primer pairs and for specifying the threshold value of majority rule consensus, respectively. If 95 was given for --consensus, 95% majority rule consensus-degenerate sequence will be generated and used for primer picking. The --sizerange and --tmrange are arguments for specifying ranges of target amplicon length and target Tm value, respectively. Part of the results of this command is shown in Fig. 3.1. The first primer pair WTGGCGGTATTTTAGTTYAT and GACGGGCAATATGTACATAT is used as universal primer pair in the following.

If taxonomically specific primer pairs which are universal within the taxon are needed, ecoPrimers (Riaz et al. 2011) is suitable for such primer picking, and ecotaxstat and ecotaxspecificity in OBITools (Boyer et al. 2016) are recommended for evaluating taxonomic specificity of such primer pairs.

3.4 Evaluation of Primer Universality

To evaluate primer universality, i.e., taxonomic coverage, ecoPCR (Ficetola et al. 2010) is used in this text. This program can perform in silico PCR amplification using template sequence database and forward and reverse primer sequences. If the universal primer pair is targeting nuclear ribosomal DNA or bacterial ribosomal DNA, TestPrime (https://www.arb-silva.de/search/testprime/) provided by SILVA Project or ProbeMatch (http://rdp.cme.msu.edu/probematch/) provided by Ribosomal Database Project is also strongly recommended (Klindworth et al. 2013; Cole et al. 2014).

```
consensus  TAADW--TGGCGGTATTTTAGTTYATTTAGAGGAATCTGTYTADTAA-TTGATADTCCACGA--------
F1_Tm52.8  ----W--TGGCGGTATTTTAGTTYAT--------------------------------------------
R1_Tm52.6  ----------------------------------------------------------------------
F2_Tm52.8  ----W--TGGCGGTATTTTAGTTYAT--------------------------------------------
R2_Tm52.4  ----------------------------------------------------------------------
F3_Tm52.8  ----W--TGGCGGTATTTTAGTTYAT--------------------------------------------
R3_Tm52.3  ----------------------------------------------------------------------
F4_Tm52.8  ----W--TGGCGGTATTTTAGTTYAT--------------------------------------------
R4_Tm52.9  ----------------------------------------------------------------------
F5_Tm52.8  ----W--TGGCGGTATTTTAGTTYAT--------------------------------------------
R5_Tm52.9  ----------------------------------------------------------------------
F6_Tm52.8  ----W--TGGCGGTATTTTAGTTYAT--------------------------------------------
R6_Tm52.1  ----------------------------------------------------------------------
           1.......10........20........30........40........50........60........70

           snip

consensus  TTTTWWWTADWWTW-WWAAAATGATT-WWWRWT-TRAAATATGTACATA-TTGCCCGTCRCTTTCAT
F1_Tm52.8  -------------------------------------------------------------------
R1_Tm52.6  --------------------------------------TATACATGTAT-AACGGGCAG--------
F2_Tm52.8  -------------------------------------------------------------------
R2_Tm52.4  ----------------------------------------TACATGTAT-AACGGGCAGYG------
F3_Tm52.8  -------------------------------------------------------------------
R3_Tm52.3  ------------------------------------------CATGTAT-AACGGGCAGYGAA----
F4_Tm52.8  -------------------------------------------------------------------
R4_Tm52.9  --------------------------------------------TGTAT-AACGGGCAGYGAAAG--
F5_Tm52.8  -------------------------------------------------------------------
R5_Tm52.9  ---------------------------------------------GTAT-AACGGGCAGYGAAAGT-
F6_Tm52.8  -------------------------------------------------------------------
R6_Tm52.1  -------------------------------------TTATACATGTAT-AACGGGCA---------
           .......430.......440.......450.......460.......470.......480.......
```

Fig. 3.1 An example of picking universal primer pairs based on degenerate consensus sequence

3.4.1 Clustering Template Sequences and Construction of Template Sequence Databases

Before performing ecoPCR, sequence clustering and extraction of centroid sequences of clusters are strongly recommended because the number of sequences is often highly varied among subtaxa and such bias may cause overestimation or underestimation of taxonomic coverage of primer pairs. The following commands cluster template sequences contained in Lepidoptera_12S_fulllength_blastn.fasta using 100% and 90% sequence identity cutoffs and extract centroid sequences to output files. After centroid sequence extraction, strand of the sequences is standardized in the following commands:

```
# Cluster amplicon sequences using 100%-cutoff
vsearch \
 --notrunclabels \
 --id 1.0 \
 --strand both \
 --threads 4 \
 --cluster_fast Lepidoptera_12S_fulllength_blastn.fasta \
 --centroids Lepidoptera_12S_fulllength_blastn_reduced100.fasta
# Standardize strand
pgstanstrand \
 Lepidoptera_12S_fulllength_blastn_reduced100.fasta \
 Lepidoptera_12S_fulllength_blastn_reduced100_stanstrand.fasta
```

```
# Cluster template sequences using 90%-cutoff
vsearch \
 --notrunclabels \
 --id 0.9 \
 --strand both \
 --threads 4 \
 --cluster_fast Lepidoptera_12S_fulllength_blastn.fasta \
 --centroids Lepidoptera_12S_fulllength_blastn_reduced90.fasta
# Standardize strand
pgstanstrand \
 Lepidoptera_12S_fulllength_blastn_reduced90.fasta \
 Lepidoptera_12S_fulllength_blastn_reduced90_stanstrand.fasta
```

Note that clustering template sequences using 95% sequence identity cutoff has been already conducted as written before (see Sect. 3.3.2). Thus, there are three reduced template sequence files at this time.

Reduced template sequence databases can be constructed by the following commands:

```
# Construct reduced template databases
# p = sequence percent identity cutoff
for p in 100 95 90; do
 perl addlabel2fasta.pl \
  Lepidoptera_12S_fulllength_blastn_reduced$p\_stanstrand.
    fasta \
  > \
  Lepidoptera_12S_fulllength_blastn_reduced$p\_stanstrand.
    ecoPCR
 python ecoPCRFormat.py \
  -f -t ~/taxonomy \
  -n Lepidoptera_12S_fulllength_blastn_reduced$p\_stanstrand \
  Lepidoptera_12S_fulllength_blastn_reduced$p\_stanstrand.ecoPCR
done
```

Perl script addlabel2fasta.pl modifies the style of sequence definition lines of input sequences as required by ecoPCRFormat.py in the above commands. ecoPCRFormat.py constructs template sequence databases for ecoPCR.

3.4.2 Performing In Silico PCR Using ecoPCR and Summarizing In Silico PCR Results

The following commands conduct in silico PCR using ecoPCR:

```
# e = allowed number of mismatches
# p = sequence percent identity cutoff
for e in 0 1 2 3; do
 for p in 100 95 90; do
  # Run ecoPCR using reduced databases
  ecoPCR \
   -d Lepidoptera_12S_fulllength_blastn_reduced$p\_stanstrand \
   -e $e \
```

```
    -l 200 \
    -L 600 \
    WTGGCGGTATTTTAGTTYAT \
    GACGGGCAATATGTACATAT \
    | perl removepolyNentry.pl \
    | perl remove3primemismatch.pl \
       -fp=WTGGCGGTATTTTAGTTYAT \
       -rp=GACGGGCAATATGTACATAT \
    > Lepidoptera_12S_fulllength_blastn_reduced$p\_stanstrand_
      e$e.txt
  done
done
```

The arguments of -d, -e, -l, and -L for ecoPCR specify template sequence database, maximum acceptable mismatches of primer pair, minimum length of amplicons without primer annealing positions, and maximum length of amplicons without primer annealing positions, respectively. The Perl scripts removepolyNentry.pl and remove3primemismatch.pl filter out amplicons with primer annealing positions containing 5 bp or longer contiguous N and amplicons containing 1 or more mismatches at the 2 bp of 3'-tail of primer annealing positions, respectively.

Next, the following commands read template sequence files and in silico PCR results and output summary table as comma-delimited file:

```
# Make summary table
# Output header line
echo "primer,cutoff,error,all,success" \
 > coveragetable.txt
# p = sequence percent identity cutoff
# e = allowed number of mismatches
for p in 100 95 90; do
 for e in 0 1 2 3; do
  # Output data line
  perl maketableofcoverage.pl \
   primer1 \
   $p \
   $e \
   Lepidoptera_12S_fulllength_blastn_reduced$p\_stanstrand.
     ecoPCR \
   Lepidoptera_12S_fulllength_blastn_reduced$p\_stanstrand_e$e.
     txt \
   >> coveragetable.txt
 done
done
```

Finally, a figure of the PCR success rate, i.e., primer universality, can be generated by the following command:

```
# Make figure
R \
 --vanilla \
 --slave \
 < makefigofcoverage.R
```

An example of the figure of the PCR success rate is shown in Fig. 3.2.

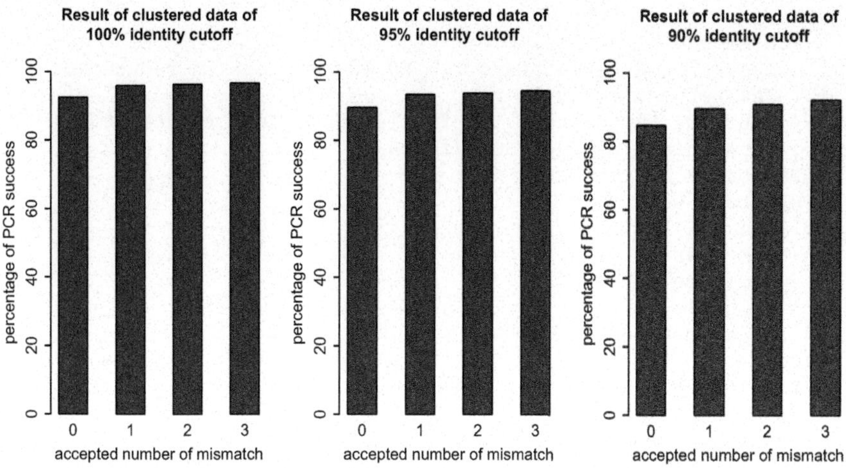

Fig. 3.2 An example of the figure of PCR success rates

3.5 Estimation of Identification Power of Amplicon Sequences Based on Database Sequences

3.5.1 Estimation of Identification Power Based on Effective Number of Sequences Registered to INSD

Effective number of sequences registered to INSD is one of the important indices for evaluating identification power of amplicon sequences because highly universal primer pair and highly variable amplicon sequences are spoiled by the small number of reference sequences. In order to estimate effective number of sequences registered to INSD, homologous sequences to amplicon sequences need to be retrieved. For example, the following commands make amplicon sequences from the results of in silico PCR, reduce the number of amplicon sequences by sequence clustering, and retrieve homologous sequences whose alignment length is equal to or longer than 200 bp by BLAST search using amplicon sequences as queries:

```
# Make amplicon sequences from the results of in silico PCR
perl extractamplicon.pl \
 Lepidoptera_12S_fulllength_blastn_reduced100_stanstrand_e3.
   txt \
 > \
 Lepidoptera_12S_fulllength_blastn_reduced100_stanstrand_e3.
   fasta
# Reduce the number of sequences using VSEARCH
vsearch \
 --notrunclabels \
 --id 0.95 \
```

```
  --strand plus \
  --threads 4 \
  --cluster_fast Lepidoptera_12S_fulllength_blastn_reduced100_
     stanstrand_e3.fasta \
  --centroids Lepidoptera_12S_fulllength_blastn_reduced100_
     stanstrand_e3_reduced95.fasta
# Perform BLAST search
clblastseq \
 blastn \
  -task blastn \
  -word_size 11 \
  -evalue 1e-5 \
  -max_target_seqs 1000000 \
 end \
 --blastdb ~/blastdb/Lepidoptera_genus \
 --minalnlen=200 \
 --output=FASTA \
 --numthreads=4 \
 Lepidoptera_12S_fulllength_blastn_reduced100_stanstrand_e3_
   reduced95.fasta \
 Lepidoptera_12S_amplicon_blastn.fasta
```

Because the sequences retrieved by BLAST may contain duplicate sequences and the number of registered sequences is often highly varied among subtaxa, sequence clustering and extraction of centroid sequences of clusters are strongly recommended to deduplicate sequences, to reduce taxonomic bias of the number of registered sequences, and to obtain the "effective" number of registered sequences:

```
# Cluster amplicon sequences using 100%-cutoff
vsearch \
 --notrunclabels \
 --id 1.0 \
 --strand plus \
 --threads 4 \
 --cluster_fast Lepidoptera_12S_amplicon_blastn.fasta \
 --centroids Lepidoptera_12S_amplicon_blastn_reduced100.fasta
# Cluster amplicon sequences using 95%-cutoff
vsearch \
 --notrunclabels \
 --id 0.95 \
 --strand plus \
 --threads 4 \
 --cluster_fast Lepidoptera_12S_amplicon_blastn.fasta \
 --centroids Lepidoptera_12S_amplicon_blastn_reduced95.fasta
# Cluster amplicon sequences using 90%-cutoff
vsearch \
 --notrunclabels \
 --id 0.9 \
 --strand plus \
 --threads 4 \
 --cluster_fast Lepidoptera_12S_amplicon_blastn.fasta \
 --centroids Lepidoptera_12S_amplicon_blastn_reduced90.fasta
```

Then, the number of sequence clusters is counted, and summary table is generated by the following commands:

```
# Make summary table
# Output header line
echo "primer,cutoff,seq" \
 > effectivenumbertable.txt
# p = sequence percent identity cutoff
for p in 100 95 90; do
 # Output data line
 perl maketableofeffectivenumber.pl \
  primer1 \
  $p \
  Lepidoptera_12S_amplicon_blastn_reduced$p.fasta \
  >> effectivenumbertable.txt
done
```

A figure of the number of sequence clusters, i.e., effective number of registered sequences, can be generated by the following command:

```
# Make figure
R \
 --vanilla \
 --slave \
 < makefigofeffectivenumber.R
```

An example of the figure of the effective number of registered sequences is shown in Fig. Fig. 3.3.

Fig. 3.3 An example of the figure of effective number of registered sequences

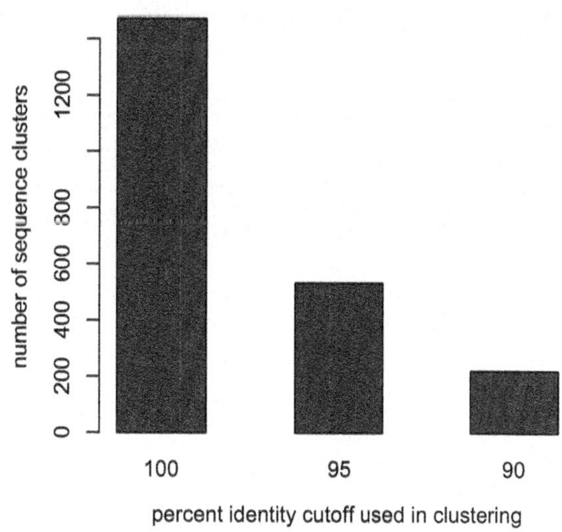

3.5.2 Estimation of Identification Power Based on Edit Distances Among Amplicon Sequences

Another important index for evaluating identification power of amplicon sequences is variability of amplicon sequences. More variable sequence should enable to distinguish closely related species more easily and clearly. To evaluate sequence variability, edit distance, i.e., the total number of substitution, insertion, and deletion, is suitable.

The following commands cluster amplicon sequences and extract centroid sequences of clusters to reduce the bias of the number of registered sequences among subtaxa:

```
# Cluster amplicon sequences using 100%-cutoff
vsearch \
 --notrunclabels \
 --id 1.0 \
 --strand plus \
 --threads 4 \
 --cluster_fast Lepidoptera_12S_fulllength_blastn_reduced100_
     stanstrand_e3.fasta \
 --centroids Lepidoptera_12S_fulllength_blastn_reduced100_
     stanstrand_e3_reduced100.fasta
# Cluster amplicon sequences using 90%-cutoff
vsearch \
 --notrunclabels \
 --id 0.9 \
 --strand plus \
 --threads 4 \
 --cluster_fast Lepidoptera_12S_fulllength_blastn_reduced100_
    stanstrand_e3.fasta \
 --centroids Lepidoptera_12S_fulllength_blastn_reduced100_
    stanstrand_e3_reduced90.fasta
```

Note that clustering amplicon sequences using 95% sequence identity cutoff has been already conducted as written before (see Sect. 3.5.1).

Then, all-against-all pairwise edit distances can be calculated by the following commands:

```
# Calculate all-against-all pairwise edit distances
# p = sequence percent identity cutoff
for p in 100 95 90; do
 perl pairdist_n.pl \
  -n=4 \
  Lepidoptera_12S_fulllength_blastn_reduced100\_stanstrand_e3_
    reduced$p.fasta \
  Lepidoptera_12S_fulllength_blastn_reduced100\_stanstrand_e3_
    reduced$p.dist
done
```

Note that this may take a while.

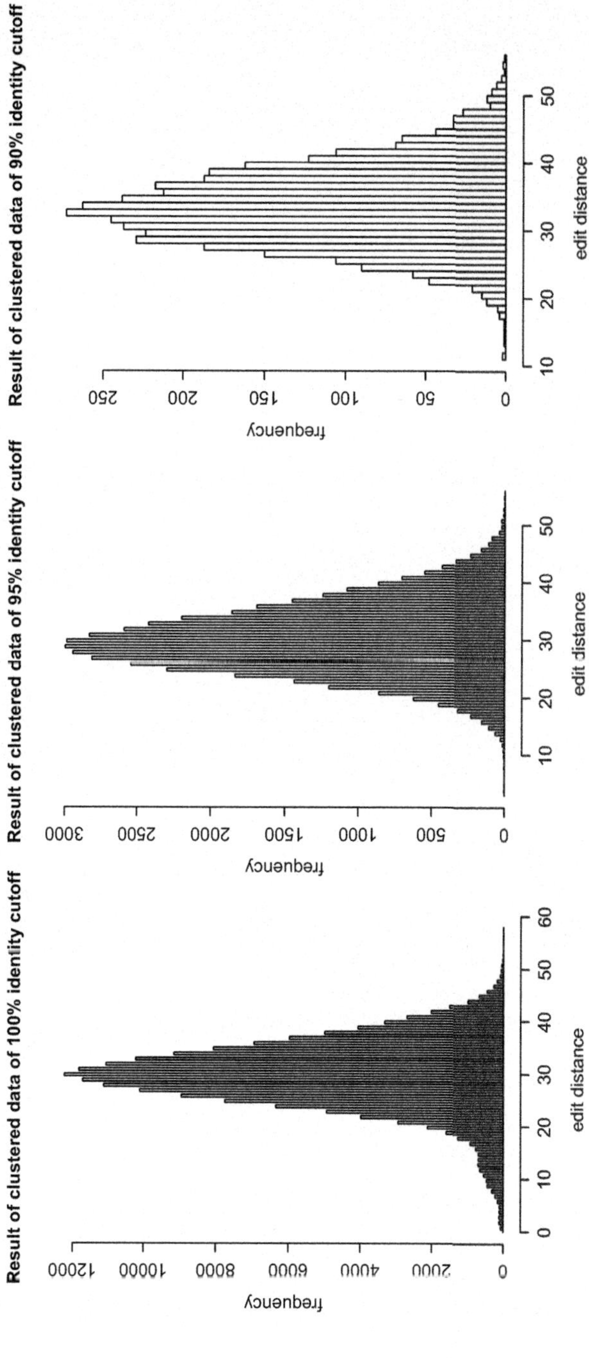

Fig. 3.4 An example of the histogram of edit distances

After calculation of edit distances, summary table can be generated like below:

```
# Make summary table
# Output header line
echo "primer,cutoff,dist" \
 > editdistancetable.txt
# p = sequence percent identity cutoff
for p in 100 95 90; do
 # Output data line
 perl maketableofeditdistance.pl \
  primer1 \
  $p \
  Lepidoptera_12S_fulllength_blastn_reduced100\_stanstrand_e3_
    reduced$p.dist \
  >> editdistancetable.txt
done
```

A histogram of edit distances can be drawn by the following command:

```
# Make figure
R \
 --vanilla \
 --slave \
 < makefigofeditdistance.R
```

An example of the histogram of edit distances is shown in Fig. 3.4.

References

Boyer F, Mercier C, Bonin A, Le Bras Y, Taberlet P, Coissac E (2016) Obitools: a unix-inspired software package for DNA metabarcoding. Mol Ecol Resour 16(1):176–182. (en). External Links: ISSN 1755-0998, Link, Document Cited by: 3.2.

Cole JR, Wang Q, Fish JA, Chai B, M D, Sun Y, Brown CT, Porras-Alfaro, Kuske CR, Tiedje JM (2014) Ribosomal Database Project: data and tools for high throughput rRNA analysis. Nucleic Acids Res 42(D1):D633–D642. External Links: ISSN 0305-1048, Link, Document Cited by: 3, 4

Ficetola GF, Coissac E, Zundel S, Tiayyba R, Shehzad W, Bessière J, Taberlet P, François P (2010) An In silico approach for the evaluation of DNA barcodes. BMC Genomics 11:434. External Links: ISSN 1471-2164, Link, Document Cited by: 4

Klindworth A, Pruesse E, Schweer T, Peplies J, Quast C, Horn M, Glöckner FO (2013) Evaluation of general 16s ribosomal RNA gene PCR primers for classical and next-generation sequencing-based diversity studies. Nucleic Acids Res 41(1):e1. External Links: ISSN 0305-1048, Link, Document Cited by: 4

Quast C, Pruesse E, Yilmaz P, Gerken J, Schweer T, Yarza P, Peplies J, Glöckner FO (2013) The SILVA ribosomal RNA gene database project: improved data processing and web-based tools. Nucleic Acids Res 41(D1):D590–D596. External Links: ISSN 0305-1048, Link, Document Cited by: 3

Riaz T, Shehzad W, Viari A, Pompanon F, Taberlet P, Coissac E (2011) ecoPrimers: inference of new DNA barcode markers from whole genome sequence analysis. Nucleic Acids Res 39(21):e145. External Links: ISSN 0305-1048, Link, Document Cited by: 3.2

Tanabe AS, Nagai S, Hida K, Yasuike M, Fujiwara A, Nakamura Y, Takano Y, Katakura S (2016) Comparative study of the validity of three regions of the 18s-rRNA gene for massively parallel sequencing-based monitoring of the planktonic eukaryote community. Mol Ecol Resour 16(2):402–414. (en). External Links: ISSN 1755-0998, Link, Document Cited by: 1

High Coverage Expression Profiling (HiCEP) of Microbial Community Genomes in the Ocean

4

Reiko Fujimura, Harunobu Yunokawa, and Koji Hamasaki

Abstract

How microbes respond to variable environmental conditions? This is a fundamental question to understand regulating mechanisms of biogeochemical cycles in the ocean. High coverage expression profiling (HiCEP) is an RNA fingerprinting method able to analyze microbial metabolic responses to various environmental conditions. Although HiCEP has been applied to various types of single organisms and revealed their responses to targeted conditions, there have been no reports on the application of this method to gene expression profiling of microbial communities. The HiCEP method can provide a unique way of analysis in metatranscriptomics. In this chapter, we provide a guidance of the HiCEP method for applying it to the omics-study of natural ecosystems. We experimentally obtained HiCEP data of prokaryotic communities in coastal surface seawater to show the feasibility of this application, indicating its methodological advantages in the analysis of metatranscriptome. In addition to high accuracy and reproducibility, this method can evaluate gene expression using either peak patterns or compositions of sequence clusters without annotation information, which enables us to highlight previously undescribed but showing distinctive expression in particular environmental conditions. The HiCEP-sequencing method can be a powerful tool for meta-omics studies in various types of environmental settings.

R. Fujimura
Department of Applied Biological Chemistry, Graduate School of Agricultural and Life Sciences, The University of Tokyo, Bunkyo-ku, Tokyo, Japan

H. Yunokawa
Messenger Scape Inc., Hachioji, Tokyo, Japan

K. Hamasaki (✉)
Atmosphere and Ocean Research Institute, The University of Tokyo, Kashiwa, Chiba, Japan
e-mail: hamasaki@aori.u-tokyo.ac.jp

T. Gojobori et al. (eds.), *Marine Metagenomics*,
https://doi.org/10.1007/978-981-13-8134-8_4

Keywords
HiCEP · Metatranscriptome · RNA fingerprinting · Microbial community

4.1 Introduction

Microbial communities play a central role in biogeochemical cycles of the ocean. Physiological reactions of microbes transform inorganic elements into organic compounds and vice versa. Their metabolic products are not only consumed for sustaining their own cells but also used for interacting with other organisms such as releasing nutrient sources, antibiotics, and signaling molecules. Furthermore, they actively release or consume climate-related gases (e.g., carbon dioxide, nitrous oxide, methane, and dimethylsulfide) in the surface ocean, suggesting the importance of marine microbes in global-scale variability in the climate.

Microbial metabolic activities are controlled by physicochemical conditions of their surrounding environments. For instance, oxygen concentration regulates the release of nitrous oxide and nitrogen dioxide in the processes of nitrification and denitrification (Bollmann and Conrad 1998). To clarify how microbes respond to variable environmental conditions is an important subject to understand the mechanisms of regulating biogeochemical cycles. The culture-based analysis is an ordinary way to study microbial activity and metabolic responses to various environmental parameters. However, it is often difficult to extrapolate those results to in situ microbial processes, because more than 90% of microbes living in nature are known to be uncultivated in vitro. It should be required to measure in situ microbial activity and metabolic responses with the use of culture-independent methodologies.

Sequence-based metagenomics provides a list of genes of a microbial community in the environment, which is a powerful culture-independent approach to predict metabolic pathways and functional potentials possessed by in situ microbial communities. Recently, gene expression profiles of microbial communities have been used to measure microbial metabolic responses to variable environmental conditions in the ocean (e.g., Frias-Lopez et al. 2008; McCarren et al. 2010; Shi et al. 2011). The expressed gene (transcriptome) analysis of a whole microbial community, so-called metatranscriptomics, supplements the information obtained by metagenomics to understand how microbial communities work and what functions are the key in ocean biogeochemical cycles.

High coverage expression profiling (HiCEP) is an RNA fingerprinting method able to analyze microbial metabolic responses to various environmental conditions. HiCEP has been developed on the basis of amplified-fragment-length polymorphism (AFLP) to analyze gene expression profiles of the organisms (Fukumura et al. 2003; Araki et al. 2006). The transcriptome fingerprinting analysis is a simple and cost-efficient way to compare the patterns of functional gene expression. The RNA fingerprinting methods such as microarray and AFLP are widely used for analyzing metabolic responses of microbial single strains or communities to the various environmental conditions among the sample sets (e.g., Booijink et al.

2010; Gunasekera et al. 2017). However, those methods have some problems on the data accuracy, sensitivity, and reproducibility. Microarray analysis is often difficult to detect low-abundant transcripts because of the background noise of fluorescence, and its reproducibility is low. It is based on hybridization reaction of RNA with designed gene probes; hence, it cannot analyze undescribed gene sequences. Also, the results of PCR-based AFLP analyses need to consider the PCR bias and mis-priming leading to the increase of the false positive rate and decrease of reproducibility. HiCEP analysis overcame these problems avoiding the use of specific primers or probes. It was succeeded to show the results with the low false positive rate (less than 4%) and differentiation of 1.2-fold difference in gene expression intensities.

In previous studies, HiCEP was applied to various types of single organisms (i.e., mouse, soybean, coral, soil microarthropod, and bacteria), and revealed their responses to targeted conditions such as exposure to radiation and symbiotic relationship (Mitani et al. 2006; Araki et al. 2006; Nakamori et al. 2008; Yuyama et al. 2011; Komatsu et al. 2015). However, there are no reports on the application of the HiCEP method to the gene expression profiling of microbial communities (i.e., metatranscriptome analysis). Currently, many studies on metatranscriptome analysis have been done by means of comprehensive RNA sequences of microbial community transcripts. The HiCEP method can provide an alternative and unique way of analysis in metatranscriptomics.

In this chapter, we provide a guidance of the HiCEP method for applying it to the omics-study of natural ecosystems. We experimentally obtained HiCEP data of prokaryotic communities in coastal surface seawater to show the feasibility of this application, indicating its methodological advantages in the analysis of metatranscriptome.

4.2 Method of HiCEP Analysis

4.2.1 Overview of the Procedures

HiCEP analysis consists of the following steps.

1. Synthesis of cDNA and fragmentation
2. Pre-PCR amplification of the fragments
3. Profiling of gene expression

The third step has two types of procedures, the gel electrophoresis method and the high-throughput sequencing method. In the former method, the gene expression profiles are analyzed by gel electrophoresis using the capillary sequencer after the selective PCR. The profiles are shown as peak patterns, and identity of each peak is determined by cloning and sequencing if necessary. In the latter method, the pre-PCR product is directly sequenced using a high-throughput sequencer and the gene expression profiles are obtained by informatic analyses. In the following sections, we briefly explain the procedures of each step. Details of original procedures and workflow were described by Fukumura et al. (2003) and patent# WO 2012/157778 A1.

4.2.2 Synthesis of cDNA and Fragmentation

Sample preparation for the cDNA fragmentation from total RNA follows the method of AFLP with some modifications. Firstly, cDNAs are synthesized using biotinylated oligo(dT) primer that fixed on avidin magnetic beads. Prokaryotic RNA needs to be treated for polyadenylation before this step (Mitani et al. 2006). A restriction enzyme, *Msp*I (or *Mse*I), is used for the first treatment of the fragmentation followed by the collection of fixed fragments on magnetic beads (Fig. 4.1a). After *Msp*I (or *Mse*I) adapters are ligated at the end of the fragments, another restriction enzyme, *Mse*I (or *Msp*I), is applied for the 2nd fragmentation and for the release of fragments from the beads. *Mse*I (or *Msp*I) adapters are ligated at the restriction site of the fragments. Those cDNA fragments are subjected to PCR to selectively amplify the fragments ligated with adapters of *Msp*I and *Mse*I at both ends.

4.2.3 Pre-PCR Amplification of the Fragments

If the quantity of template RNA was low (less than 1 μg), PCR amplification is needed to obtain enough amount of template cDNA for library preparation step (pre-PCR). PCR cycles were chosen appropriate number according to template RNA concentration.

4.2.4 Profiling of Gene Expression by the Gel Electrophoresis Method

Gel electrophoresis method by capillary sequencer provides the profiles of distribution patterns of fragment length and numbers as peak patterns (Fukumura et al. 2003). Prior to the electrophoresis, the selective PCR is required to have enough resolution for each transcript sequence. The PCR uses the adapter primers adding two nucleotides (NN, selection sequence) at the 3′ end. Owing to the combination of 4 bases (A, T, G, C) as the selection sequence, 16 (4×4) types of forward primer and 16 types of reverse primer are designed, and 256 (16×16) types of primer sets are generated. Using all combinations of primer sets, the PCR amplification is performed with small number of cycles as much as possible to avoid PCR bias. Subsequently, gel electrophoresis by capillary sequencer is performed for all PCR products, providing 256 peak profiles for each sample. Theoretically, each peak corresponds to one original transcript and the peak height reflects the abundance

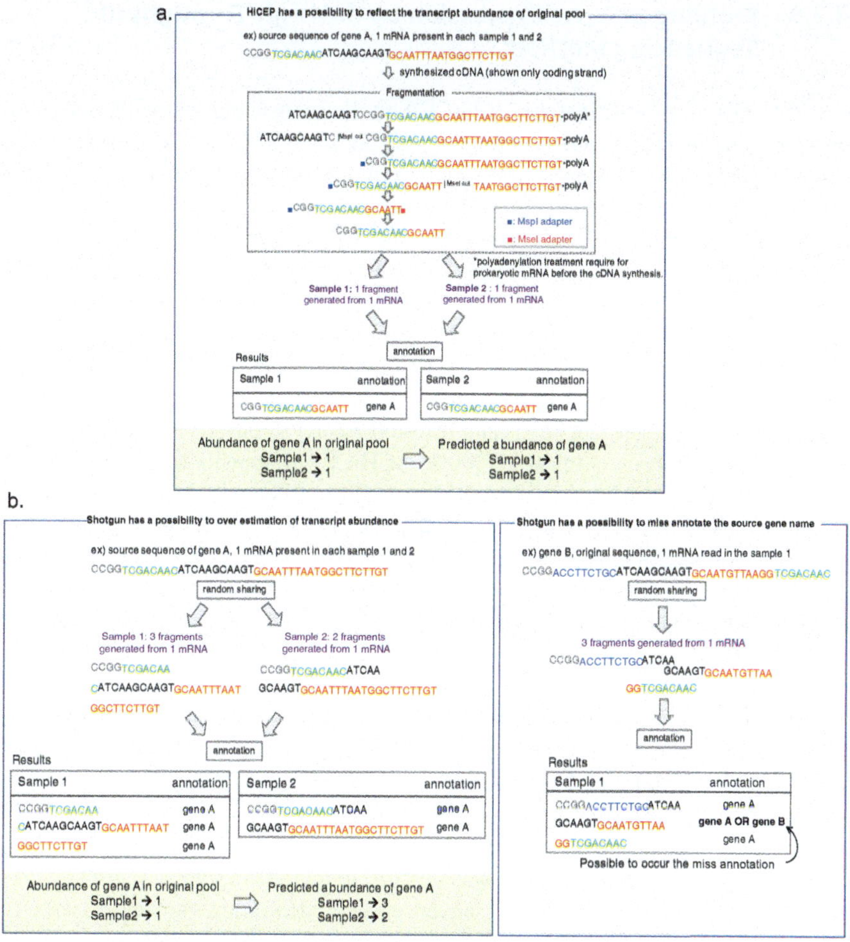

Fig. 4.1 Methodological difference between (**a**) HiCEP and (**b**) shotgun sequencing analyses on cDNA fragmentation step. (**a**)AFLP-based process is used in HiCEP. This method theoretically generates one fragment from one transcript. If same sequence with same numbers existed in sample 1 and 2, the results also show same source gene name and count of the transcript sequence read. (**b**) Several numbers of sequence fragments generated from one transcript by random sharing process. As a result, the duplicated annotation (left) or miss annotation (right) occurs when the source gene has same sequence regions with the other source genes

of each transcript in the samples. In this way, we can detect gene expression profiles by peak patterns without annotation information of each transcript. If we need annotation information of focused peaks (transcripts), we additionally sequence targeted fragments in the selective PCR products by the clone library analysis.

4.2.5 Profiling of Gene Expression by the High-Throughput Sequencing Method

Profiling of transcript fragments of HiCEP by the sequencing method is based on the clustering of sequence reads (patent# WO 2012/157778 A1). Consequently, the profiles can be shown using the information of read length and read count of each clustered sequence without the information of sequence annotation.

Products of the pre-PCR (generated in step 2) are used as the template for sequencing analysis. Sequencing analysis is performed by the Ion PGM™ sequencer (Thermo Fisher Scientific, Waltham, MA) according to the manufacturer's protocols. Our earlier experiments confirmed that Ion PGM™ was the best choice for sequencing with HiCEP among currently available sequencers. Firstly, Ion PGM™ has a potential to produce high-quality sequences around 400 bases, which satisfies the need of read length (>400 bases) to ensure the data coverage in the HiCEP analysis. Almost all (90%) of HiCEP fragments from the selective PCR are less than 400 bases, although some fragments show longer read. Secondly, frequency distribution of the fragment length found in the Ion PGM™ data indicated similar patterns to that of the electrophoresis method. Thirdly, the test by using the Illumina Hiseq sequencer (Illumina Inc., San Diego, CA) resulted in dissimilar frequency distribution of the fragment length to either the electrophoresis method or the Ion PGM™.

The next step for the profiling of the transcript pools is an informatic analysis of sequence reads. Firstly, high-quality HiCEP sequences are selected from the output sequences. Output sequences consist of three types of structure: (1) sequence reads containing the adapter sequences at both the 5′ and 3′ ends, (2) sequence reads containing the adapter sequences at either the 5′ or 3′ end, (3) sequence reads containing neither adapter sequence. Those sequence reads are screened by Cross_match (http://bozeman.mbt.washington.edu/phrap.docs/general.html) which is used for marking the adapter sequences and identifying the three types of sequence reads. The 1st type of sequence reads are classified as the high-quality sequence and used for the next clustering analysis (Fig. 4.2).

The high-quality sequence reads are assembled to generate the clusters based on the sequence length and similarity using TGICL tool (Pertea et al. 2003). The parameters for clustering were set at strict conditions: maximum length of unmatched overhangs is two bases and minimum identity for overlaps is 93%. In this step, the adapter sequences play an important role for increasing the accuracy of sequence alignment. Our previous test data showed that adapter primer sequences successfully reduced the singleton read count compared with alignment without the adapters, and 95% of the total reads were clustered.

The cluster information is used for the construction of database (HiCEP database) for the comparison of transcript profiles among samples (Fig. 4.2). The HiCEP database contains the information of cluster name, length, selection sequences, and number of reads in the cluster for each analyzed sample. Annotation information of clustered sequences are added to the database, if the sequences are identified

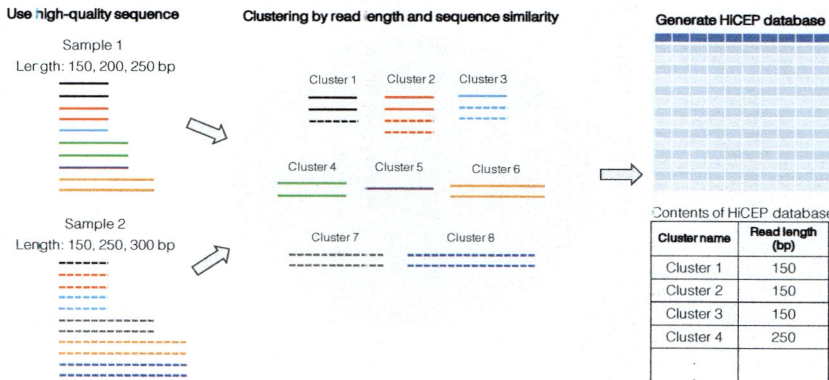

Cluster name	Read length (bp)	Selective sequence 5'	Selective sequence 3'	Read count Sample 1	Read count Sample 2	Read count Sample 3	… continue
Cluster 1	150	AA	AT	2	1	0	
Cluster 2	150	TT	GC	2	2	0	
Cluster 3	150	AA	AT	1	2	0	
Cluster 4	250	CA	AT	2	0	0	
⋮ continue							

Fig. 4.2 Workflow of data processing of HiCEP method. Figure shows an example of clustering process for two sample dataset. High-quality sequences are combined in all sample dataset, then clustered by read length and sequence similarity

by homology search in public databases. The 2nd type of the sequence reads are classified to the clusters in the HiCEP database by homology search using BLASTn program (Camacho et al. 2009). The read count of each cluster obtained from the 2nd type of the sequence reads is added to the read count of the original cluster.

4.2.6 Comparison of Two Methods for Gene Expression Profiling

The methods of gel electrophoresis and high-throughput sequencing have advantages and disadvantages. The gel electrophoresis method is a simple way to profile the transcript pools, in which gene expression is indicated by peak position (fragment size) and peak height (abundance). The peaks mostly originate from one specific gene, and thus the peak patterns can be directly used for comparing the profiles. However, this method requires intensive labors to perform the PCR using 256 primer sets and subsequent electrophoreses of 256 PCR products. By contrast, transcript profiling by the high-throughput sequencing method is technically simpler than the gel electrophoresis method. In this analysis, the selective PCR using 256 primer sets and subsequent electrophoreses are unnecessary because these steps are replaced by informatic analyses. Once we get sequences, clustering analysis shows the distribution of fragment size and abundance as gene expression profiles.

Moreover, the peak shown in the gel electrophoresis method does not always correspond to one specific gene but occasionally originates from multiple gene transcripts. If we want to identify such overlapped genes contained in one peak, we need more labors to separate those sequences by means of *E. coli* cloning or any other methods. Also, the presence of overlapped genes may be problematic when the method is applied to analyze complex microbial communities such as a metatranscriptome analysis. For instance, the result of human cells presented that 70% of peaks contained a fragment from one source gene (Fig. 4.3a). However, our experimentally obtained data of seawater microbial communities (described below) presented that most of the peaks contained fragments derived from some different source genes (Fig. 4.3b). The peak originating in one source gene accounted for only less than 20% of the total number of peaks. Furthermore, our experiment showed 42,358 peaks by the electrophoresis method, while 548,672 clusters by the sequencing method. This result suggested that the electrophoresis method detect only 7.7% of total fragments, and most of the fragments could not be distinguished by peak pattern analysis relying on only length of each fragment. The HiCEP-sequencing method relies on not only the fragment length but also the sequence identity for clustering, leading to the higher resolution of distinguishing transcript variability. Therefore, the HiCEP-sequencing method is suitable for meta-omics analysis.

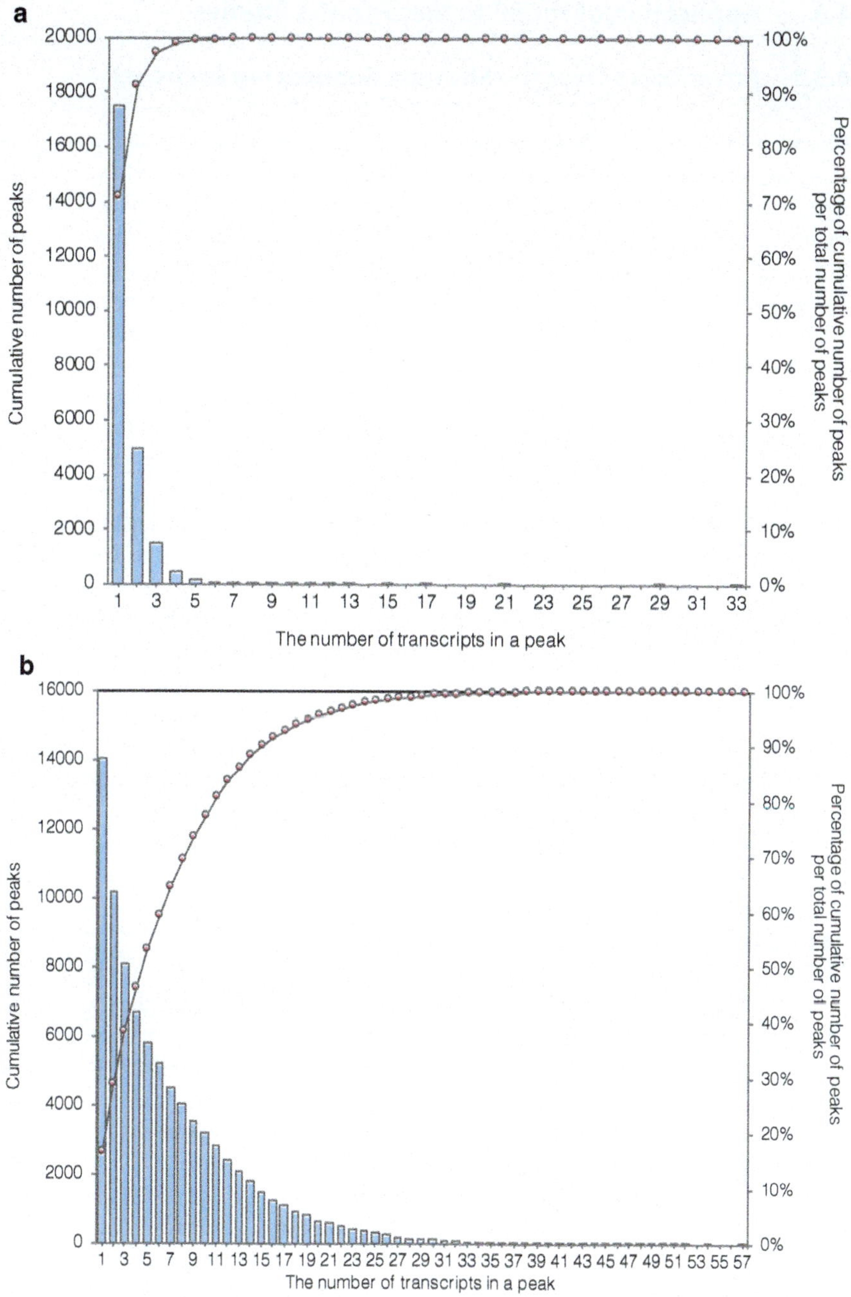

Fig. 4.3 Frequency distribution of the peak containing several number of source genes in one peak. X-axis indicates the number of source transcripts in one peak. Bar plot indicates cumulative number of peaks (first Y-axis), and circular plot indicates the percentage of cumulative number of peaks per total count of generated peaks (second Y-axis). (**a**) HiCEP clustering result of a human cell sample. (**b**) HiCEP clustering result of our test samples

4.3 Application of HiCEP to Meta-Omics Studies

4.3.1 Advantages Over the Shotgun Sequencing Approach

Recent mainstream of RNA meta-omics (metatranscriptomics) analysis uses shotgun sequencing technique with high-throughput sequencers. However, we expect that HiCEP analysis with the high-throughput sequencing method (HiCEP-sequencing) will be another useful tool for metatranscriptomic studies. In the following sections, we describe advantages of the HiCEP-sequencing over the shotgun sequencing method when we apply those methods to metatranscriptomic analysis. Furthermore, we also present the results of applying the method to surface seawater samples to show its feasibility.

An advantage of the HiCEP method over the shotgun sequencing method is high accuracy and reproducibility, because HiCEP has lower probability of methodological bias occurring in sample preparation and data processing. We expect that the transcript profiles of HiCEP are closer to actual compositions of transcript pools in the environment than those of the shotgun sequencing method. The most significant difference in sample preparation is the fragmentation step of synthesized cDNA (Fig. 4.1). Generally, total cDNA is randomly shared with physical or chemical techniques in the fragmentation procedure of the shotgun sequencing method. As a result, multiple fragments are generated from one cDNA fragment, and the longer cDNA gives more fragments (transcript length bias) (Fig. 4.1b). To avoid such bias, the sequence read counts of transcripts are generally normalized by unified numbers of read count and transcript gene length. RPKM/FPKM (reads/fragments per kilobase of gene per million reads mapped) and TPM (transcripts per million) are widely used for the calculation of mRNA abundances (Wagner et al. 2012). However, the information of gene length used for the calculation is based on gene annotations of known organisms. Hence, we should be cautious when applying the method for metatranscriptome samples which is supposed to contain many undescribed organisms.

On the other hand, the fragmentation method of HiCEP theoretically generates one cDNA fragment from one original transcript (Fig. 4.1a). Unlike the shotgun sequencing method, it is not necessarily to normalize the read count with gene length. Transcript profiles of the HiCEP method are expected to reflect the original transcript pools more precisely than the shotgun sequencing method. In fact, our preliminary experiment confirmed that relative abundance of transcripts positively correlated with the quantity data of each mRNA (data not shown).

The HiCEP-sequencing also has an advantage in reducing time of data processing for informatic analysis. Clustering and database construction take 1 day for one sample when using 12 thread CPU and 32 GB memory. By contrast, the sequence annotation of shotgun sequences of metatranscriptome analysis (e.g., 150 bp, ten million read count) may take more than a week or a month depending on the database size.

4.3.2 Potential to Discover Unknown Gene Functions

The HiCEP method has a potential to identify the genes encoding previously undescribed functions (e.g., hypothetical protein). As we mentioned above, this method can evaluate gene expression using either peak patterns or compositions of sequence clusters without annotation information. Hence, we can evaluate unknown genes by picking up transcripts which are previously undescribed but showing distinctive expression in particular environmental conditions. By contrast, shotgun sequencing method has lower resolution for profiling the transcript pools. Most of the analysis of shotgun sequences depends on annotation information of the sequence reads. In the annotation-based analysis, sequence reads are generally grouped by annotation information such as gene names or ortholog IDs. If the sequences are not annotated, it will be combined to unknown groups. Hence, annotation-based approach is difficult to follow the change of individual transcript abundance. Although genome mapping can compare gene expression patterns of individual transcripts including hypothetical proteins, it is only applicable to the organisms from which genome sequence information has already been archived.

Annotation-based analysis has another problem on low ratio of assigned sequences when applied to metatranscriptome analysis. Unfortunately, over 50% of sequence reads derived from environmental samples are annotated to hypothetical proteins or unassigned to any sequences in the databases. The sequences assigned to hypothetical proteins are not used for the prediction of transcript pools in most cases. Data analysis of shotgun sequences reportedly showed the ratios of assigned sequences were only approximately 10–50% of total read counts, although it depends on the database size and diversity of the community (Ganesh et al. 2014; Fujimura et al. 2016).

4.3.3 HiCEP of Prokaryotic Community in Coastal Seawater

We applied HiCEP-sequencing to compare gene expression patterns of prokaryotic communities in coastal surface seawater. Seawater samples (8 L) were filtered on the 0.2 μm pore-size Sterivex cartridge filter (Merck KGaA, Darmstadt, Germany), followed by total RNA extraction from the filter. Ribosomal proteins were removed from the extracts, and RNAs were polyadenylated before the cDNA synthesis. The following procedures were according to the general protocol described above. HiCEP database was constructed using information of sequences and clusters successfully obtained from our samples.

Subio platform (https://www.subioplatform.com) was used to show the overview of the difference of gene expression profiles. A comparison of gene expression profiles between two samples is shown in Fig. 4.4a. Sequence counts of each cluster from the samples set on the x and y axes were plotted to show the difference of each expression level and its distribution patterns. According to the HiCEP database, we also obtained the description of each sequence (e.g., cluster ID and read name;

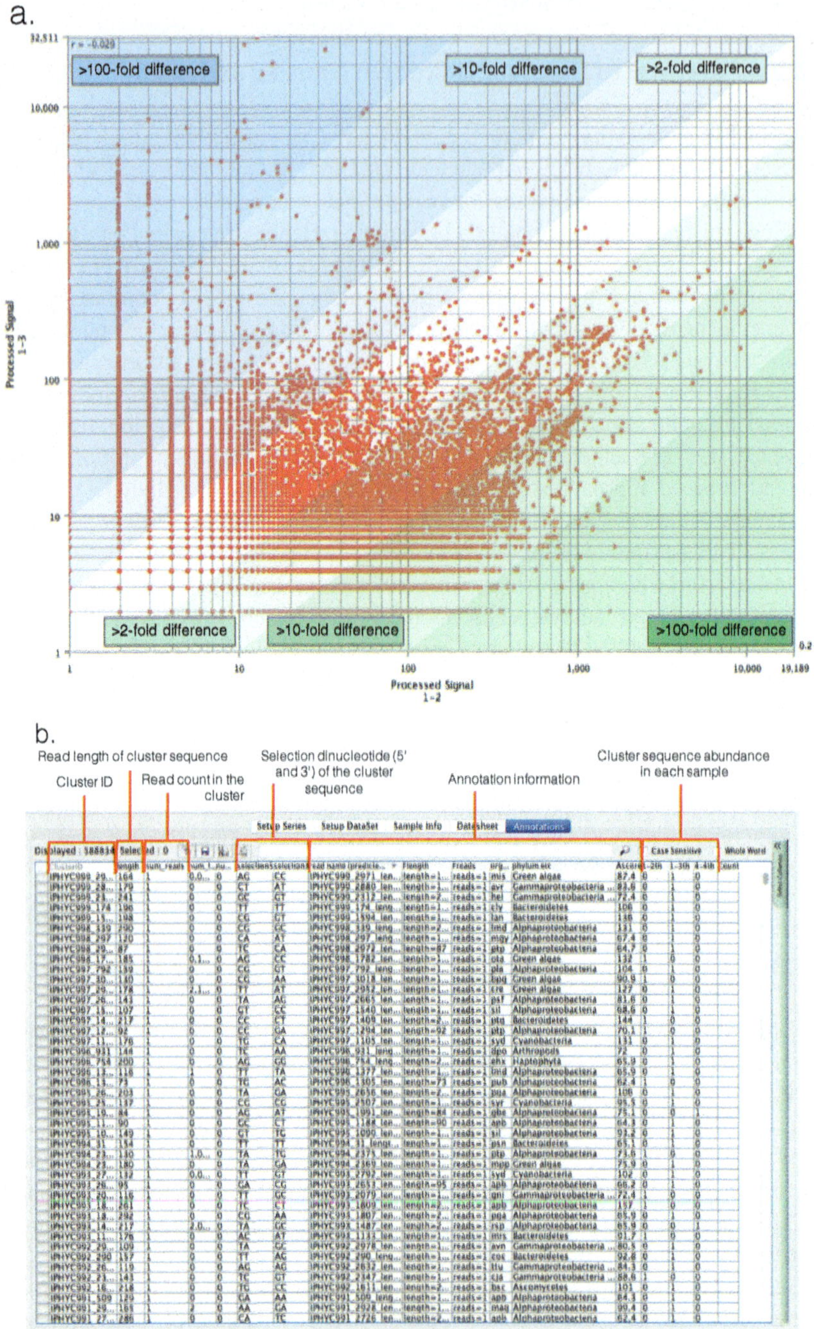

Fig. 4.4 Results of the HiCEP-sequencing analysis of coastal surface seawater samples. (**a**) Scatter plot of cluster compositions of two tested samples (1–2 and 1–3 on x- and y-axis, respectively). The number on the axes indicates the cluster size (read count in a cluster). Green and blue regions indicate abundant clusters in samples 1–2 and 1–3, respectively. Color gradient shows the degree of cluster size difference, e.g., clusters plotted in the deeper green show the 100-fold higher abundance than 1–3. (**b**) Contents of HiCEP database. Table shows the description of example dataset. Table data will be highlighted when we select the plots from the graph (**a**). The figures were created by Subio platform v.1.19

Fig. 4.4b). The gene name with the closest match and the source organism were shown based on the assignment information. For instance, when we select the plots showing 100-fold higher abundance at x-axis than that at y-axis from the graph, Subio platform highlights the sequence descriptions of these plots in the database table. Even if we don't have annotation information of some sequences in the database, we can extract them from the FASTA files of cluster sequence by using read names. The extracted sequences are annotated by homology search such as BLASTn program (Camacho et al. 2009). It is usually hard to predict the function of the sequences when they are assigned to hypothetical proteins. However, the comparison of gene expression patterns using HiCEP provides us the link between genes and their working environmental conditions, which may help us to find out key genes that have never been described previously.

4.4 Summary

Culture-independent molecular approaches are useful for illustrating prokaryotic dynamics in natural ecosystems, which is inaccessible by the monoculture experiment. For instance, gene expression profiling with time-sequential change of environmental parameters offers a deep insight into the mechanism for microbes to maintain ecosystem redundancy and robustness in the ocean. We expect that the HiCEP-sequencing method is a powerful tool for meta-omics studies in various types of environmental settings. Transcript profiling by such a database-independent method have a potential to find undescribed genetic responses, which lead to fill out the missing link of microbial processes in natural ecosystems.

Acknowledgments We are very grateful to R. Araki and M. Sunayama for HiCEP analysis and their helpful comments regarding the manuscript and data analysis. We thank S. Asakawa and E. Tan for performing IonPGM sequencing analysis. We also thank K. Kogure and the Tohoku Ecosystem Associated Marine Sciences (TEAMS) for sea water samples obtained on the research cruise. Additionally, we thank the captains, crew members, and participants on the R/V Daisankaiyo-maru cruises for their cooperation. Computational analyses were partially performed by maze Inc. The Japan Science and Technology Agency (CREST) and JST-SENTAN Program supported this research.

References

Araki R, Fukumura R, Sasaki N et al (2006) More than 40,000 transcripts, Including novel and noncoding transcripts, in mouse embryonic stem cells. Stem Cells 24:2522–2528. https://doi.org/10.1634/stemcells.2006-0005

Bollmann A, Conrad R (1998) Influence of O_2 availability on NO and N_2O release by nitrification and denitrification in soils. Glob Chang Biol 4:387–396. https://doi.org/10.1046/j.1365-2486.1998.00161.x

Booijink CCGM, Boekhorst J, Zoetendal EG et al (2010) Metatranscriptome analysis of the human fecal microbiota reveals subject-specific expression profiles, with genes encoding proteins involved in carbohydrate metabolism being dominantly expressed. Appl Environ Microbiol 76:5533–5540. https://doi.org/10.1128/AEM.00502-10

Camacho C, Coulouris G, Avagyan V et al (2009) BLAST+: architecture and applications. BMC Bioinformatics 10:421. https://doi.org/10.1186/1471-2105-10-421

Frias-Lopez J, Shi Y, Tyson GW et al (2008) Microbial community gene expression in ocean surface waters. Proc Natl Acad Sci U S A 105:3805–3810. https://doi.org/10.1073/pnas.0708897105

Fujimura R, Kim S-W, Sato Y et al (2016) Unique pioneer microbial communities exposed to volcanic sulfur dioxide. Sci Rep 6:19687. https://doi.org/10.1038/srep19687

Fukumura R, Takahashi H, Saito T et al (2003) A sensitive transcriptome analysis method that can detect unknown transcripts. Nucleic Acids Res 31:e94

Ganesh S, Parris DJ, DeLong EF, Stewart FJ (2014) Metagenomic analysis of size-fractionated picoplankton in a marine oxygen minimum zone. ISME J 8:187–211. https://doi.org/10.1038/ismej.2013.144

Gunasekera TS, Bowen LL, Zhou CE et al (2017) Transcriptomic analyses elucidate adaptive differences of closely related strains of Pseudomonas aeruginosa in fuel. Appl Environ Microbiol 83:e03249–e03216. https://doi.org/10.1128/AEM.03249-16

Komatsu S, Sakata K, Nanjo Y (2015) ' Omics ' techniques and their use to identify how soybean responds to flooding. J Anal Sci Technol 6:1–8. https://doi.org/10.1186/s40543-015-0052-7

McCarren J, Becker JW, Repeta DJ et al (2010) Microbial community transcriptomes reveal microbes and metabolic pathways associated with dissolved organic matter turnover in the sea. Proc Natl Acad Sci U S A 107:16420–16427. https://doi.org/10.1073/pnas.1010732107

Mitani Y, Suzuki K, Kondo K et al (2006) Gene expression analysis using a modified HiCEP method applicable to prokaryotes: a study of the response of Rhodococcus to isoniazid and ethambutol. J Biotechnol 123:259–272. https://doi.org/10.1016/j.jbiotec.2005.11.004

Nakamori T, Fujimori A, Kinoshita K et al (2008) Application of HiCEP to screening of radiation stress-responsive genes in the soil microarthropod Folsomia candida (Collembola). Environ Sci Technol 42:6997–7002

Pertea G, Huang X, Liang F et al (2003) TIGR gene indices clustering tools (TGICL): a software system for fast clustering of large EST datasets. Bioinformatics 19:651–652. https://doi.org/10.1093/bioinformatics/btg034

Shi Y, Tyson GW, Eppley JM, DeLong EF (2011) Integrated metatranscriptomic and metagenomic analyses of stratified microbial assemblages in the open ocean. ISME J 5:999–1013. https://doi.org/10.1038/ismej.2010.189

Wagner GP, Kin K, Lynch VJ (2012) Measurement of mRNA abundance using RNA-seq data: RPKM measure is inconsistent among samples. Theory Biosci 131:281–285. https://doi.org/10.1007/s12064-012-0162-3

Yuyama I, Watanabe T, Takei Y (2011) Profiling differential gene expression of symbiotic and aposymbiotic corals using a high coverage gene expression profiling (HiCEP) analysis. Mar Biotechnol 13:32–40. https://doi.org/10.1007/s10126-010-9265-3

Part II

Technological Aspects of Marine Metagenomics: Metagenome Data Analysis

The Application of DDCA to Metagenomic Analysis

5

Kazutoshi Yoshitake, Kyohei Matsuno, and Atsumi Tsujimoto

Abstract

With the advent of the next-generation sequencer (NGS), sequencing costs have dramatically declined, but data analysis still costs as much as before or more. One of the reasons is that for NGS data analysis, it is necessary to key in commands on Linux which is a less common operating system. Therefore, we have developed the digital DNA chip analysis (DDCA) system which processes NGS data to produce results similar to microarray that are familiar to wet researchers. With this system, it is not necessary to have a high-performance computer in the laboratory, and analysis can be performed by using simple mouse operations on a web browser. After uploading sequence data to our server, users can click on some options to perform an analysis called digital hybridization on DDCA. In this paper, after describing the DDCA system, we introduce the procedure to analyze metagenome data using actual sequence data.

Keywords

DDCA · Digital hybridization · Digital DNA chip · Visualization

With the advent of the next-generation sequencer (NGS), sequencing costs have dramatically declined, but data analysis still costs as much as before or more. One of the reasons is that for NGS data analysis, it is necessary to key in commands on Linux which is a less common operating system. It is difficult for wet researchers who are not specialized in bioinformatics to input commands, so it is inevitable that data analysis is outsourced. Another reason is that according to Moore's Law, sequencers have evolved faster than computers. As a result, a single computer has

K. Yoshitake · K. Matsuno · A. Tsujimoto (✉)
Japan Software Management Co., Ltd., Kanagawa-ku Yokohama-shi, Kanagawa, Japan
e-mail: tsujimoto@jsm.co.jp

T. Gojobori et al. (eds.), *Marine Metagenomics*,
https://doi.org/10.1007/978-981-13-8134-8_5

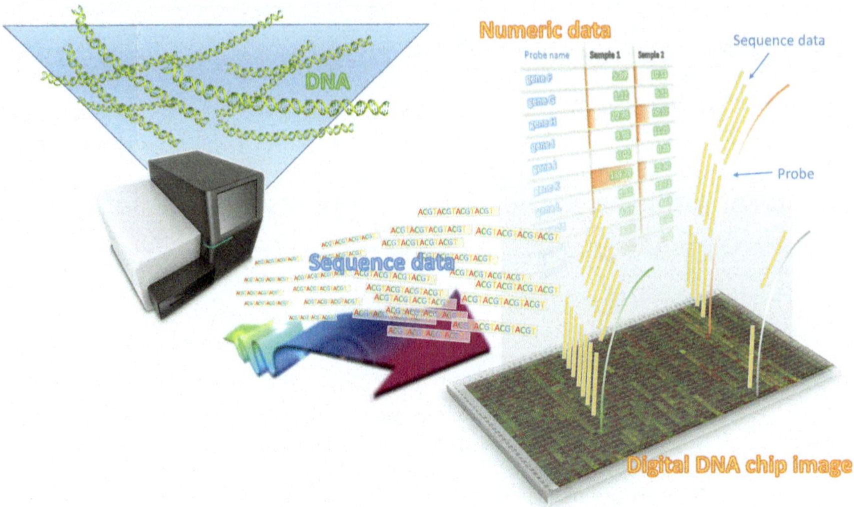

Fig. 5.1 Image of DDCA
Analysis results like microarray can be obtained from NGS data

insufficient processing power, and large-scale parallel computing is necessary. Thus, it is becoming more difficult for a laboratory to operate an NGS data analysis computer.

Therefore, we have developed the digital DNA chip analysis (DDCA) system which processes NGS data to produce results similar to microarray that are familiar to wet researchers. With this system, it is not necessary to have a high-performance computer in the laboratory, and analysis can be performed by using simple mouse operations on a web browser. After uploading sequence data to our server, users can click on some options to perform an analysis called digital hybridization on DDCA. Digital hybridization is an analysis tool that virtualizes hybridization on a computer and is a tool to virtually hybridize sequence data to a digital DNA chip created on a computer. In digital hybridization, a base sequence highly homologous to the probe sequence on the digital DNA chip is virtually hybridized to the probe sequence (Fig. 5.1). As a result of digital hybridization, it is possible to obtain graphical digital DNA chip images similar to microarrays and hybridized numerical data.

The advantages of visualizing data in a digital DNA chip is that even if there are tens of thousands of reference sequences, it is possible to visualize abundance or expression level at once and is suitable to have an overview of large amount of data. Furthermore, in situations like case control study, it is possible to display results similar to two-color microarrays which is easy to understand.

Henceforth, this document will explain how to operate the DDCA, and then examples of analysis using actual sequence data will be provided.

5.1 How to Use DDCA

Although DDCA can be used in various ways depending on the design of the digital DNA chip, it can be easily utilized for data analysis in fields where microarray is used (e.g., analysis of expression level of transcriptome and exhaustive detection of microorganisms by metagenome). As a quintessential usage of DDCA, a procedure for digitally hybridizing the sequence data of RNA-seq to the human cDNA chip prepared on DDCA and calculation of gene expression level are introduced.

5.1.1 Login

Go to http://bio.jsm.co.jp/ddca and log in. General users who plan to use public data can use the public user account: login ID, publicuser, and password, publicuser (Fig. 5.2).

If you plan to use private data, please contact bio@jsm.co.jp to request for a DDCA account. An individual account will be issued, and files uploaded by you will remain private.

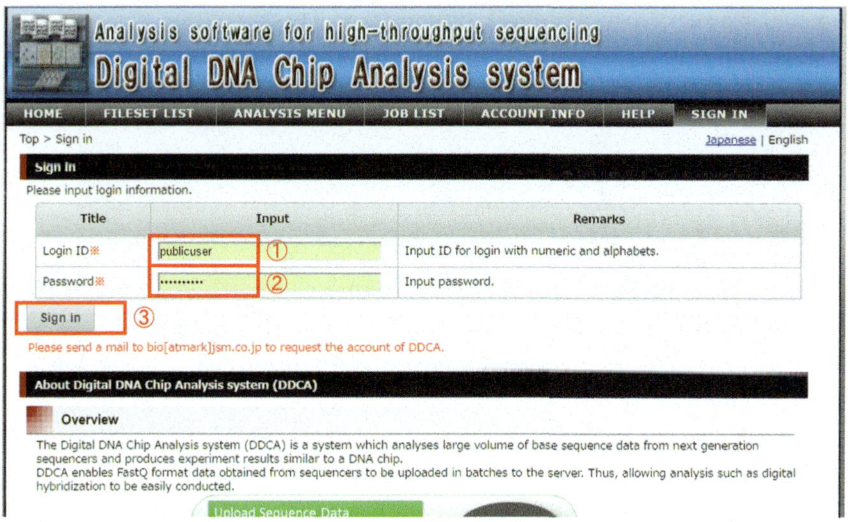

Fig. 5.2 DDCA login page
Go to http://bio.jsm.co.jp/ddca to log in. Please fill (*1*) Login ID and (*2*) Password, and then click (*3*) "Sign in" button
If you use public data, you can login with account ID, publicuser, and password, publicuser

5.1.2 File Upload

Once you are logged in, first, upload the sequence data to the server. At this time, it is possible to upload multiple sequence data at once. The procedure is as follows: from the menu bar, select the "FILESET LIST," and then select the "File Upload." After that, a Windows user can upload a file to the server by opening Explorer and dragging and dropping the paired-end FASTQ files to the web browser (Fig. 5.3). A Mac user will open Finder instead of Explorer. If drag and drop does not work, you can click "Select File", open a file selection dialog, and select a FASTQ file.

If appropriate public data is not available, you may use the public data of RNA-seq shown below. (In the case of paired-end FASTQ format, two sequence data files for forward and reverse reads are required.)

ftp://ftp.ddbj.nig.ac.jp/ddbj_database/dra/fastq/SRA040/SRA040091/SRX083302/SRR309282_1.fastq.bz2
ftp://ftp.ddbj.nig.ac.jp/ddbj_database/dra/fastq/SRA040/SRA040091/SRX083302/SRR309282_2.fastq.bz2

The above files are compressed to bz2 format. Windows users should download and install a software such as lhaplus to decompress the bz2 files. If you are a Mac user, double click on the downloaded file to decompress them.

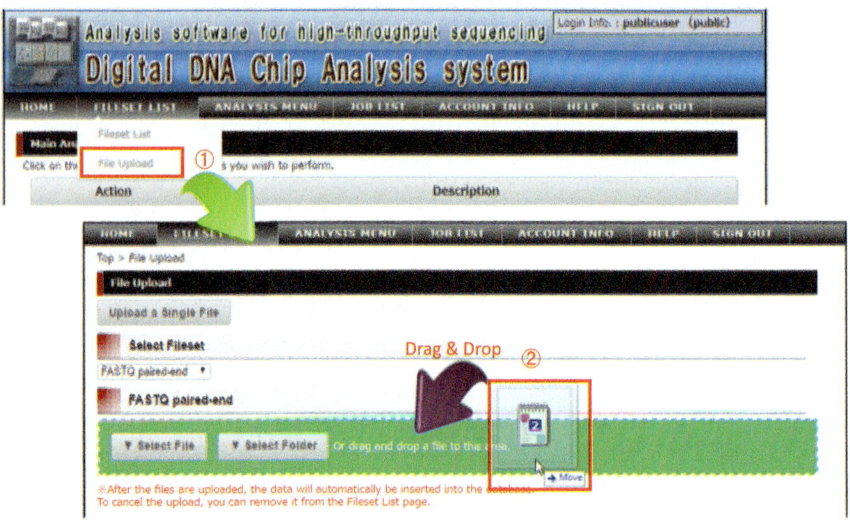

Fig. 5.3 File upload page
Multiple paired-end FASTQ can be easily uploaded at once by clicking (*1*) "File Upload" from the menu and moving to the file upload page and dragging and dropping the files to the web browser (*2*)

5.1.3 Performing Digital Hybridization

In order to perform digital hybridization, it is necessary to select the fileset to be used for analysis. Click "FILESET LIST" from the menu page to display the fileset list page (Fig. 5.4). Select the fileset uploaded earlier from among the displayed filesets. Select the human cDNA sequence Fileset ID: 30357 "Homo_sapiens.GRCh38.cdna.all.fasta" as a reference sequence. To use another reference sequence, you can upload the file in FASTA format in advance and select the uploaded FASTA format fileset (concrete example will be provided later). Next, click "Select Fileset for Analysis" to go to the next page. Then, go to the "Analysis Menu" page and the selected fileset will be displayed. After confirming the contents, click the "Execute" button next to "DDCA hybridization (v2.0)." Check the values of the options on the "Option List" page, and change it as necessary. When you click the "Execute Analysis" button, the job will be registered automatically in the analysis server. When the registration is completed, you will be redirected to the job list page (Fig. 5.5). After this, it takes a while for the job to complete, so you may close the web browser.

To view the progress of the job, click "JOB LIST" from the menu bar. A list of the jobs will be displayed, and you should click the ID of the job to view the logs. Next, the detailed information page of the job will be opened, and it is possible to see the progress status and analysis log.

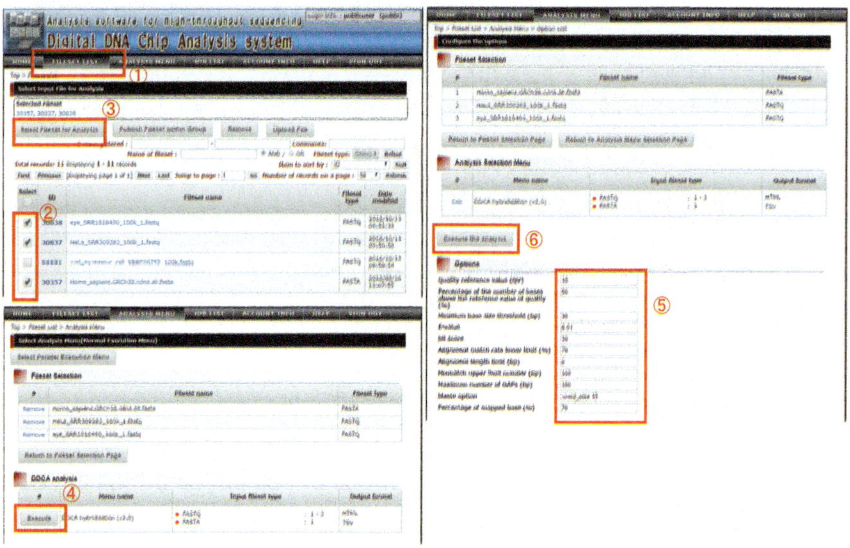

Fig. 5.4 Analysis execution procedure
(*1*) Click "FILESET LIST" from the menu to display the file set list page. (*2*) Select the file set to be used for analysis from the displayed file set. (*3*) Click "Select Fileset for Analysis" to proceed to the next page. (*4*) Click the "Execute" button next to "DDCA hybridization (v2.0)." (*5*) Check options. (*6*) Click "Execute Analysis" to execute analysis

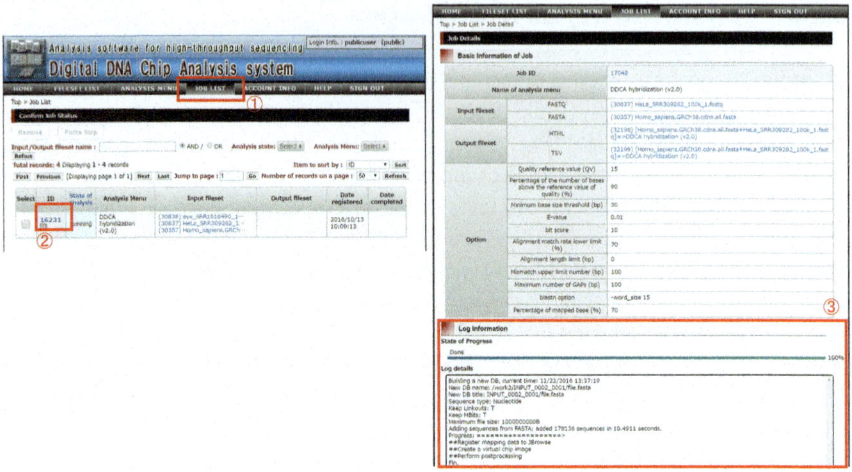

Fig. 5.5 Execution status check
(*1*) Click "JOB LIST" from the menu to display the job list page. (*2*) Select the job you want to view logs from among the displayed jobs. (*3*) It is possible to see progress status and logs in "Log Information"

5.1.4 Viewing Results of the Digital Hybridization Analysis

To view the analysis result, click on "JOB LIST" from the menu bar to display the job list page. Once the job has completed, the Analysis state will indicate "Finished," and the fileset name will be displayed in the column of the output fileset of the job. Then click on the fileset name to view the detailed information page of the output fileset (Fig. 5.6). Clicking the html file name in the detailed information page will display the digital DNA chip image as a result of the digital hybridization.

First, an image similar to a microarray is displayed, and it is possible to have an overview of the abundance of each spot by looking at the intensity of the spots. With your mouse cursor hovering over the digital DNA chip, scroll up or down to zoom in or out (Fig. 5.7). Furthermore, by dragging and dropping on the digital DNA chip, it is possible to move the display range of the digital DNA chip. In Fig. 5.7, sample 1 is displayed in red, and sample 2 is displayed in green. When you zoom in to the digital DNA chip with your mouse, information on where the sequence data is hybridized in the probe is displayed in a vertical line. Sample 1 is mRNA RNA-seq and sample 2 is SAGE data. You can verify that sample 1 is hybridized over the entire length of the probe and sample 2 is hybridized only to the end of the probe. Furthermore, double clicking on the spots will display the hybridization result for each base on the JBrowse genome browser (Skinner et al. 2009).

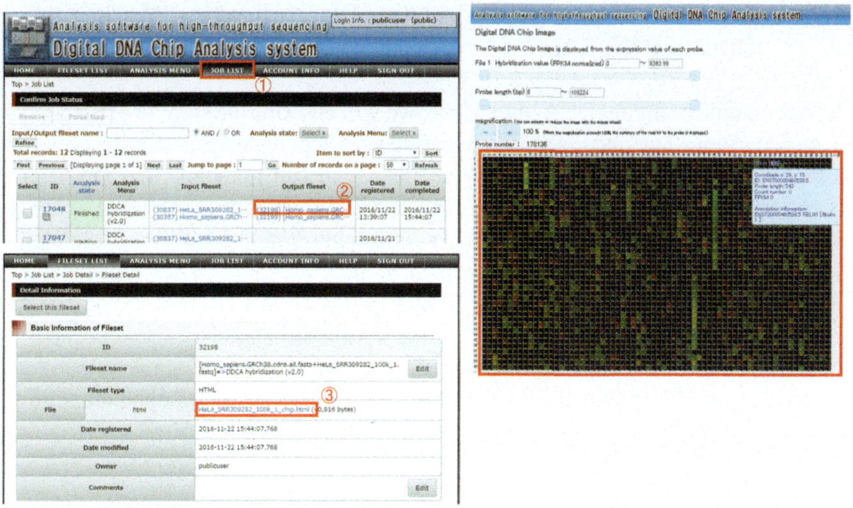

Fig. 5.6 Display analysis result
(*1*) Click "JOB LIST" from the menu to display the job list page. (*2*) When the job is finished, the output file set is displayed, and when clicking it, the detailed information page is displayed. (*3*) By clicking on the analysis result file, the result can be downloaded or displayed on the browser. (*4*) Mouse scrolling on the digital DNA chip enables enlargement and reduction

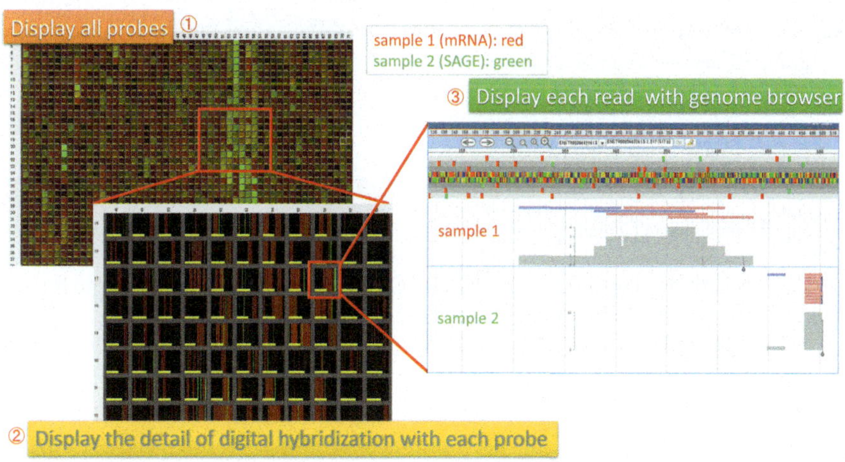

Fig. 5.7 Digital DNA chip operation
(*1*) Display screen with magnification factor of 100%. It is possible to view all probes in bird's eye view. (*2*) Display screen with magnification factor of 200%. Multiple probes can be browsed at the same time while checking the hybridization position for each probe. A yellow bar shows probe length. (*3*) By double clicking on the probe, the genome browser is activated, and detailed hybridization results can be viewed

5.2 Example of Using Fish Species Discrimination Marker as Digital DNA Chip

As an example of digital DNA chip usage, we will describe the steps to extract DNA from aquarium water and discriminate fish species kept in an aquarium. As a species discrimination marker of fish, rDNA sequences in fish mitochondrial DNA have been published (Iwasaki et al. 2013). To use this mitochondrial sequence as a digital DNA chip, download mitogenomes.zip from http://mitofish.aori.u-tokyo.ac.jp/. Since this file is compressed into zip, it is necessary to decompress the data. Thousands of FASTA files are created when you unzip the zip file, so you need to combine them into one file. If you are using Windows, open the extracted folder in Explorer, click "File" from Explorer menu, and then click "Open command prompt" to open command prompt. Then, enter the following command to combine all files into one file (mitogenomes.fasta).

```
copy /b *.fa mitogenomes.fasta
```

If you are using Mac, open a terminal, navigate to the directory you have extracted the files, and enter the following command to merge all the files into one file.

```
cat *.fa > mitogenomes.fasta
```

Next, we use the data collected from Kuroshio Tank of Churaumi Aquarium as metagenomic data we want to distinguish (Miya et al. 2015). In the paper, multiple samples are used, but we will just download one of the paired-end FASTQ sequence data, DRR030411. Data can be downloaded from the following URLs.

ftp://ftp.ddbj.nig.ac.jp/ddbj_database/dra/fastq/DRA003/DRA003066/DRX027427/DRR030411_1.fastq.bz2
ftp://ftp.ddbj.nig.ac.jp/ddbj_database/dra/fastq/DRA003/DRA003066/DRX027427/DRR030411_2.fastq.bz2

Since the above files are compressed bz2 files, it is necessary to decompress them using software such as lhaplus as described above.

Next, upload both the FASTA file and the FASTQ files prepared above to the DDCA. The FASTQ files "DRR030411_1.fastq" and "DRR030411_2.fastq" are uploaded as FASTQ (paired-end) in the file upload procedure introduced earlier. When you upload the FASTA file, "mitogenomes.fasta," select "FASTA" as the fileset format.

When the file is uploaded, there are two filesets, "DRR030411" and "mitogenomes.fasta" in the file set list page. Select these two filesets and perform digital hybridization as described above.

The analysis will be completed after some time. If you open the HTML result file, you will see the digital DNA chip image as in Fig. 5.8. DRR030411 is metagenomic data taken from Kuroshio Tank of Churaumi Aquarium. It was shown that the Kuroshio water tank of the Churaumi Aquarium is dominated by the DNA

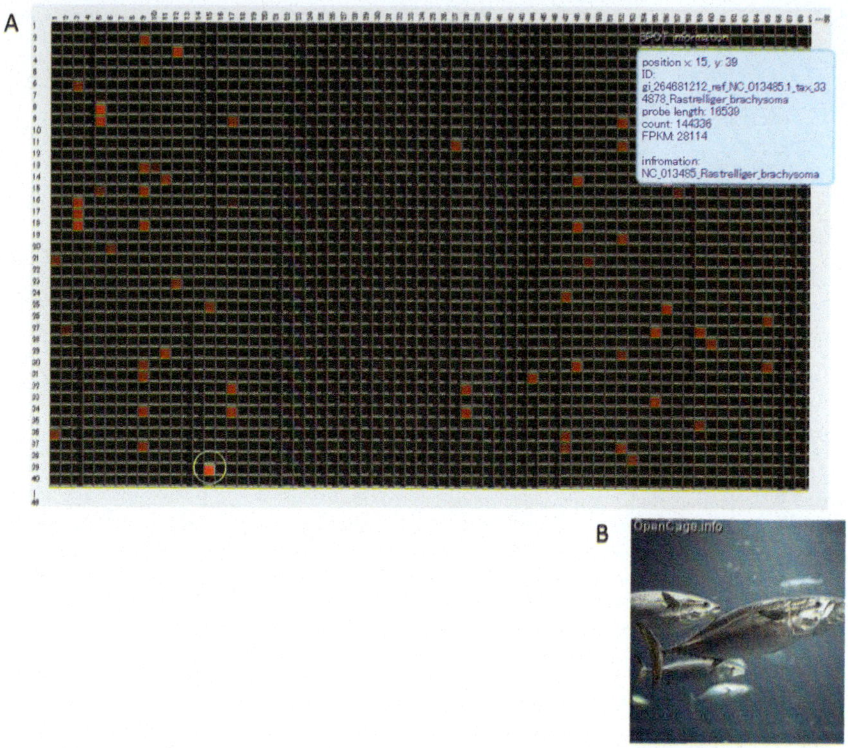

Fig. 5.8 Identification of fish in aquarium
(**a**) In the Kuroshio Tank of the Churaumi Aquarium, the most common DNA of fish is that of *Rastrelliger brachysoma* (the probe of the *Rastrelliger brachysoma* surrounded by a yellow circle is displayed brightest.) (**b**) The image of *Rastrelliger brachysoma*

of a fish called *Rastrelliger brachysoma* (Miya et al. 2015). The DNA marker of *Rastrelliger brachysoma* was shown to be the brightest even in the result of the digital hybridization (Fig. 5.8), and it could be confirmed that *Rastrelliger brachysoma* is a dominant species.

5.3 Extraction and Visualization of Red Tide Predictive Marker by Shotgun Metagenome Analysis

Red tide has caused major damage to the aquaculture industry in Japan. By using the digital DNA chip, we thought that we could warn the fishermen before the red tide occurs. Therefore we extracted DNA markers which increased specifically before red tide from the shotgun metagenome sequence data of seawater and visualized them with digital DNA chips (Fig. 5.9a). Shotgun metagenome data of Ariake Seawater in Japan from 2012 to 2013 was used as data. Ariake Sea is a sea area

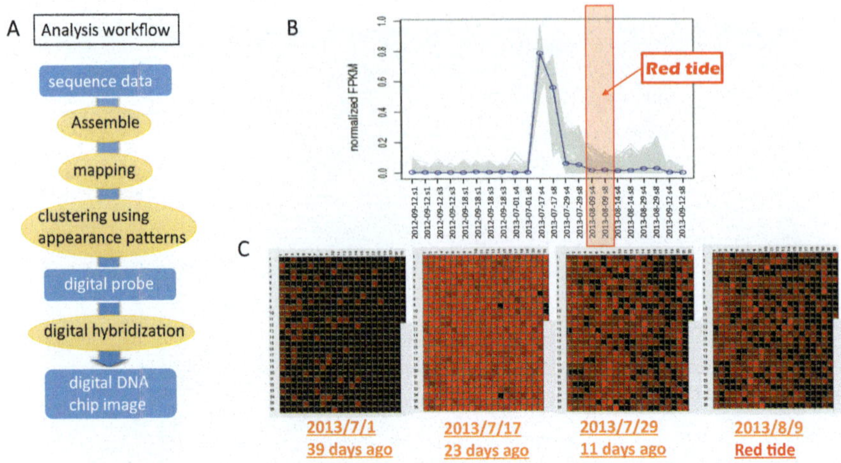

Fig. 5.9 Detection of increasing sequence before red tide
(**a**) Analysis flow for creating a digital DNA chip. (**b**) Appearance pattern of 557 DNA sequences specifically increased before the red tide that occurred on August 9, 2013, in Ariake Sea. (**c**) Digital DNA chip image created using DNA probe increased before red tide occurrence

where red tide occurs frequently in summer, and it occurred in both 2012 and 2013. We sampled sea water about once a week in the summer of this period, extracted DNA from size fraction of 0.2–1 μm, and obtained shotgun metagenome data. After assembling according to the analysis flow shown in Fig. 5.9a, clustering was carried out so that contigs having similar appearance patterns with a correlation coefficient of 0.95 or more were clustered into one group (Nielsen et al. 2014). Analysis of clusters that increased by 10 times or more before the red tide occurs revealed the presence of clusters that increased specifically 3 weeks before the red tide occurred (Fig. 5.9b). From the result of DDCA visualization of the amount of contig present in the cluster, it was confirmed that abundance abruptly increased before red tide occurrence (Fig. 5.9c). It was confirmed that the increased contigs belong to the cluster of Synechococcus phage.

5.4 Conclusion: Advantages of DDCA

We have designed a friendly GUI that is easy for wet researchers to use, and microarray-like results can be obtained from NGS data with a few clicks of a mouse.

It is possible to have an overview of the data with microarray-like images in digital DNA chips, compare two colors, and view the detailed mapping result of the reads.

And then expression level information can be downloaded in Excel file format.

References

Iwasaki W et al (2013 Nov) Mol Biol Evol 30(11):2531–2540. https://doi.org/10.1093/molbev/mst141

Miya M et al (2015) R Soc Open Sci 2(7):150088. https://doi.org/10.1098/rsos.150088

Nielsen HB et al (2014) Nat Biotechnol 32(8):822–828. https://doi.org/10.1038/nbt.2939

Skinner ME et al (2009) Genome Res 19(9):1630–1638. https://doi.org/10.1101/gr.094607.109

Horizontal Gene Transfer in Marine Environment: A Technical Perspective on Metagenomics

6

Yoji Nakamura

Abstract

In the environment, unicellular organisms such as prokaryotes are exposed to direct invasion of viruses and its consequent transduction. In addition, some of the prokaryotic species can uptake naked DNA molecules outside or transfer their own DNA to other species through conjugative plasmids. Hence, prokaryotic genomes could be often mosaic: they may have the extrinsic genes which are not vertically transmitted from the ancestor but horizontally transferred from other organisms. Such a phenomenon, namely, "horizontal (or lateral) gene transfer," is the main issue of this chapter. Horizontal gene transfer can rapidly cause genotypic/phenotypic changes in the recipient organisms, apparently beyond the theory of traditional population genetics based on mutation. Thus, it has been considered that horizontal gene transfer has influenced very much on the evolution of prokaryotes. In response to the accumulation of genomic data, the amount of horizontally transferred genes has been estimated at the large scale, but the significance of horizontal gene transfer in real environment has not been fully assessed. How often does horizontal gene transfer occur among taxa? How much does it affect the gene pool in environment? The challenging studies have just started. Metagenomic approaches have a great potential for this purpose, but many methodological limitations for treating the data remains unsolved. In this chapter, traditional genomics methods for estimating horizontally transferred genes are first reviewed. In the latter part, technical perspectives on prediction of horizontal gene transfer from the metagenomics data are discussed.

Y. Nakamura (✉)
National Research Institute of Fisheries Science, Japan Fisheries Research and Education Agency, Yokohama, Kanagawa, Japan
e-mail: yojnakam@affrc.go.jp

T. Gojobori et al. (eds.), *Marine Metagenomics*,
https://doi.org/10.1007/978-981-13-8134-8_6

Keywords
Horizontal gene transfer · Phylogenetic tree · Nucleotide composition · Chimeric sequence · Mobilome

6.1 Introduction

6.1.1 Horizontal Gene Transfer

In principle, the genetic material of organism is inherited from one generation to the next within the species population. In some cases, however, the genetic material can be naturally introgressed from another species. Such a phenomenon is called horizontal transfer or lateral transfer, and particularly when the genetic material encodes genes, it is called *horizontal gene transfer* (*HGT*) or *lateral gene transfer* (*LGT*). The gene transferred is called *xenologous gene* or *xenolog* (Koonin et al. 2001) to the vertically transmitted homologs in the recipient species. HGT is often observed in prokaryotes, because the prokaryotic chromosome is not packed by nucleus in the cell, and exogenous DNA is relatively easily integrated into the genomic DNA. The exogenous DNA is derived from another organism in the environment and can invade the recipient prokaryotic cell in several ways (see Sect. 6.1.2). If any gene is encoded in the DNA, a new allele may occur in the recipient. Particularly, prokaryotes are unicellular, and hence the newly introgressed allele is directly inherited to the next generation. In addition, if the new allele is advantageous for the recipient, it is possibly spread and fixed in the population. Contrastingly, multicellular eukaryotes, particularly higher ones, partition the inheritable genetic materials into germline cells, which are in general isolated from the environment. Thus, HGT in eukaryotes is considered to be relatively rare, although some cases of HGT have been reported (Nikoh and Nakabachi 2009). Exceptionally, HGT from organelle (e.g., mitochondrion or chloroplast) to nuclear chromosome has often occurred in eukaryotes. This may be a consequence of ancient endosymbiosis between the ancestral host and organelle, where the donor and recipient genomes have long been coexistent in the cells. It is widely believed that horizontal gene transfer is likely to occur between partner organisms in symbiosis or between organisms in the same environmental niche (Mongodin et al. 2005).

Much attention has been paid to the evolutionary significance of HGT (Bushman 2001). In prokaryotes, a well-known case is the relationship with evolution of pathogenicity. The virulence-related genes are often of xenologous origin in pathogenic bacteria. It is known that some serotypes of *Vibrio cholerae*, a Gram-negative aquatic bacterium, have caused pandemics in the past. They carry a toxin, named "cholera toxin," which is encoded by two genes corresponding to six subunits (an A subunit and five B subunits). These genes are located in a temperate bacteriophage (i.e., a virus infecting bacteria) region in the *V. cholerae* genome (Waldor and Mekalanos 1996), indicating that such genes have been originally present in another organism and carried through the virus which had been released from the organism (see Sect. 6.1.2). The virulence-related genes are often adjacently located as a cluster in the genome, and it is called **pathogenicity island**. The pathogenicity islands are often horizontally transferred among species (Hacker et al. 1997). Enteropathogenic *Escherichia coli* has a locus for enterocyte effacement (LEE) required for invasion into host animal cells (McDaniel et al. 1995). The LEE is xenologous in virulent *E. coli* like cholera toxin genes in *V. cholerae*. Recently, it was reported that a fish pathogenic aquatic bacterium, *Edwardsiella tarda*, possessed homologous genes to those in LEE (Nakamura et al. 2013). In that study, nucleotide and phylogenetic analysis suggested that a direct or indirect flow of LEE genes has occurred between fish and zoological pathogens. In marine environments, it is also known that gene flows between cyanobacteria and bacteriophages have often occurred. Cyanobacteria are photosynthetic bacteria and play a crucial role for the production of oxygen in the oceans as well as photosynthetic picoeukaryotes. Cyanophages are the bacteriophages that infect cyanobacteria, and the virulent host ranges can affect the population sizes of cyanobacterium. Photosynthesis genes in cyanobacteria are often found in such cyanophage genomes (Lindell et al. 2004; Mann et al. 2003). Therefore, these genes have been transferred among cyanobacteria mediated through cyanophages, suggesting that the host-phage gene flow influences on photosynthetic metabolism in the marine environment.

Thus, HGT is now considered one of the driving forces of evolution in prokaryotes. In particular, HGT allows gene pools to be formed beyond species barrier in real environments. In the next section, I briefly introduce the mechanisms which enable the interspecies transfer and acquisition of genes.

6.1.2 Mechanisms of Horizontal Gene Transfer in Prokaryotes

In general, there are three ways for invasion of exogenous DNA into the recipient cells: conjugation, transduction, and transformation (Bushman 2001). **Conjugation** is involved in the transfer of DNA through a cell-to-cell contact. In general, the genes responsible for this mechanism are encoded in plasmids. A well-known example is F factor encoded in a plasmid of *Escherichia coli* (Lederberg et al. 1952). The F plasmid encodes pilus genes, hence host *E. coli* with the plasmid (F^+) have pili (or sex pili) on the cell surface. By the function of pili, a cell of F^+ *E. coli* can attach to an F^- recipient cell, and then the DNA of F plasmid is injected into the recipient cell, resulting in F^+ by the expression of transferred pilus genes.

Transduction is the transfer of DNA through viruses. Temperate viruses have the ability to insert their DNA into the host genome by themselves. The inserted virus can be activated by some stimuli, and then the DNA of viral genome is packed into the particle and released to the environment. At this moment, a part of host genome region, neighboring the virus-inserted region, is accidentally integrated together into viral particles. Thus, when the released virus again invades into another organism and inserts its DNA into the host genome, an HGT between two organisms is indirectly established: DNA of an organism has been moved to another organism.

Transformation is the uptake of naked DNA from the surrounding environment (and the consequent alteration of genotype or phenotype). It is well known that some bacteria naturally have an ability of DNA competence. The most historic case is a transformation in *Streptococcus pneumoniae* (Griffith 1928), which was reported before the genetic material of organism turned out to be DNA. Injection of DNA in transformation is in general carried out by type IV pili.

In each of the mechanisms, specific recombination systems further mediate integration of exogenous DNA into the recipient host genome. For example, in transduction, integrases are encoded in the bacteriophage genome, and thereby the bacteriophage is lysogenized in the host genome. Regarding conjugation or transformation, the invading DNA often persists as an episome, but it is possible to be integrated into the host genome by host's recombination system (e.g., homologous recombination). Plasmids sometimes carry transposons, which can mediate the insertion of episomal DNA into the host chromosome, independently from the host system.

6.1.3 Mobilome

As mentioned above, genetic elements such as plasmids, viruses, and transposons are deeply involved in the occurrence of horizontal gene transfer. These elements are particularly called **mobile genetic elements** (MGEs). MGE can be a kind of "vehicle" for genes. If resident genes in an organism are integrated into an episome (virus or plasmid) genome through recombination by chance, such genes may be possibly transferred into another organism together with the MGE. Actually, viruses or plasmids carrying prokaryotic genes have been sometimes isolated. Such genes are seemingly irrelevant to the amplification or maintenance of MGE, although the gene interaction may happen in the recipient cell. Prokaryotic genes encoded in MGE genomes are possibly in a transition state, on the way to moving to another organism.

The entirety of MGEs is called **mobilome**, which is the gene pool of plasmid, virus, and transposon in environments. Recently, high-throughput DNA sequencers have been developed, which produce a large amount of shotgun sequence data. Therefore, it is now in principle possible to collect MGE sequences in metagenome data. Understanding the content of mobilome in environmental DNA samples might contribute to understanding the potential of gene flow in the target environment. For example, one can imagine that quantity of MGE might be positively correlated with frequency of HGT in real environments. This perspective is still unclear, because regarding MGEs detected in mobilome data, the actual activity is unknown. Some MGEs may be responsible for interspecies gene traffic as vehicles, but others may not have such a leeway. Regarding infecting viruses, some may rather influence the host population sizes by their own virulence, but others may be rather dormant in the hosts.

It should be noted that focusing on only MGE as a vehicle of genes may not be enough for assessing the evolutionary significance of HGT. It may be more important to know how often host-derived genes are horizontally transferred among species and how much species are evolutionarily affected by such HGTs. For this purpose, we have to conduct prediction of horizontally transferred genes in host nucleotide sequences. In the next section, I introduce the methods for estimating the frequency of HGT in actual nucleotide sequence data.

6.2 Prediction Methods of Horizontally Transferred Genes

6.2.1 Overview

HGT is a gain of genes which are not inherited from the ancestral genome. Therefore, detecting horizontally transferred genes in a genome is equivalent to detecting the phylogenetic inconsistency of target genes with vertically transmitted genes. For this purpose, mainly two methods are available. One is a straightforward method, which is based on the construction of molecular phylogenetic tree from nucleotide or amino acid sequences. Another is a roundabout method, which measures the deviation from sequence patterns common to vertically transmitted genes. First, I explain these two methods and then introduce other clues implying HGT, such as co-occurrence with MGE, and orphan genes in pan-genome analysis. It should be noted that these methods are originally for completed genome sequences while also applied to metagenomic sequences. Finally, I discuss the application of HGT detection methods to metagenomic sequences. Although the principle is invariant, there are some limitations in metagenomic data analysis.

6.2.2 Phylogenetic Analysis

A straightforward method for predicting HGT is to construct molecular phylogenetic trees for the genes encoded in the genomes. The topology of horizontally transferred gene should be incongruent with the topology of species mostly supported by vertically transmitted genes. This method is in general the best, because it directly reflects the evolutionary scenario of xenolog in which such a gene is not inherited from the parent but derived from another organism. Stated in another way, the donor organism might be inferred from the topology if the valid tree is obtained. The timing of transfer may also be estimated if the outgroup is appropriately chosen. The first step of this method is collection and selection of homologs to target genes. This is performed using BLAST or BLAST-like programs against nucleotide/protein sequence databases. The second step is the multiple alignment using the homologous sequences selected. The third step is construction of molecular phylogenetic tree, and finally based on the topology of tree obtained, the occurrence of HGT is assessed, and, if possible, the donor is estimated.

Although the phylogenetic method is the best for predicting horizontal gene transfer, there is still an ambiguity about the quality. Particularly, when the method is applied to a massive amount of data, such as more than 1000 or 10,000 genomes, the usefulness is not clear. When all the genes are thoroughly examined step by step, sampling of homologous genes would be at first a pain in the neck. It is not easy to select homologs appropriate for tree construction from a massive amount of genomic data. If paralogs are included in the sample and outgroup orthologs are uncertain, the rooted topology obtained can be incongruent with the correct one, resulting in an artifact of HGT. In tree construction, multiple alignment is also complicated in a large amount of sequence analysis. After that, for each of the sequence alignments, appropriate substitution model and tree construction method need to be selected for the robust topology output. Since we eventually have no choice but to automate these steps by computer programs for massive gene sequences, it is very difficult to identify error factors affecting the final output. Thus, unlike careful case studies for a small number of genes, phylogenetic HGT estimation using a large amount of genome data may not be so reliable as expected. Rather, comprehensive phylogenetic analysis of genome-scale data may be useful for visualizing rough inconsistency of topology. To follow rapid accumulation of genome data, further technical improvements are necessary.

6.2.3 Nucleotide Composition

The second method is based on nucleotide frequency of genomic region or gene. In general, particularly in prokaryotes, nucleotide composition such as G + C content is homogenous across the genome sequence. That is often called "genome signature" (Abe et al. 2003) and explained by point mutational pressure uniformly acting on the genome (Lawrence and Ochman 1997; Medigue et al. 1991). Therefore, exogenous sequences due to horizontal transfer from another organism can be distinguished by heterogeneity of the nucleotide composition compared to the average composition. For this purpose, nucleotide tuple frequency or codon usage bias has been used as the indicator of horizontal transfer. Codon usage is, for example, represented as a frequency vector of $4^3 = 64$ codons, that is, a 64-dimension vector (to be strict, termination codons should not be counted). Clustering methods based on frequency vectors may be applicable for detecting HGT. If the nucleotide composition of donor species is also homogeneous, it is a clue for origin of transferred genes. As a method using four or five tuples, for example, an index based on Markov model was proposed for detecting HGT (Nakamura et al. 2004), where donor species was also predictable from the index.

The advantage of nucleotide composition method is its quickness. In nucleotide composition methods, it does not always have to refer the sequence data in other organisms. This is a big difference between nucleotide composition method and phylogenetic method: in the latter, the search for homologous sequences in other species is required for tree construction. Thus, it does not take more time for the nucleotide method analysis than for phylogenetic method. In addition, the error factor is not affected by the other genomes, while phylogenetic method is affected by inadequate homologs (e.g., paralogs) in databases. The point of nucleotide composition method is the assumption that nucleotide composition of genome sequence is homogenous, but such an assumption may not sometimes be the case. Positively or negatively selected genes can have abnormal nucleotide compositions. For example, highly expressed genes such as ribosomal protein genes are often biased (Karlin et al. 1998; Sharp and Li 1987). Attention must be paid to pseudogenes, too. If pseudogenes go through frameshift mutation, which may shorten them, abnormal codon frequencies will be observed. Even if the assumption of homogeneity in nucleotide composition is satisfied to native genes, there is another limitation: HGT from closely related species might be immune from the detection by nucleotide composition, because the transferred genes have similar nucleotide compositions to native genes.

Another methodological problem is in gene length. Biases of nucleotide composition are usually measured by some kinds of statistics, such as GC% with standard deviation. These statistics are apparently influenced by sequence length. In other words, the statistics for shorter genes tend to be more deviated than for longer genes, or vice versa. When the bias of codon usage is validated by chi-square goodness-of-fit test, for example, longer genes will be preferentially predicted as outliers (i.e., xenologous genes) than shorter genes. Therefore, chi-square statistic is usually applied to the frequency of nucleotide tuples in a fixed window size (Heidelberg et al. 2000). In this case, however, horizontal transfer may not be predicted at single gene level, because the fixed window sometimes includes multiple short genes or only a part of single long gene. It should be noted that gene length may also be influential in phylogenetic analysis, where sampling bias in computation of multiple sequence alignments and genetic distances may be deviated particularly in short genes. However, it is not easy to obtain a test statistic correlated directly with gene length in phylogenetic HGT analysis, and practically we may have no choice but to look at the tree topology itself for judging HGT. In that sense, it may be easy to evaluate the stochastic prediction bias depending on gene length in nucleotide composition methods.

6.2.4 Co-occurrence with Mobilome Genes

As mentioned before, MGEs can be vehicles of gene for horizontal transfer. Therefore, simply checking the genes located within or beside MGEs may also be an approach for HGT detection. It is possible that such genes might be horizontally transferred ones integrated together with the MGEs. In the *V. cholerae* genome, for example, cholera toxin genes are encoded in a temperate bacteriophage region, suggesting that the genes may have been carried by the bacteriophage. The genomic region into which many transposons are inserted might be a hotspot of HGT. In virus or plasmid genomes, genes irrelevant to the maintenance of such MGEs are sometimes found. Some of those might be in a kind of "transit" state, in which the genes might have been clipped from the host genomes.

6.2.5 Pan-Genome

In response to the rise of high-throughput DNA sequencing technology, which has enabled rapid determination of complete genome sequences at low cost, comparative genomics studies on multiple strains in a species or a group of closely related species have been popular. In such studies, the entirety of genes present in a clade, called **pan-genome** (Medini et al. 2005), is of interest to biologists. The genes conserved among all the members in the clade are called "core genes," and the genes present in only a specific member are called "orphan genes." Orphan genes are often considered to be of xenologous origin; otherwise the presence of such genes might be explained by parallel gene losses in other members in the clade. Based on parsimony, it is more likely that a single event of gene gain has occurred than parallel gene losses.

HGT can also be estimated with reference to core genes or common genes to some members in the clade. Gene exchanges within the clade can be inferred by sequence comparison. Here, it should be noted that identification of orthologous genes and the following analysis are overlapped with those in the abovementioned phylogenetic analysis. In the case of pan-genome data, such steps are performed among strains in a species or closely related species. Therefore, it is expected that errors like misidentification of orthologs are fewer than in the analysis using distantly related species. Since accurate sequence alignments are obtained among strains in a species or closely related species, the aligned sequences can be further checked in detail. In the case of orphan genes or the genes present in limited members in the clade, the possibility of HGT is estimated by comparing the sequences around DNA insertion sites (Homma et al. 2007). Thus, estimation of HGT in pan-genome data is relatively easy, although the scope of study is limited to the closely related species.

6.2.6 HGT Detection Methods for Metagenomic Data

In the previous sections, I discussed the methods for HGT detection in the complete genome data. The complete genome data are static: in many cases, the samples are prepared from culture stock, the genomic DNAs are amplified by cell divisions, and the purified sequenced data are finally obtained. In the case of metagenomics data, the situation is more complicated than as in genomic data. The DNAs are sampled from environments, and we don't understand what types of or how many species exist in the samples: to understand those is exactly the purpose of metagenomics. Thus, we will face metagenomic-specific problems. The biggest problems are shortness and massiveness of sequences to be treated. Currently, metagenomics sequencing is performed using high-throughput parallel DNA sequencers, which usually output the sequences of very short DNA fragments. Some of the high-throughput sequencing platforms can treat relatively longer DNAs, but instead the reads obtained are of lower-quality and the number is smaller. Rather, long-read sequencing platforms are applicable for basic metagenome analyses, in which the DNA sequenced can be derived from a specific target gene, such as the (PCR-amplified) ribosomal RNA gene. However, such an amplicon sequencing is not suitable for metagenomic HGT analysis. Because, to examine HGT from metagenomics data, the sequencing must be performed using a random-shotgun library to cover the whole genomes. Thus, first, we need to treat a huge amount of short reads, which are fragmented genome sequences derived from a variety of species. Usually, to reduce the number of reads, joining of the reads overlapped to each other, namely, "assembly," is performed using software (Namiki et al. 2012), and thereby somewhat longer sequences, namely, **contigs**, than read sequences are obtained.

The next step is taxon identification of assembled contigs. Taxon identification itself is applicable also for read sequences, although the accuracy is low because of short sequences. A simple way for taxon identification is to search for similar sequences in nucleotide sequence databases such as the GenBank/EMBL/DDBJ. If there are any significant matches in the databases, we can say that the query sequences are derived from the species or closely related species. The sequences matched in databases are sometimes from unidentified organisms in environmental DNA samples. To avoid this, expansion of reference genome data derived from identified organisms is very important. When the homologous sequences are not found in databases, we can try nucleotide composition methods for taxon identification. If homogeneity of genome sequence in each species is guaranteed, such as prokaryotic metagenome data, taxonomic classification of each contig (or read) sequence is possible to some extent. The resolution is, however, at the class or phylum level and inferior to the best performance of homologous sequence search (species or genus level).

It should be noted that these methods are in general used for detecting one-to-one correspondence between each of the reads or contigs and a single taxon in the traditional metagenomic analysis. Indeed, this is not enough for HGT detection in metagenomic data. Reads/contigs are a pool of sequence fragments derived from a variety of taxa, and each of those has little information on the location in genome. In HGT analysis, we must find insertions of exogenous nucleotide sequences into recipient genomic regions. The genomic region around the inserted site is a "chimera" of two types of sequences, one of which is intrinsic in the recipient organism and the other is transferred from another organism. Therefore, detection of chimeric regions is crucial for prediction of horizontal transfer in metagenomic data (Fig. 6.1).

A simple way is to split the read/contig obtained into two in the middle and perform the abovementioned taxon identification methods (e.g., nucleotide composition method) for those. If the former and latter sequences are classified into different taxa, respectively, such a read/contig can be considered chimeric due to horizontal transfer. In the case of paired-end or mate-pair libraries, pairs of reads may be used in a similar way: if two reads in each pair are mapped to different reference genomes, such a pair may be chimeric due to horizontal transfer. Figure 6.2 shows an example in which paired-end read mapping has detected

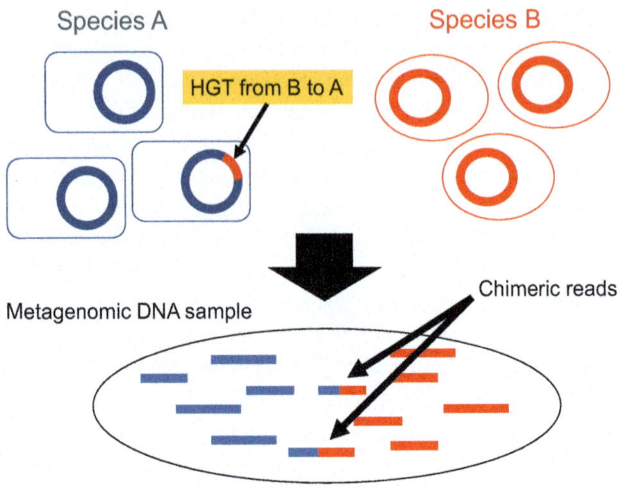

Fig. 6.1 HGT detection from metagenomic data
Here, let us suppose that there are two species, A and B, in the target environment. When HGT has occurred between species A and B (from B to A), the metagenomic DNA sample should contain chimeric reads, in which one is of A and the other is of B

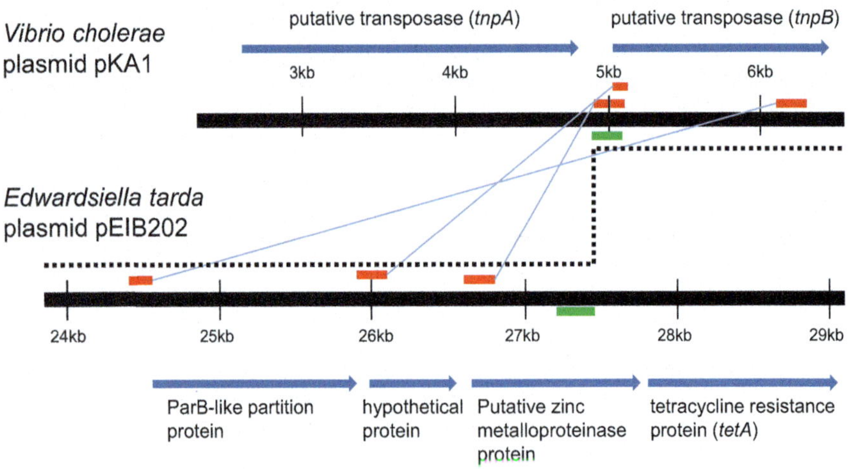

Fig. 6.2 HGT detection by surveying chimeric reads
The partial regions of *Vibrio cholera* plasmid pKA1 and *Edwardsiella tarda* plasmid pEIB202 are shown, respectively. Blue arrows indicate protein-coding regions. Read sequences from a metagenomic sample (unpublished) were mapped to these reference regions by BLAST. The former/latter sequences of a single-end read (green) are mapped to the two plasmid sequences, respectively. Regarding paired-end reads, forward/reverse reads in three pairs (red) are mapped to the two plasmid sequences, respectively. Thus, in this metagenomic sample, a plasmid may have undergone a recombination between two regions (dashed line)

a chimeric structure of nucleotide sequence in a plasmid: a region including a transposase gene in *V. cholerae* plasmid pKA1 is linked to a region including different genes in *E. tarda* plasmid pEIB202, suggesting that either of the regions has been transferred into the other by recombination.

To effectively find chimeric reads/contigs, we need to obtain long ones as possible. For very short reads, it is better to construct contigs using assembly software, although we have to be careful about artifact chimeric contigs due to mis-assembly. In particular, it is desirable that several genes (at least two) are encoded in the reads/contigs, because gene sequences are more useful for homology search than noncoding sequences. In the case of protein-coding genes, nucleotide composition methods such as codon usage comparison are also available for HGT prediction. In prokaryotes, roughly estimating, each gene is located in a ratio of one per 1 kb in the genome. Therefore, to obtain the information of genomic regions containing at least two genes, where one is intrinsic and the other is of xenologous origin, more than 2 kb single-end reads or paired-end/mate-pair reads with >2 kb in insert length would be necessary in high-throughput DNA sequencing. For single-end reads, long-read sequencers such as PacBio may be useful, although the read accuracy is not enough. For paired-end/mate-pair reads, unfortunately, short-read libraries with >2 kb insert are little used in metagenomics studies, because long-insert libraries are more difficult to be prepared than short-insert or single-end libraries. In the case of short-read libraries, therefore, de novo assembly would be necessary, whereby contigs with >2 kb might be obtained.

After long reads/contigs are prepared, homology search of the sequences will be performed against the nucleotide databases such as the GenBank/EMBL/DDBJ. If the former and latter parts in a query sequence are significantly matched to sequences of different taxa, respectively, the query sequence may possibly include a horizontally transferred region. For the purpose of accuracy, phylogenetic analysis will be further necessary, but as mentioned above, the entire process including homology search may be time-consuming. As a sequence similarity search program, BLAST (basic local alignment search tool) (Altschul et al. 1997; Altschul et al. 1990) has been a de facto standard for two decades, but expansion of metagenomic data produced from high-throughput sequencers is going beyond the processing capability. Recently, the programs faster than BLAST have been developed (Buchfink et al. 2015; Suzuki et al. 2014), raising new challenges for big data analysis. Even if the reads/contigs have little homologous sequences in databases, the former and latter sequences in each of the reads/contigs can be classified into taxa by nucleotide composition methods. Here, gene prediction is not always required for further analysis, because comparison of tuple frequency (3 bp, 4 bp, or larger) is available. Instead, resolution of taxon identification may not be good.

6.2.7 Mobilome in Metagenome Data

In metagenomic sample preparation, viral particles can be separated from microbe cells by appropriately sized filters. Therefore, metagenomic shotgun sequencing of only viral DNAs is theoretically possible. Actually, some viral metagenomic data from size-fractionated samples are available at the GenBank/EMBL/DDBJ. The viral metagenomic data often contain non-viral (e.g., bacterial) sequences. Although these sequences might be derived from contamination of DNAs released from dead microbe cells, which should go through the filters in sample preparation, there is a possibility that such are the horizontally transferring DNAs integrated into viral genomes. In order to clarify this, construction of long contigs by de novo assembly is necessary. In addition, enriching the contents of nucleotide database is also very important. There are still unidentified species in the environments, and the nucleotide sequences are lacking in databases. Thus, in metagenomics projects, in parallel with metagenomic sequencing, genome sequencing of species identified and cultured from the samples should also be done if possible. In red tide metagenomics project in Japan (see Sect. 6.3.2), for example, the complete genome sequences of bacterial or bacteriophage strains isolated from the sampling areas were determined and deposited into the GenBank/EMBL/DDBJ (Yasuike et al. 2017; Nishiki et al. 2016; Yasuike et al. 2015; Kawato et al. 2015; Yasuike et al. 2014; Yasuike et al. 2013b; Yasuike et al. 2013a).

6.2.8 Summary

We can utilize mainly two types of method for detecting horizontal gene transfer, namely, phylogenetic method and nucleotide composition method. In general, phylogenetic method is straightforward for proving horizontal transfer (i.e., rejecting vertical transmission) and the best if done carefully. However, many steps such as homolog choice, multiple sequence alignment, tree construction, and test of tree topology must be automated for a massive amount of genome data, thereby the results obtained might include many artifacts which are hard to be verified later. Nucleotide composition method is easily applied to a massive amount of genome data, based on a simple assumption that the native sequences in genome have homogeneous nucleotide compositions to each other by mutation pressure. Conversely, when we use the nucleotide composition method, we must pay attention to whether such an assumption is held in the genomes examined, for example, positively/negatively selected regions might be artifacts having irregular nucleotide compositions. Thus, the both methods have pros and cons, and there is no accurate method for detecting HGT unless the targets to be compared are closely related strains/species. A promising result may be possibly obtained from the extrapolation of pan-genomic analysis on each clade, in which HGT are more accurately estimated. In the analysis of metagenome data, the situation is harder, because we have to start from short fragmented sequences. Nevertheless, the methods of HGT detection are in principle common to those for complete genome data. In addition, it is necessary to prepare long reads or contigs, and the reference genome sequences are also important for identifying the taxa of reads/contigs.

6.3 HGT Analysis

6.3.1 HGT Estimated in the Complete Genomes of Prokaryotes

As mentioned above, regarding prokaryote, the frequency of HGT has been estimated at the genome-wide level. A pioneering research was first done for the *E. coli* genome (Lawrence and Ochman 1998). This was based on nucleotide composition analysis, and they estimated that 18% of the current *E. coli* K12 genome were of xenologous origin. Although this estimation may not be robust (another estimate was 13% (Ochman et al. 2000)), further studies suggested that around 10% per genome among prokaryotes were recently transferred from another species (Nakamura et al. 2004). These studies suggested that HGT is less frequently observed in the species with smaller genomes, which are reducing by gene loss in the course of evolution. In general, many of the horizontally transferred genes detected are mobilome-related genes such as transposon, plasmid, and prophage-encoding genes. Otherwise, pathogenicity, cell defense, or cell surface genes are often detected as horizontally transferred genes, though the frequencies are not so high as mobilome-related genes. Conversely, essential genes such as those involved in DNA replication, transcription, or translation tend to be rarely transferred between taxa (Rivera et al. 1998). These genes are strictly regulated for maintaining life cycle: for example, the codon usage is optimized for keeping the high level of expression, and the timing of expression is controlled by proper promoters. Therefore, coexistence of the essential gene's xenolog derived from another species or replacement by those in the recipient genome is probably undesirable for the organism, which may disturb the balance in gene regulation.

6.3.2 HGT Analysis Using Metagenomic Data: A Test Case in Red Tides in Japan

Algal broom has been a serious issue in coastal aquaculture worldwide. In Japan, for example, red tides by *Chattonella antiqua* occurred in the Yatsushiro sea and caused the mass mortality of cultured yellowtail in 2009 and 2010. Therefore, a metagenome project has been launched for monitoring the microbiome during the red tide. In this project, seawater samples were collected with time at the fixed points, and the metagenomic DNAs extracted were sequenced by high-throughput sequencers. In addition, many strains of marine bacteria and bacteriophages were isolated and cultured, and the complete genomes were determined as reference data. Moreover, some of the seawater samples were fractionated according to microbial sizes (planktons, prokaryotes, and viruses), and the DNAs extracted

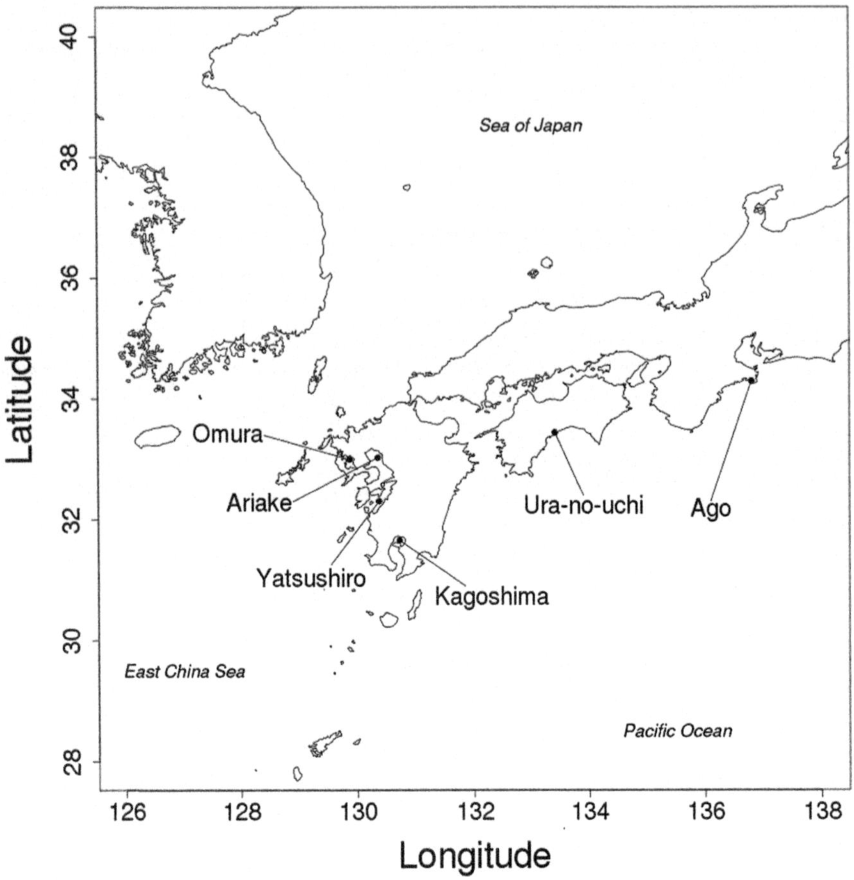

Fig. 6.3 Sampling areas in the red tide metagenome project in Japan

were sequenced separately by shotgun method, providing test data for examining horizontal DNA exchanges between host microbes and viruses. From 2013 to 2015, DNAs of prokaryotes and viruses were sequenced from a total of 70 seawater samples collected in 6 areas of Yatsushiro sea, Ariake sea, Omura bay, Kagoshima bay, Uranouchi bay, and Ago bay (Fig. 6.3) (70 × 2 = 140 runs) (unpublished). Using these data, the evaluation of nucleotide composition was performed. The read sequences were processed by PEAR (Zhang et al. 2014) and assembled by MetaPlatanus (URL: http://platanus.bio.titech.ac.jp/?page_id=174). Then, as nucleotide composition data, frequencies of 4-bp tuples (136 patterns) in the contigs obtained (> = 500 bp) were computed. The data of 136-dimension vectors were reduced by t-SNE (van der Maaten and Hinton 2008), and the 2D data obtained were plotted (Figs. 6.4 and 6.5). In the graphs, the contigs with similar nucleotide composition are plotted near to each other. The probability density function was estimated by the kernel density estimation.

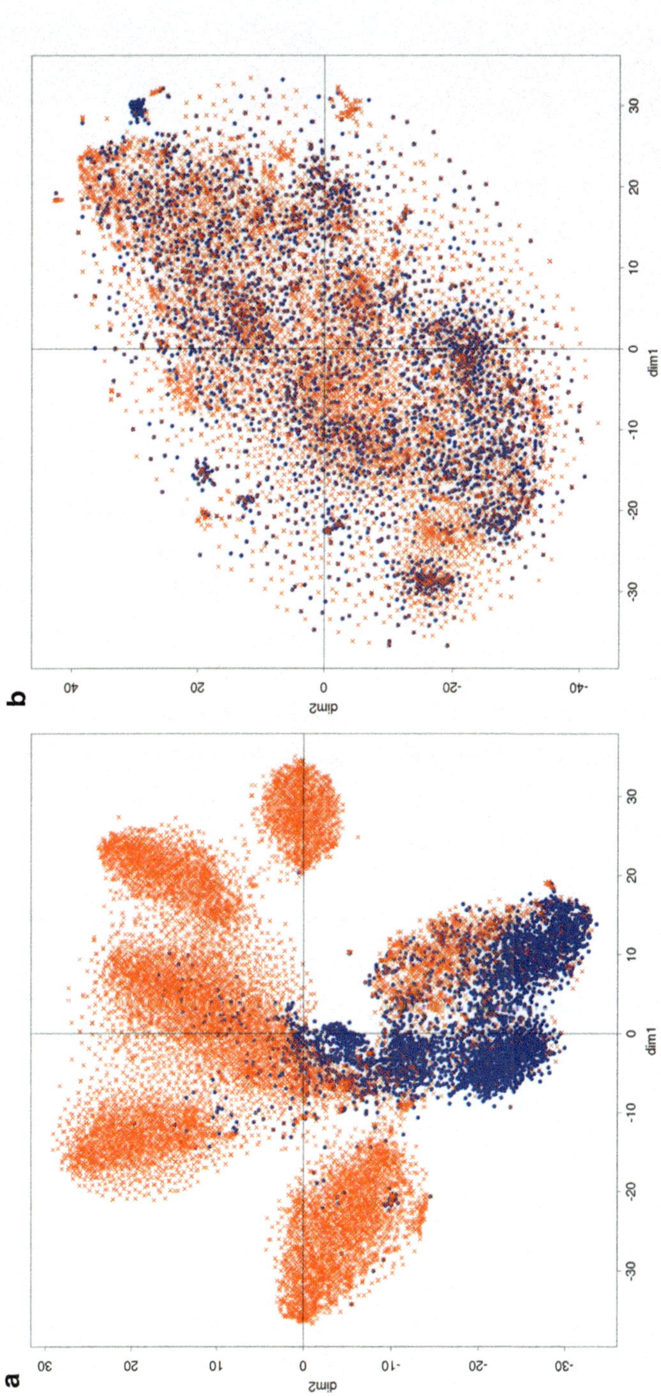

Fig. 6.4 Plot of contigs based on 4-bp-tuple frequency
Each dot corresponds to one of contigs assembled from prokaryotic (blue) or viral (red) DNA samples in Kagoshima bay on April 20, 2015 (**a**), and on June 25, 2015 (**b**), respectively

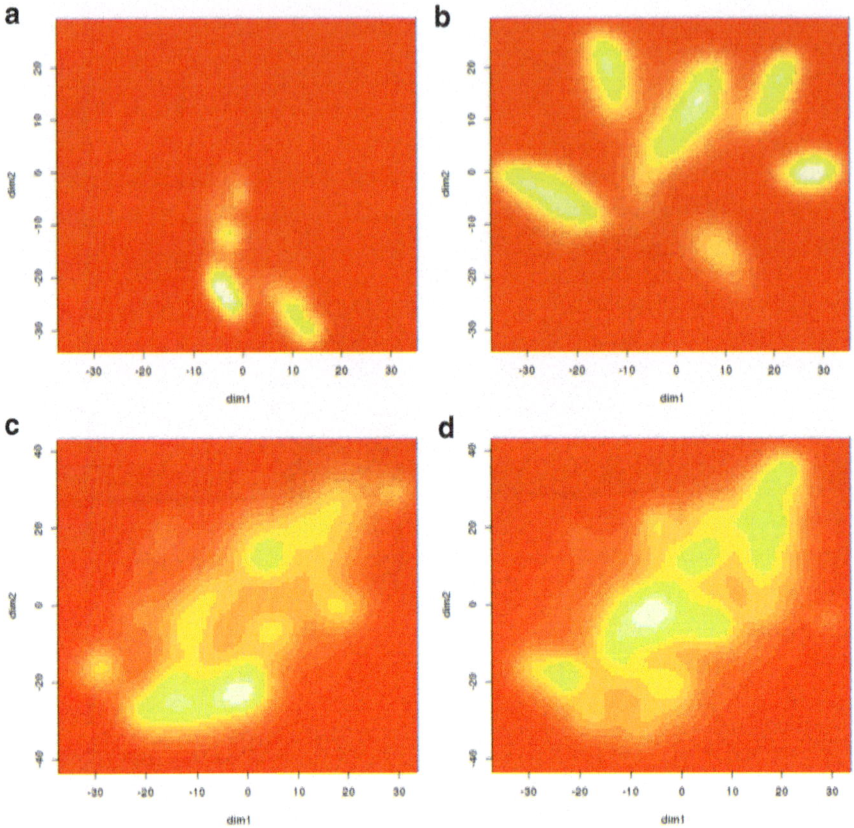

Fig. 6.5 Distribution of contigs based on 4-bp-tuple frequency
(**a** and **b**) Distributions of prokaryotic (**a**) and viral (**b**) contigs in Fig. 6.4a. (**c** and **d**) Distributions of prokaryotic (**c**) and viral (**d**) contigs in Fig. 6.4b. Distributions are computed by KDE, and contig-dense regions are shown in yellow

In some cases, the distribution of prokaryote contigs is clearly different from that of virus contigs (Figs. 6.4a, 6.5a, b). In others, however, those seem to be heavily overlapped with each other (Figs. 6.4b, 6.5c, d). Ideally, the existence of overlapped regions would indicate that prokaryotes and bacteriophages have the genomic regions with similar nucleotide composition. Therefore, the gene flow between prokaryotic and bacteriophage genomes should be estimated from the computation of overlapped region. However, things are not so simple. A concern is the experimental bias in DNA sequencing. In some of high-throughput sequencers, PCR steps are required in the protocols. Therefore, there might be some biases in the amplification of DNA: for example, high-GC or high-AT DNAs might not be fully amplified. The most difficult problem is the accuracy of DNA fractionation. Viral fractions may contain degraded prokaryotic DNA, or prokaryotic fractions may contain viral particles attaching to other materials (e.g., host cells). Actually, prokaryotic rRNA gene sequences are often detected in the contigs assembled from viral DNA samples.

Thus, measuring overlapping regions between prokaryotic and viral plots perhaps leads to overestimation of HGT. Rather, it is better to think that this analysis may provide implication about a maximum ratio of HGT in prokaryotic genomes mediated by viruses in the target environment. In order to accurately measure the ratio of HGT, upgrading of reference genome data, particularly of uncultured species, is important, which allows us to accurately identify the taxon of read/contig. Rather than DNA fractionation for prokaryotes and bacteriophages, which contains many experimental weaknesses as mentioned above, simply obtaining longer reads/contigs and detecting chimeric regions within may be better.

6.3.3 Ultra-Long Read Sequencers

At present, there are still challenges in HGT analysis using metagenomic data, but technological advancement of high-throughput sequencers might solve the problems. In particular, real-time single-molecule sequencers such as Oxford Nanopore have allowed us to quickly obtain very long reads (more than 100 kb). The basic idea of nanopore sequencing was published before 2010, and it has been recently put to practical use (Pennisi 2014). When millions of reads with hundreds of kb in size are sequenced in metagenomics, the HGT prediction (Fig. 6.1) becomes much easy and accurate. In addition, the complete reference genomes are possibly obtained by de novo assembly. Actually, nanopore technology has started to be used in metagenomics study (Brown et al. 2017). It is no long before the promising achievement about HGT estimation from metagenomic data will be reported.

6.4 Conclusion

Metagenomics technology has enabled us to monitor the real-time fluctuations of gene pool in environments and has started to be applied to the studies on HGT. It is not long before the quantification and evaluation of HGT between microbes in real environments are fully done. There are currently still limitations for accurate performance, but further technical improvements will alter the situation. The processing of big data output from high-throughput DNA sequencers may be more complicated in HGT detection than in traditional metagenomic analyses. In general, HGT is predicted by phylogenetic or nucleotide composition analyses, and these methods are applicable for metagenomic data as well as for complete gnome data. In particular, detection of chimera sequences is crucial for predicting HGT, which depends on whether longer and more accurate reads (or contigs in assembly) are obtained from high-throughput sequencers. Reference genome sequences are also very important for taxon identification of reads/contigs, although many microbes in environments cannot be easily cultured for genome sequencing.

References

Abe T, Kanaya S, Kinouchi M, Ichiba Y, Kozuki T, Ikemura T (2003) Informatics for unveiling hidden genome signatures. Genome Res 13(4):693–702. https://doi.org/10.1101/gr.634603

Altschul SF, Gish W, Miller W, Myers EW, Lipman DJ (1990) Basic local alignment search tool. J Mol Biol 215(3):403–410. https://doi.org/10.1016/S0022-2836(05)80360-2

Altschul SF, Madden TL, Schaffer AA, Zhang J, Zhang Z, Miller W, Lipman DJ (1997) Gapped BLAST and PSI-BLAST: a new generation of protein database search programs. Nucleic Acids Res 25(17):3389–3402

Brown BL, Watson M, Minot SS, Rivera MC, Franklin RB (2017) MinION nanopore sequencing of environmental metagenomes: a synthetic approach. Gigascience. https://doi.org/10.1093/gigascience/gix007

Buchfink B, Xie C, Huson DH (2015) Fast and sensitive protein alignment using DIAMOND. Nat Methods 12(1):59–60. https://doi.org/10.1038/nmeth.3176

Bushman F (2001) Lateral DNA transfer: mechanisms and consequences. Cold Spring Harbor Laboratory Press. doi:citeulike-article-id:703582

Griffith F (1928) The significance of pneumococcal types. J Hyg (London) 27(2):113–159

Hacker J, Blum-Oehler G, Muhldorfer I, Tschape H (1997) Pathogenicity islands of virulent bacteria: structure, function and impact on microbial evolution. Mol Microbiol 23(6):1089–1097

Heidelberg JF, Eisen JA, Nelson WC, Clayton RA, Gwinn ML, Dodson RJ, Haft DH, Hickey EK, Peterson JD, Umayam L, Gill SR, Nelson KE, Read TD, Tettelin H, Richardson D, Ermolaeva MD, Vamathevan J, Bass S, Qin H, Dragoi I, Sellers P, McDonald L, Utterback T, Fleishmann RD, Nierman WC, White O, Salzberg SL, Smith HO, Colwell RR, Mekalanos JJ, Venter JC, Fraser CM (2000) DNA sequence of both chromosomes of the cholera pathogen Vibrio cholerae. Nature 406(6795):477–483. https://doi.org/10.1038/35020000

Homma K, Fukuchi S, Nakamura Y, Gojobori T, Nishikawa K (2007) Gene cluster analysis method identifies horizontally transferred genes with high reliability and indicates that they provide the main mechanism of operon gain in 8 species of gamma-Proteobacteria. Mol Biol Evol 24(3):805–813. https://doi.org/10.1093/molbev/msl206

Karlin S, Mrazek J, Campbell AM (1998) Codon usages in different gene classes of the *Escherichia coli* genome. Mol Microbiol 29(6):1341–1355

Kawato Y, Yasuike M, Nakamura Y, Shigenobu Y, Fujiwara A, Sano M, Nakai T (2015) Complete genome sequence analysis of two Pseudomonas plecoglossicida phages, potential therapeutic agents. Appl Environ Microbiol 81(3):874–881. https://doi.org/10.1128/AEM.03038-14

Koonin EV, Makarova KS, Aravind L (2001) Horizontal gene transfer in prokaryotes: quantification and classification. Annu Rev Microbiol 55:709–742. https://doi.org/10.1146/annurev.micro.55.1.709

Lawrence JG, Ochman H (1997) Amelioration of bacterial genomes: rates of change and exchange. J Mol Evol 44(4):383–397

Lawrence JG, Ochman H (1998) Molecular archaeology of the *Escherichia coli* genome. Proc Natl Acad Sci U S A 95(16):9413–9417

Lederberg J, Cavalli LL, Lederberg EM (1952) Sex compatibility in *Escherichia Coli*. Genetics 37(6):720–730

Lindell D, Sullivan MB, Johnson ZI, Tolonen AC, Rohwer F, Chisholm SW (2004) Transfer of photosynthesis genes to and from Prochlorococcus viruses. Proc Natl Acad Sci U S A 101(30):11013–11,018. https://doi.org/10.1073/pnas.0401526101

Mann NH, Cook A, Millard A, Bailey S, Clokie M (2003) Marine ecosystems: bacterial photosynthesis genes in a virus. Nature 424(6950):741. https://doi.org/10.1038/424741a

McDaniel TK, Jarvis KG, Donnenberg MS, Kaper JB (1995) A genetic locus of enterocyte effacement conserved among diverse enterobacterial pathogens. Proc Natl Acad Sci U S A 92(5):1664–1668

Medigue C, Rouxel T, Vigier P, Henaut A, Danchin A (1991) Evidence for horizontal gene transfer in *Escherichia coli* speciation. J Mol Biol 222(4):851–856

Medini D, Donati C, Tettelin H, Masignani V, Rappuoli R (2005) The microbial pan-genome. Curr Opin Genet Dev 15(6):589–594. https://doi.org/10.1016/j.gde.2005.09.006

Mongodin EF, Nelson KE, Daugherty S, Deboy RT, Wister J, Khouri H, Weidman J, Walsh DA, Papke RT, Sanchez Perez G, Sharma AK, Nesbo CL, MacLeod D, Bapteste E, Doolittle WF, Charlebois RL, Legault B, Rodriguez-Valera F (2005) The genome of Salinibacter ruber: convergence and gene exchange among hyperhalophilic bacteria and archaea. Proc Natl Acad Sci U S A 102(50):18147–18,152. https://doi.org/10.1073/pnas.0509073102

Nakamura Y, Itoh T, Matsuda H, Gojobori T (2004) Biased biological functions of horizontally transferred genes in prokaryotic genomes. Nat Genet 36(7):760–766. https://doi.org/10.1038/ng1381

Nakamura Y, Takano T, Yasuike M, Sakai T, Matsuyama T, Sano M (2013) Comparative genomics reveals that a fish pathogenic bacterium Edwardsiella tarda has acquired the locus of enterocyte effacement (LEE) through horizontal gene transfer. BMC Genomics 14:642. https://doi.org/10.1186/1471-2164-14-642

Namiki T, Hachiya T, Tanaka H, Sakakibara Y (2012) MetaVelvet: an extension of Velvet assembler to de novo metagenome assembly from short sequence reads. Nucleic Acids Res 40(20):e155. https://doi.org/10.1093/nar/gks678

Nikoh N, Nakabachi A (2009) Aphids acquired symbiotic genes via lateral gene transfer. BMC Biol 7:12. https://doi.org/10.1186/1741-7007-7-12

Nishiki I, Oinaka D, Iwasaki Y, Yasuike M, Nakamura Y, Yoshida T, Fujiwara A, Nagai S, Katoh M, Kobayashi T (2016) Complete genome sequence of nonagglutinating lactococcus garvieae Strain 122061 Isolated from Yellowtail in Japan. Genome Announc 4(4). https://doi.org/10.1128/genomeA.00592-16

Ochman H, Lawrence JG, Groisman EA (2000) Lateral gene transfer and the nature of bacterial innovation. Nature 405(6784):299–304. https://doi.org/10.1038/35012500

Pennisi E (2014) Genomics. DNA sequencers still waiting for the nanopore revolution. Science 343(6173):829–830. https://doi.org/10.1126/science.343.6173.829

Rivera MC, Jain R, Moore JE, Lake JA (1998) Genomic evidence for two functionally distinct gene classes. Proc Natl Acad Sci U S A 95(11):6239–6244

Sharp PM, Li WH (1987) The codon Adaptation Index – a measure of directional synonymous codon usage bias, and its potential applications. Nucleic Acids Res 15(3):1281–1295

Suzuki S, Kakuta M, Ishida T, Akiyama Y (2014) GHOSTX: an improved sequence homology search algorithm using a query suffix array and a database suffix array. PLoS One 9(8):e103833. https://doi.org/10.1371/journal.pone.0103833

van der Maaten LJP, Hinton GE (2008) Visualizing data using t-SNE. J Mach Learn Res 9(Nov):2579–2605

Waldor MK, Mekalanos JJ (1996) Lysogenic conversion by a filamentous phage encoding cholera toxin. Science 272(5270):1910–1914

Yasuike M, Sugaya E, Nakamura Y, Shigenobu Y, Kawato Y, Kai W, Fujiwara A, Sano M, Kobayashi T, Nakai T (2013a) Complete genome sequences of Edwardsiella tarda-Lytic Bacteriophages KF-1 and IW-1. Genome Announc 1(1). https://doi.org/10.1128/genomeA.00089-12

Yasuike M, Sugaya E, Nakamura Y, Shigenobu Y, Kawato Y, Kai W, Nagai S, Fujiwara A, Sano M, Kobayashi T, Nakai T (2013b) Complete genome sequence of a novel myovirus which infects atypical strains of Edwardsiella tarda. Genome Announc 1(1). https://doi.org/10.1128/genomeA.00248-12

Yasuike M, Kai W, Nakamura Y, Fujiwara A, Kawato Y, Hassan ES, Mahmoud MM, Nagai S, Kobayashi T, Ototake M, Nakai T (2014) Complete genome sequence of the Edwardsiella ictaluri-Specific Bacteriophage PEi21, isolated from river water in Japan. Genome Announc 2(2). https://doi.org/10.1128/genomeA.00228-14

Yasuike M, Nishiki I, Iwasaki Y, Nakamura Y, Fujiwara A, Sugaya E, Kawato Y, Nagai S, Kobayashi T, Ototake M, Nakai T (2015) Full-genome sequence of a novel myovirus, GF-2, infecting Edwardsiella tarda: comparison with other Edwardsiella myoviral genomes. Arch Virol 160(8):2129–2133. https://doi.org/10.1007/s00705-015-2472-5

Yasuike M, Nishiki I, Iwasaki Y, Nakamura Y, Fujiwara A, Shimahara Y, Kamaishi T, Yoshida T, Nagai S, Kobayashi T, Katoh M (2017) Analysis of the complete genome sequence of Nocardia seriolae UTF1, the causative agent of fish nocardiosis: The first reference genome sequence of the fish pathogenic Nocardia species. PLoS One 12(3):e0173198. https://doi.org/10.1371/journal.pone.0173198

Zhang J, Kobert K, Flouri T, Stamatakis A (2014) PEAR: a fast and accurate Illumina Paired-End reAd mergeR. Bioinformatics 30(5):614–620. https://doi.org/10.1093/bioinformatics/btt593

MAPLE Enables Functional Assessment of Microbiota in Various Environments

7

Hideto Takami

Abstract

A main goal of metagenomic analysis is to elucidate comprehensive functions (i.e., the functionome) of entire communities across various environments, but only PCR amplicon analysis of 16S rDNA has been performed in most metagenomic projects so far. One reason for this is that a standard evaluation method for discerning the functional potentials harbored within genomes or metagenomes has not yet been established. To break this deadlock, a new method was developed to infer the potential functionome based on the completion ratio of the individual Kyoto Encyclopedia of Genes and Genomes functional modules. A prototype system of the MAPLE (**M**etabolic **A**nd **P**hysiological potentia**L** **E**valuator) to automate all processes used in this new method was then launched in December 2013. The MAPLE system was further improved to increase its usability, and the latest version, MAPLE 2.3.1, is now available through a web interface (https://maple.jamstec.go.jp/maple/maple-2.3.1/).

Keywords

Functional metagenomics · MAPLE system · KEGG module · Marine metagenomics

---

H. Takami (✉)
Yokohama Institute, Japan Agency for Marine-Earth Science and Technology (JAMSTEC), Yokohama, Japan
e-mail: takamih@jamstec.go.jp

T. Gojobori et al. (eds.), *Marine Metagenomics*,
https://doi.org/10.1007/978-981-13-8134-8_7

7.1 Introduction

Considering that culturable microbes constitute less than 1–10% of all microbes thriving in natural environments, metagenomic analysis is one of the most powerful ways to understand species and functional diversity in whole communities (WCs) of microbes. Since the first report of a comprehensive metagenomic analysis of a marine environment was published (Venter et al. 2004), many metagenomic analyses focused on microbial communities ranging from natural environments to human gut microbiota have been reported. However, the diversity estimates based on 16S rDNA and some other key genes, which enable the characterization of the habitats, have just been discussed in most metagenomic papers. However, a main goal of metagenomics is the actual deduction of not only community structure but also potential comprehensive functions (the functionome) harbored by entire communities across various environments. We define the functionome as the comprehensive functions occurring through combinations of individual functions, such as carbon fixation, nitrogen fixation, nitrification, denitrification, and amino acid metabolism, encoded by multiple genes. This goal remains poorly addressed because the evaluation of the potential functionomes still remains difficult compared with the functional annotation of individual genes or proteins, mainly because there is no established standard methodology for extracting functional category information such as data pertaining to metabolism, energy generation, and membrane transport systems.

Until the late 2000s, comprehensive functional categories presented in detail in the Kyoto Encyclopedia of Genes and Genomes (KEGG) (Kanehisa and Goto 2000) and SEED (Overbeek et al. 2005) databases have been used to identify the functional features in comparative genomics and metagenomics represented by the KEGG Automatic Annotation Server (KAAS) (Moriya et al. 2007), Metagenomics Rapid Annotation using Subsystem Technology (MG-RAST) (Meyer et al. 2008), and Metagenome Analyzer (MEGAN) (Huson et al. 2011) systems. Each system employs a similarity-based method for functional annotations but uses different databases for protein sequences, default threshold values, and ortholog identification numbers for mapping annotated sequences to functional categories depending on the desired outputs (specifically, pathways in KEGG and subsystems in SEED). However, these functional categories still remain too broad to distinguish metabolic or physiological features. Thus, the evaluation of current tools and the development of new tools are required to characterize potential physiological and metabolic pathways in actual ecosystems (Filippo et al. 2012). To resolve this problem, a new method for evaluating the potential functionome was developed based on calculating the completion ratio of four types of KEGG modules: pathways, molecular complexes, functional sets, and signatures (Kanehisa et al. 2008). This is represented as the percentage of a module component filled with the input KEGG Orthology (KO)-assigned genes by KAAS (Takami et al. 2012). A prototype system of MAPLE (Metabolic And Physiological potentiaL Evaluator) for automating the newly developed method was then launched in December 2013. MAPLE first assigns KO to the query gene, maps the KO-assigned genes to the corresponding KEGG functional modules, and then calculates the module completion ratio (MCR) of each functional module to characterize the potential functionome of the user's own genomic and metagenomic data. Afterward, two new useful functions for calculating module abundance and *Q*-value, which indicate the functional abundance and statistical significance of the MCR results, respectively, were added to the new MAPLE ver. 2.1.0 to enable more detailed comparative genomic and metagenomic analyses (Takami et al. 2016).

Although this system was very useful at the time, especially for metagenomic data analysis, the high computational time associated with analyzing the often massive amino acid sequence datasets (with 1–3 million sequences typically employed in metagenomic research) reduced its utility. Thus, the MAPLE system was further developed to reduce calculation time by adapting KAAS to use GHOSTX (Suzuki et al. 2014), a much faster homology search program, instead of BLAST. The latest MAPLE system 2.3.1 is now available through a web interface (https://maple.jamstec.go.jp/maple/maple-2.3.1/).

Although next-generation sequencing (NGS) can easily produce massive sequence datasets, these raw data cannot be directly applied to the system because the data submitted to MAPLE must be a multi-FASTA file of amino acids. Unfortunately, it is difficult for researchers who are unfamiliar with bioinformatic tools to process such massive raw datasets properly. To increase user convenience, we developed MAPLE Submission Data Maker, which can convert raw NGS data into multi-FASTA files of amino acid sequences. This useful software can be installed on personal computers that run MacOS X or Windows OS.

MAPLE results, such as MCR, *Q*-value, and module abundance, can be easily downloaded as an Excel file. However, MAPLE results are difficult to judge visually as they consist of rows of numerical values. Thus, we developed the downloadable program that is available through the website to draw histograms based on the MAPLE results (MAPLE Graph Maker). With this program, users can easily visualize MAPLE results by importing resulting Excel files. This chapter presents the methodology for functional genomics and metagenomics to reveal the functional diversity of individual microbes and WCs of microbes. An example of functional metagenomics for marine environments using MAPLE is also provided in this chapter.

7.2 Development of a New Method for Evaluating the Potential Functionome

7.2.1 KEGG Module

KEGG MODULE (Kanehisa et al. 2008) is a collection of pathway modules and other functional units designed for automatic functional annotation and pathway enrichment analysis. As of July 13, 2018, a total of 305 pathway modules have been defined for energy, carbohydrate, lipid, nucleotide, and amino acid metabolisms,

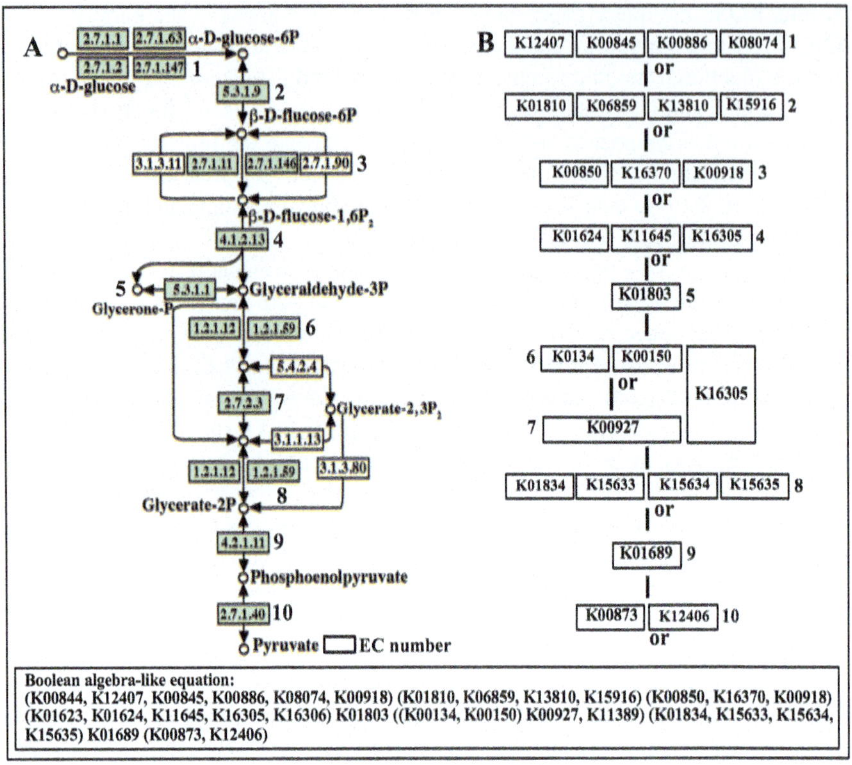

Fig. 7.1 Glycolysis reactions registered in the Kyoto Encyclopedia of Genes and Genomes (KEGG) pathway database. (**a**) KEGG reaction map for glycolysis. (**b**) KEGG functional module for glycolysis corresponding to the reaction map. The module M00001 comprising 10 reactions is defined for the glycolysis module and represented as a Boolean algebra-like equation of KEGG Orthology identifiers or K numbers for computational applications. The relationship between this module and the corresponding KEGG pathway map is also indicated by the corresponding K number sets in the module and Enzyme Commission (EC) numbers in the pathway map using the same index. In each K number set, horizontally arranged K numbers indicate alternatives, related to each other with "Or" or "," in the equation

including genetic and environmental information processing pathways. A total of 796 KEGG modules (305 pathways, 294 structure complexes, 157 functional sets, and 40 signatures) can be accessed through the website (http://www.genome.jp/kegg-bin/get_htext?ko00002.keg). Pathway modules in KEGG MODULE correspond to smaller portions of subpathways (Fig. 7.1a), manually defined as consecutive reaction steps, operons, or other regulatory units, and phylogenetic units obtained by genome comparisons (Fig. 7.1b). Other functional units include (1) structural complexes representing sets of protein subunits for molecular machineries such as ATPase, (2) functional sets representing other types of essential sets such as aminoacyl-tRNA synthases, and (3) signature modules representing markers of phenotypes such as the enterohemorrhagic *Escherichia coli* pathogenicity signature for Shiga toxin. The latest KEGG MODULE is available from the KEGG FTP site (http://www.kegg.jp/kegg/download). Each module is defined by a combination of KO identifiers (IDs) such that it can be used for annotation and interpretation purposes in individual genomes or metagenomes. Notations of the Boolean algebra-like equation (Fig. 7.1) for this definition include space-delimited items indicating pathway elements, comma-separated items in parentheses indicating alternatives, plus signs indicating complexes, and minus signs indicating optional items.

7.2.2 Calculation of the Module Completion Ratio Based on the Boolean Algebra-Like Equation

The completion ratio of all KEGG functional modules in each organism was calculated on the basis of a Boolean algebra-like equation. For this analysis, 1 genome was selected from each of the 5257 available prokaryotic and 457 eukaryotic species genomes, and a reference genome set was constructed to cover all completely sequenced organisms, excluding draft genomes as of July 13, 2018 (3622 total genomes, 3186 prokaryote genomes, and 436 eukaryotic genomes).

For example, M00001 is a pathway module for glycolysis, comprising 10 reactions as shown in Fig. 7.1. In each KO number set depicted, the horizontally arranged rows of K numbers indicate alternatives, which are related to each other by “Or” or “,” in the equation (Takami et al. 2012). When genes are assigned to all KO IDs in each reaction according to the Boolean algebra-like equation, the module completion ratio becomes 100%. For example, when genes are not assigned to KO IDs in two reactions, the module completion ratio is calculated to be 80% ($8/10 \times 100 = 80$).

7.2.3 Assignment of KO IDs to the Query Sequences

The functional annotation of rapidly growing sequence data from complete genomes and metagenomes requires efficient and accurate computational methods. KAAS is an efficient tool for assigning KO IDs to genes from complete genomes based on a BLAST or GHOSTX search of the KEGG GENES database combined with a bidirectional or single-directional best-hit method (Moriya et al. 2007). Because of its efficiency, KAAS is used to assign KO IDs to protein sequences from individual genome or metagenome projects. The latest stand-alone KAAS system is available from the KAAS HELP website (http://www.genome.jp/tools/kaas/help.html).

We used KAAS to estimate the database dependency on accuracy of KO assignments. First, we selected *E. coli* as a representative prokaryotic species and constructed four different types of datasets: without *E. coli* and closely related species (1239 species), without all species within family *Enterobacteriales* (1200 species), without all species within class *Gammaproteobacteria* (1040 species), and without all species within phylum *Proteobacteria* (755 species). In addition, we created artificially fragmented protein sequences from *E. coli* to confirm the

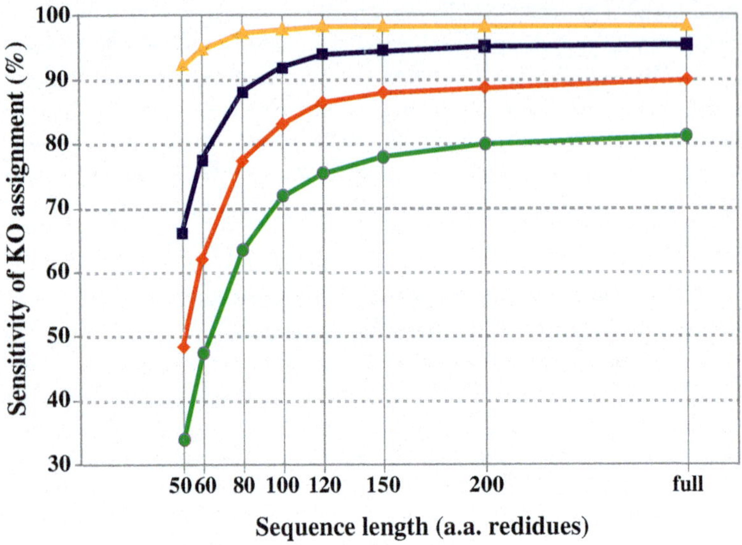

Fig. 7.2 Effect of database dependency on accuracy of Kyoto Encyclopedia of Genes and Genomes Orthology (KO) assignment. The sampled *Escherichia coli* was isolated from a Norwegian infant (draft genome sequenced using the 454 GS FLX Titanium platform). Red diamonds show the results using the dataset without proteins from the genera *Escherichia*, *Salmonella*, *Shigella*, and *Yersinia* (1239 species). Similarly, blue squares, yellowish-brown triangles, and green dots show the results without proteins from the order *Enterobacteriales* (1200 species), class *Gammaproteobacteria* (1040 species), and phylum *Proteobacteria* (755 species), respectively. KO IDs specific to the genera *Escherichia*, *Salmonella*, *Shigella*, and *Yersinia* (16 KO IDs), order *Enterobacteriales* (90), class *Gammaproteobacteria* (203), or phylum *Proteobacteria* (370) were removed in advance from the protein dataset. Here, the accuracy is defined by the sensitivity TP/(TP + FN), where TP and FN are the numbers of true positives and false negatives, respectively. We also used truncated proteins to confirm the effect of amino acid (a.a.) sequence lengths on the accuracy of KO assignments. The 4410 proteins from the *Escherichia coli* isolates were randomly fragmented into fragments of 50, 60, 80, 100, 120, 150, and 200 a.a.s in length, and each length of a.a. sequences was used for verification of the accuracy of KO assignment

accuracy of KO assignment for the truncated proteins. This analysis examined gene products predicted from the end of assembled contigs and singletons, which were often truncated in the sequences produced by NGS. The amino acid sequences of complete CDSs were randomly fragmented into fragments of 50, 60, 80, 100, 120, 150, and 200 residues in length, and each fragment was subjected to verification of database dependency based on the accuracy of KO ID assignment (Fig. 7.2).

In general, because most microbes thriving in natural environments are unculturable, many genes in environmental metagenomic data do not show significant similarity to those from known species in public genome databases. Especially when microbial genomes belonging to the same phylum as the query microbe are missing from the genome database utilized, the accuracy rate of KO assignment to proteins phylogenetically distant from known phyla is expected to be low. Indeed, when all species within phylum *Proteobacteria* were not included in the dataset, the accuracy rate of KO assignment for full *E. coli* proteins decreased to 80%, but an accuracy rate of approximately 70% was maintained even for proteins fragmented to lengths of about 100 residues (Fig. 7.2). Considering these results, even if genes from unidentified phyla of the so-called candidate division are included in the metagenomes, the KAAS system can presumably assign KO IDs to genes longer than 300 bp (100 amino acids) with an accuracy rate of approximately 70% (Takami et al. 2012).

7.2.4 Distribution Patterns of the Module Completion Ratio for 3186 Prokaryotes

Each KEGG module is designed for automatic functional annotation by a Boolean algebra-like equation of KEGG Orthology IDs. However, it remains unclear which species possess common modules or whether certain modules demonstrate universality or rareness among specific taxa. Specific information regarding the phylogenetic profiles of each module holder would be especially useful for annotating metagenomes (Takami et al. 2012). Thus, we first examined distribution patterns of the MCRs of the KEGG modules for the 3186 prokaryotic species whose genomic sequences have been completed. Although the distribution of the MCRs for the 3186 species varied greatly depending on the kind of module (Fig. 7.3), we found that there were essentially four patterns (i.e., universal, restricted, diversified, and nonprokaryotic) regardless of the module type (i.e., pathway, structural complex, signature, or functional set) when considering 70% of all species to represent a majority measurement for the patterns.

Pattern A (defined as "universal") comprised modules completed for more than 70% of the 3186 species (Fig. 7.3A-1), and more than 70% of the 3186 species possessed MCRs of >80%, referred to as pattern A-2 (Fig. 7.3A-2). Only 9.0% of the pathway modules were grouped into pattern A, and they mainly belonged to the categories of central carbohydrate metabolism and cofactor and vitamin biosynthesis. Although there are many species, more than 70% of the 3186 prokaryotes possessed MCRs of 80%. Species with a 100% completion ratio were very limited within the pattern A-2 group. M00096, a representative of pattern A-2 (Fig. 7.3), is a pathway module for C5 isoprenoid biosynthesis, a non-mevalonate pathway comprising eight reaction steps. Pattern B (defined as "restricted") is composed of modules completed by less than 30% of the species (Fig. 7.3b) and accounted for 24.7% of all the pathway modules, and 66 modules were rare modules, completed for less than 10% of the 3186 species.

Pattern C (defined as "diversified") accounted for 33.7% of all the pathway modules and comprised modules ranging widely in completion ratios. M00012 (the glyoxylate cycle, comprising five reactions) is one of the representatives of pattern C (Fig. 7.3c). One or several KO IDs were assigned to each reaction in this module; however, KO IDs, except for K01637 and K01638 assigned to the third and fourth reactions, were also assigned to other pathway modules such as the tricarboxylic

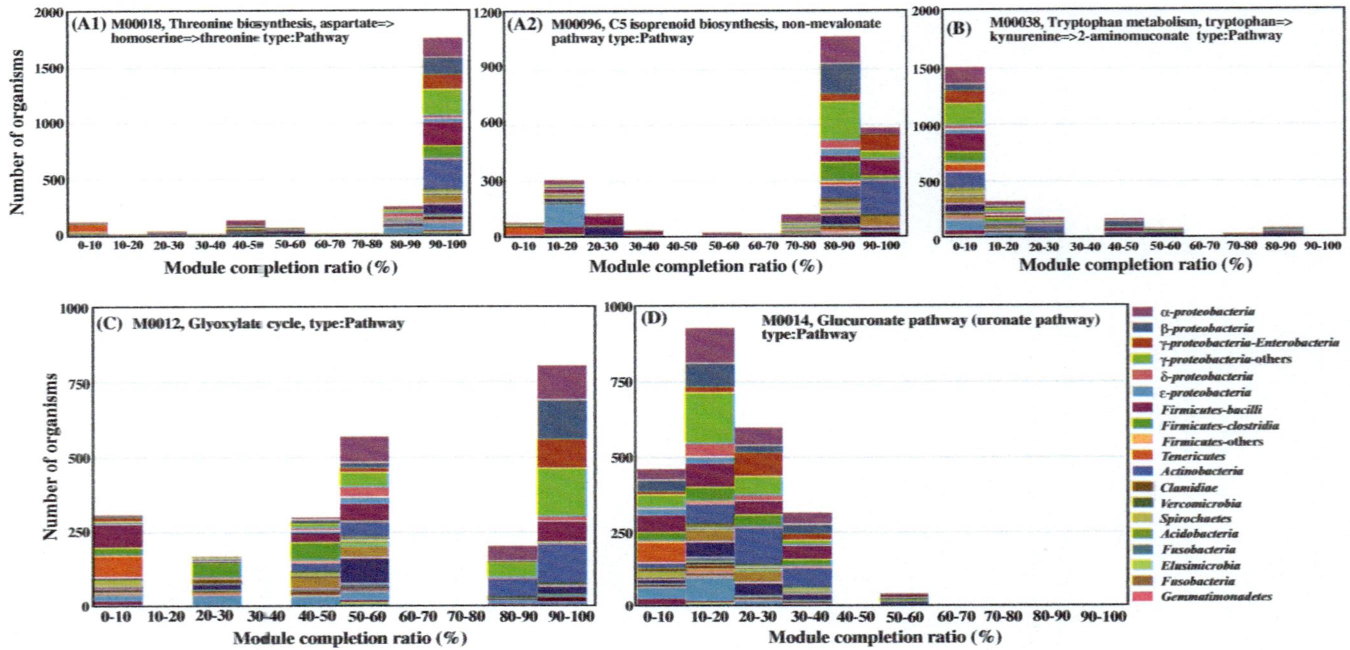

Fig. 7.3 Typical completion patterns of the Kyoto Encyclopedia of Genes and Genomes (KEGG) modules for 2367 prokaryotic species. (**a**) Universal modules. (*A-1*) Modules completed by more than 70% of 768 prokaryotic species, such as M00018, which represents threonine biosynthesis (aspartate homoserine threonine). (*A-2*) Modules for which more than 70% of 2367 prokaryotic species show a MCR of >80%, such as M00096, which represents C5 isoprenoid biosynthesis and is a non-mevalonate pathway. (**b**) Restricted modules completed by less than 30% of 768 prokaryotic species, such as M00038, which represents tryptophan metabolism (tryptophan kynurenine 2-aminomuconate). (**c**) Diversified modules, that is, modules that vary in their MCRs among 768 prokaryotic species, such as M00012, which represents the glyoxylate cycle. (**d**) Nonprokaryotic modules completed by no prokaryotic species, such as M00014, which represents the glucuronate pathway (uronate pathway). Some examples of the taxonomic variation that completes each KEGG module are shown in Fig. 7.4

acid (TCA) or Krebs cycle (M00009), reductive TCA cycle (M00173), and C4-dicarboxylate cycle (nicotinamide adenine dinucleotide $(NAD)^+$-malic enzyme type; M00171). Some KO IDs assigned to many of the modules categorized into pattern C were also assigned to several other independent modules (Takami et al. 2012). Thus, when the MCR is low, the relationship between the MCR of the targeted module and others for which the same KO IDs were assigned should be considered. Pattern D, which accounted for 32.6% of all pathway modules, comprised nonprokaryotic modules that are not completed for prokaryotic species (Fig. 7.3d).

Among structural complex modules redefined from modules with various complex patterns, 150 modules were categorized into pattern B (51.7%), and 119 were rare modules. Pattern C accounted for only 18.6% of the structural modules compared with 33.7% of the pathway modules. Thus, it was hypothesized that most of the structural complex modules, except for those conforming to pattern D, are shared only among limited prokaryotic species. Nonprokaryotic modules account for 32.6% of the pathway (e.g., M00741) and 26.6% of structural complex modules, respectively, and other modules were classified into various taxonomic patterns such as prokaryotic, Bacteria-specific, and Archaea-specific based on the module completion profiles (Fig. 7.4). These four patterns indicate the universal and unique nature of each module and also the versatility of the KO IDs mapped to each module. Thus, the four criteria and taxonomic classifications for each module should be helpful for interpretation of results based on module completion profiles.

M00012 [jump to KEGG]
Glyoxylate cycle
Pathway modules

	Prokaryote	Eukaryote
TAXONOMY	Prokaryote	Plants/Fungi/Protists
CLASS	C ---diversified---	C ---diversified---

M00264 [jump to KEGG]
DNA polymerase II complex
Complex modules

	Prokaryote	Eukaryote
TAXONOMY	Archaea	Non-eukaryote
CLASS	B --restricted--, Rare	D--non-prokaryotic/non-eukaryotic--

M00001 [jump to KEGG]
Glycolysis, (Emden-Meyerhof pathway) glucose => pyruvate
Pathway modules

	Prokaryote	Eukaryote
TAXONOMY	Prokaryote	Eukaryote
CLASS	A(1) ---universal---	A(1) ---universal---

M00630 [jump to KEGG]
D-galacturonate degradation, D-galacturonate => glycerol
Pathway modules

	Prokaryote	Eukaryote
TAXONOMY	Prokaryote	Animal/Plants/Protists
CLASS	C ---diversified---	C ---diversified---

M00004 [jump to KEGG]
Pentose phosphate pathway (Pentose phosphate cycle)
Pathway modules

	Prokaryote	Eukaryote
TAXONOMY	Bacteria	Eukaryote
CLASS	C ---diversified---	A(1) ---universal---

M00741 [jump to KEGG]
Propanoyl-CoA metabolism, propanoyl-CoA => succinyl-CoA
Pathway modules

	Prokaryote	Eukaryote
TAXONOMY	Non-prokaryote	Fungi
CLASS	D--non-prokaryotic/non-eukaryotic--	B --restricted--, Rare

Fig. 7.4 Taxonomic variation that completes Kyoto Encyclopedia of Genes and Genomes (KEGG) modules. Taxonomic variation and completion patterns of each module can be displayed in the MAPLE results page. Some typical examples are shown in this figure

7.3 The MAPLE System

7.3.1 Overview

MAPLE is an automatic system that can perform a series of steps used in the evaluation of potential comprehensive functions (i.e., functionomes) harbored by genomes and metagenomes. From April 2016 through March 2017, MAPLE was accessed 2.5 million times. However, beginners still have difficulty in processing such massive raw datasets produced by NGS prior to data submission to MAPLE and in interpreting MAPLE results, which contain many rows of numerical values. Thus, we now provide a complete system to support every step from initial data processing to final visualization of the MAPLE results (see Sect. 7.3) (Fig. 7.5).

MAPLE first assigns a KO ID to the query gene using KAAS (Fig. 7.6b), then maps the KO-assigned genes to the KEGG functional modules (Fig. 7.6c), and finally calculates the MCR of each functional module and its abundance when the module is complete (Fig. 7.6d). There are two methods for KO assignment by KAAS: bidirectional best hit (BBH) and single-directional best hit (SBH).

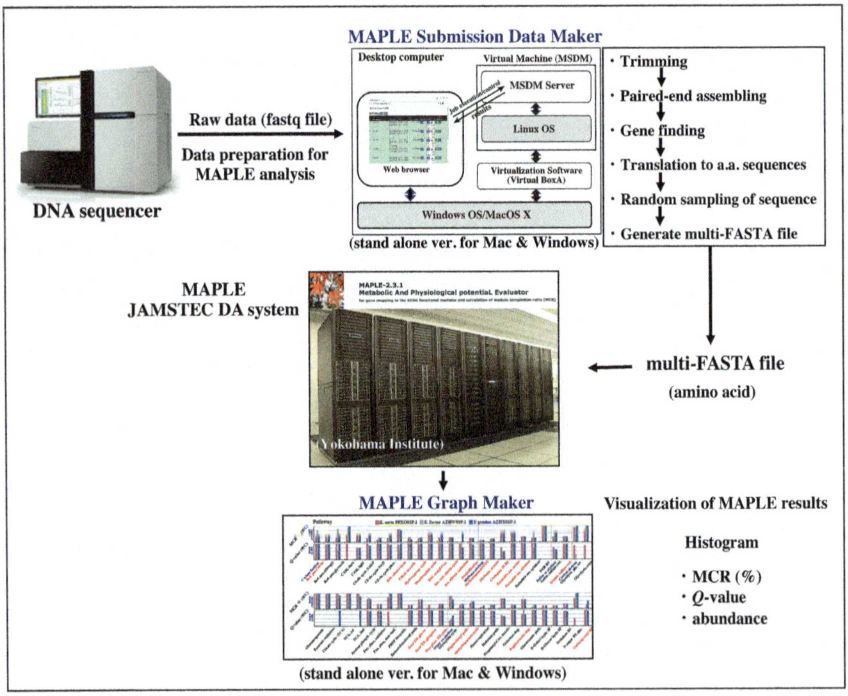

Fig. 7.5 Outline of metagenomic analysis using the MAPLE system. MAPLE Submission Data Maker (MSDM), for preparing multi-FASTA file, and MAPLE Graph Maker (MGM) for visualizing MAPLE results, are available from the website (https://maple.jamstec.go.jp/maple/maple-2.3.1/softdownload/)

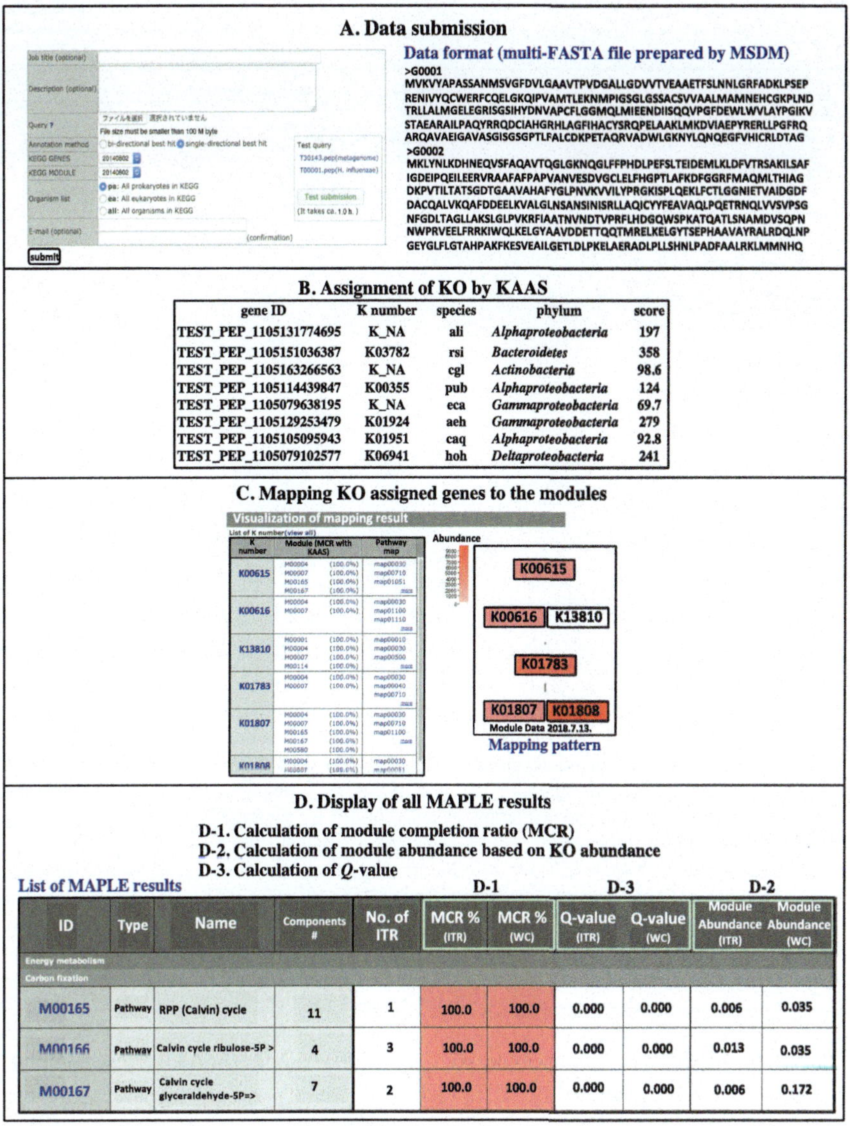

Fig. 7.6 Workflow of the MAPLE system, including the four steps and intermediate results at each step. This diagram was slightly modified from the original version with permission from *DNA Research* (Takami et al. 2016). (**a**) Query sequences submitted to the MAPLE system must be amino acid sequences containing partial genes from metagenomic sequences generated by high-throughput DNA sequencers, such as Illumina MiSeq and HiSeq. (**b**) The Kyoto Encyclopedia of Genes and Genomes (KEGG) Orthology (KO) assignments are performed by the KEGG Automatic Annotation Server (KAAS) on the basis of results from the BLASTP program. (**c**) The query sequences should be in multi-FASTA format with unique IDs, and the gene IDs must not include tabs. After the KO assignment to each query sequence is finished, mapping of the KO-assigned sequences to the KEGG functional modules starts, and (*D-1*) subsequently, the module completion ratio (MCR) is calculated. (*D-2*) After MCR calculation, KO abundance and module abundance are calculated. (*D-3*) Finally, the *Q*-value of the module is calculated based on the MCR results, the abundance, and the similarity score of each KO-assigned sequence mapped to the module. MSDM, MAPLE Submission Data Maker

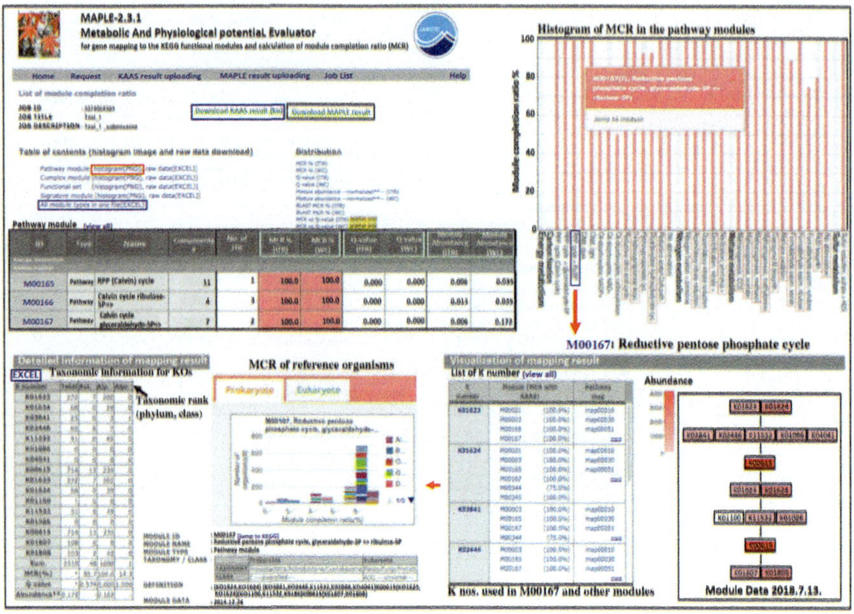

Fig. 7.7 Overview of MAPLE results. This figure was modified from the original version with permission from *DNA Research* (Takami et al. 2016). The table containing all MAPLE results (module completion ratio [MCR], Kyoto Encyclopedia of Genes and Genomes [KEGG] Orthology [KO] and module abundances, and *Q*-value) is displayed by clicking a job ID. The MCR results are displayed as both a table and histogram. Detailed results for each module, such as the mapping pattern and taxonomic information, can be displayed by clicking the module ID in the table or module name in the histogram. The list of KO-assigned genes by the KEGG Automatic Annotation Server (KAAS), MAPLE results, and taxonomic information for KO-assigned sequence are downloadable from the links highlighted by blue boxes on the first results page. An example of a downloaded file is shown in Fig. 7.8. The data package needed to redisplay all MAPLE results after the expiration of data storage on the MAPLE server is also downloadable from the same page

The BBH method is suitable for complete gene sets identified from complete genomes or contigs, while the SBH method is mainly appropriate for short-read sequences of metagenomes or incomplete genomes. When the query sequences are submitted to MAPLE (Fig. 7.6a), the MCR and abundance of each KEGG module as well as the taxonomic information for the KO-assigned genes mapped to the module are displayed along with a mapping pattern (Fig. 7.7). When a KO mapped to a module is shared by two or more modules, the module IDs sharing the same KO are listed (Fig. 7.7). The MCR calculation is performed based on a Boolean algebra-like equation defined by KEGG for each module. The results of KO assignment produced by KAAS, taxonomic information for the genes mapped to the KEGG modules, and calculated MCRs are downloadable in an Excel spreadsheet format (Fig. 7.8). MAPLE can display the results of comparative analyses of mapping patterns, MCR results, and the abundance of complete modules between the different metagenomic samples (Fig. 7.9). To evaluate the working probability of

1. KAAS result

Gene ID	K number	species	Phylum/Class/Order	Score
F2-W_4.1_A_pred_2_-_2292_3848_0_C	K08600	cha	*Firmicutes - Clostridia*	188.348
F2-W_14.1_A_pred_11_-_10682_11242_0_C	K_NA	dmi	*Firmicutes - Clostridia*	170.629
F2-W_58.1_A_pred_7_+_5544_6758_0_C	K17810	asq	*Actinobacteria*	369.777
F2-W_21.1_B_pred_12_+_9609_10586_0_C	K06889	tbe	*Spirochaetes*	244.202
F2-W_71.1_B_pred_5_+_4839_6200_0_C	K01685	puf	*Firmicutes - Others*	833.872
F2-W_33.1_A_pred_5_+_4609_6048_0_C	K01270	ova	*Firmicutes - Clostridia*	801.13
F2-W_71.1_B_pred_6_+_6197_7066_0_C	K_NA	pru	*Bacteroidetes*	317.29
F2-W_35.1_A_pred_7_+_10786_12177_0_C	K02316	rbr	*Firmicutes - Clostridia*	384.8
F2-W_43.1_B_pred_4_-_1643_2092_0_C	K_NA	cce	*Firmicutes - Clostridia*	125.861
F2-W_18.1_B_pred_6_+_4744_5772_0_C	K_NA	bth	*Bacteroidetes*	890.112

2. MAPLE result

Large category	Small category	ID	Type	Name (abbreviation)	Components #	No. of ITR	MCR % (ITR)	MCR % (WC)	Q-value (ITR)	Q-value (WC)	Module abundance (ITR, sum)	Module abundance (ITR, max)
Energy metabolism	Carbon fixation	M00377	Pathway	Reductive acetyl-CoA pathway	7	0	71.4	71.4	0.909	0.938	-	-
Energy metabolism	Carbon fixation	M00579	Pathway	Phosphate acetyltransferase-acetate kinase pathway	2	4	100	100	0	0	0.794	0.366
Energy metabolism	Carbon fixation	M00620	Pathway	Incomplete reductive citrate cycle, acetyl-CoA =>	7	0	28.6	71.4	1	0.938	-	-
Energy metabolism	Nitrogen metabolism	M00175	Pathway	Nitrogen fixation, nitrogen => ammonia	1	0	0	0	0.857	0.857	-	-
Energy metabolism	Nitrogen metabolism	M00531	Pathway	Assimilatory nitrate reduction, nitrate => ammonia	2	0	50	50	0.727	0.727	-	-
Energy metabolism	Nitrogen metabolism	M00530	Pathway	Dissimilatory nitrate reduction, nitrate => ammonia	2	0	50	50	0.96	0.934	-	-

3. Taxonomic breakdown of the genes mapped to each KO in the KEGG module

K number	Total	Actinobacteria	Bacteroidetes	Chlorobi	Deltaproteobacteria	Firmicutes - Bacilli	Firmicutes - Clostridia	Firmicutes - Others	Gammaproteobacteria - Enterobacteria	Gammaproteobacteria - Others	Spirochaetes
K00618	42	3	14	0	3	1	17	2	0	2	0
K00616	8	2	3	0	0	0	3	0	0	0	0
K13810	0	0	0	0	0	0	0	0	0	0	0
K01783	16	0	7	1	0	0	5	0	2	0	1
K01807	2	2	0	0	0	0	0	0	0	0	0
K01808	15	0	5	0	2	0	7	1	0	0	0
Sum.	83	7	29	1	5	1	32	3	2	2	1
MCR(%)	-	75	100	25	50	25	100	50	25	25	25
Q-value	-	0.750319	0	0.998722	0.926176	0.99361	0	0.926176	0.998722	0.99361	0.998722
Abundance**	0.419	-	0.209812	-	-	-	0.209812	-	-	-	-

Fig. 7.8 Examples of downloadable results. All results generated by MAPLE are downloadable in an Excel-compatible format (tab delimited). This figure was reproduced with permission from *DNA Research* (Takami et al. 2016). (1) An example of downloaded files containing the Kyoto Encyclopedia of Genes and Genomes (KEGG) Orthology (KO) assignment results for the genes performed by the KEGG Automatic Annotation Server (KAAS). (2) Module completion ratio (MCR), individual taxonomic rank (ITR), *Q*-value, and module abundance. (3) Taxonomic breakdown of the genes mapped to the KEGG modules. WC, whole community

the physiological function in the incomplete modules, we proposed the *Q*-value as a more appropriate way to interpret MCR results. The *Q*-value, which indicates the probability that a reaction module is identified by chance, is calculated based on the statistics of the sequence similarity score and KO abundance using the concept of multiple testing corrections according to Boolean algebra-like equations (Fig. 7.10).

7.3.2 Evaluation of the Module Completion Ratio According to *Q*-Values

Generally, it is expected that the MCR is linked to the likelihood that the organisms perform the physiological function corresponding to a particular module. However, when the KOs used for a module are shared with the other modules, the MCR does not necessarily reflect the working probability of each functional module (Fig. 7.10f). Thus, the MCRs of the targeted module, module completion of other modules to which the same KOs are assigned, and the contribution of specific KOs of each module to module completion should be considered when a module is incomplete. That is, even if the same MCR was observed among different modules,

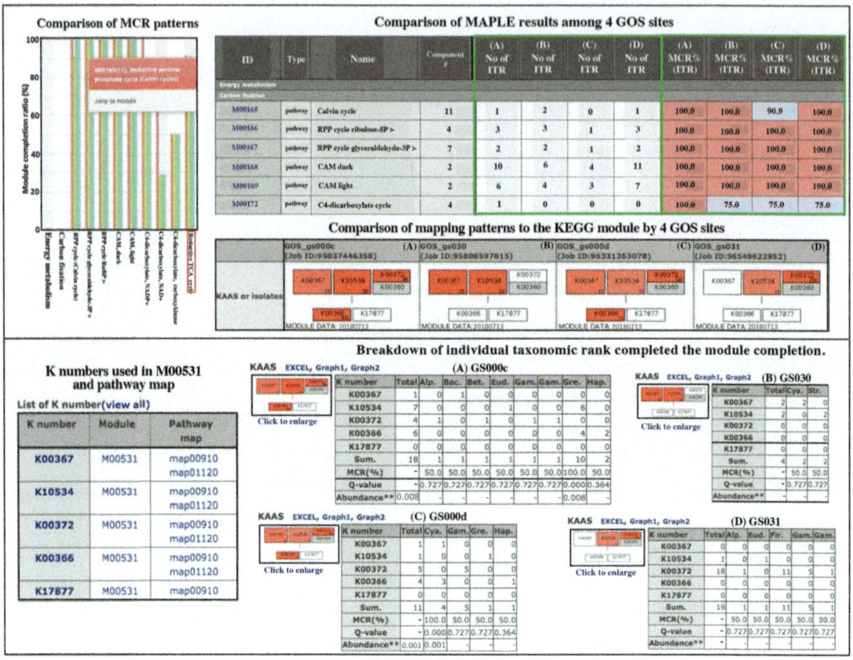

Fig. 7.9 Results from comparative analyses of different metagenomic samples. This figure was modified from the original version with permission from *DNA Research* (Takami et al. 2016). Module completion ratio (MCR), patterns of mapping to the modules, abundance of complete modules, and taxonomic information are displayed and available in Excel format. GOS, Global Ocean Sampling; ITR, individual taxonomic rank; KEGG, Kyoto Encyclopedia of Genes and Genomes; KAAS, KEGG Automatic Annotation Server

the working probability of the physiological function is not necessarily equal among these MCRs. To avoid these problems, we propose the use of the *Q*-value for determining the significance of module completion. This measure, which represents the probability that a reaction module is identified by chance, is calculated based on statistical sequence similarity scores (e.g., E-values) using the concept of multiple testing corrections, according to the definition of the KEGG reaction module (i.e., Boolean algebra-like equation) (Takami et al. 2016).

To explain the *Q*-value, we use a simple example, a reaction module (e.g., M00020: serine biosynthesis) in which each reaction step consists of only one gene (i.e., one KO). Assuming that gene i is available when at least one read (or sequence) can be detected, the probability that gene i is identified is $1 - \prod_{j \in r_i} P_j$, where P_j denotes the P-value of sequence j calculated based on the E-value, r_i is the set of reads (or sequences) of gene i, and $|r_i|$ denotes the abundance of gene i. When gene i corresponds to n_i different reaction modules, the gene must be observed in all of these modules. Thus, the probability that gene i is identified is corrected as $\left(1 - \prod_{j \in r_i} P_j\right)^{n_i}$. Taken together, the probability (i.e., Q-value) that the reaction

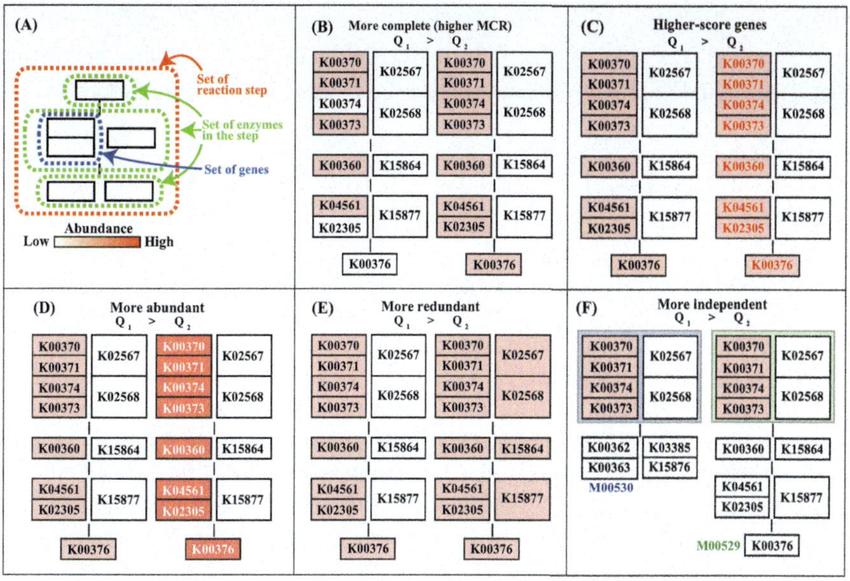

Fig. 7.10 Illustration of the *Q*-value concept. This image was slightly modified from the original version with permission from *DNA Research* (Takami et al. 2016) because the M00530 module was redefined by the Kyoto Encyclopedia of Genes and Genomes. (**a**) Schematic diagram of a reaction module. (**b–e**) As an example, the reaction module M00529 is shown. The weight of the K number (e.g., K00370) indicates the sequence similarity scores. The *Q*-value is lower in the following cases. (**b**) The module is more complete (i.e., it has a higher module completion ratio). (**c**) The module consists of genes with higher similarity scores. Red numbers indicate high similarity scores. (**d**) Genes are more abundant. (**e**) Enzymes or reactions include alternative elements (e.g., isozymes exist). (**f**) The module has less overlap with the other modules because there are fewer multiple comparisons. Note that the left-side module in (**f**) is ideal (i.e., M00529 is assumed to be independent from M00530) for the purpose of illustration. MCR, module completion ratio

module is identified by chance is calculated as

$$Q = 1 - \prod_{i \in S} \left(1 - \prod\nolimits_{j \in r_i} P_j\right)^{n_i},$$

where *S* is the set of reaction steps (genes or KOs, in this case) of a reaction module.

In general, however, each reaction step may consist of alterative genes (i.e., isozymes) and gene complexes (i.e., enzymes composed of a protein complex; Fig. 7.10a). Because the *Q*-value for a gene complex enzyme can be calculated only when all genes encoding the enzyme are identified (because of the AND relationship), the probability that the enzyme is identified is written as $\prod_{h \in K_j} \left(1 - \prod_{k \in r_h} P_k\right)^{n_i}$, where K_i is the set of genes in enzyme *j* (blue in Fig. 7.10a). Moreover, the calculation assumes the reaction step can occur when at least one enzyme is available (because of the OR relationship);

thus, the probability that the reaction step is identified is represented as $1-\prod_{j\in Z_i}\left\{\prod_{h\in K_j}\left(1-\prod_{k\in r_h}P_k\right)^{n_i}\right\}$, where Z_i is the set of enzymes in reaction step i (green in Fig. 7.10a). Taken together, the Q-value is computed as follows:

$$Q=1-\prod_{i\in S}\left[1-\prod_{j\in Z_i}\left\{\prod_{h\in K_j}\left(1-\prod_{k\in r_h}P_k\right)^{n_i}\right\}\right].$$

We found higher MCRs were generally associated with lower Q-values, and the Q-value was almost 0 when the MCR was 100%. However, the Q-value is not necessarily negatively correlated with the MCR. For example, we considered the module M00532 (photorespiration; Fig. 7.11). This module was not completed by the WC for the GS000d, GS030, and GS031 samples of metagenomic sequences from the Global Ocean Sampling (GOS) project; however, the MCR was higher than 80% (Fig. 7.11). Notably, the Q-value of this module was high for each site (0.875 for GS000d and 0.75 for both GS030 and GS031). In this module comprising ten reaction steps, the KOs assigned to seven steps (steps 2–5 and 8–10) were module-specific, while sites GS030 and GS031 lacked K14272 in step 5 and GS000d lacked K03781 and K14272 in step 4. Moreover, module M00373 (ethylmalonyl pathway related to glyoxylate and dicarboxylate metabolisms) was not completed by the WC at sites GS000c, GS000d, and GS030, although the MCR was 92.9%. Half of the 14 reaction steps included in this module were module-specific, and these three sites lacked only one KO, K14451, which was assigned to the last module-specific step in this module. According to its definition, the Q-value emphasizes the completion of module-specific KOs for evaluating the working probability; thus, even though a high MCR value is shown, the Q-value is high when the module-specific KO is missing (Takami et al. 2016).

Although module M00095 (C5 isoprenoid biosynthesis in the mevalonate pathway) was not completed in the WC, the MCR was high (85.7%) for the four GOS sites, similar to those of module M00532; however, the Q-value of this module was low (0.25), as shown in Fig. 7.11. There are four module-specific reaction steps (steps 3–6) in the module comprising seven reaction steps, and the KO assigned to step five was missing for all GOS sites, similar to the results observed for module M00532. However, a notable difference was noted in the module structure of the missing reaction step; namely, reaction step five consisted of alternative genes (KOs), that is, isozymes. According to the definition, the Q-value becomes lower when more isozymes are available; thus, it is low in such a case, even if module-specific KOs are missing (Takami et al. 2016).

7.3.3 Calculation of Module Abundance

MAPLE highlights the difference in the potential functions of organisms and environmental samples if the MCRs of various modules differ. MAPLE can also

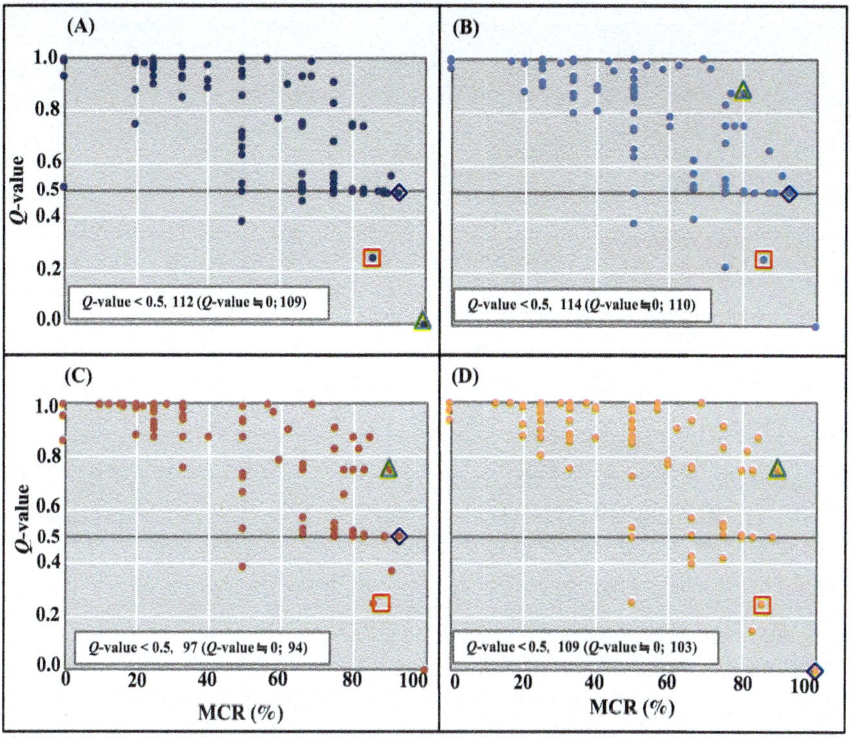

Fig. 7.11 Comparison of module completion ratio and *Q*-value patterns of pathway modules for whole communities (WCs) of microbes among the four global ocean sampling sites. (**a**) GS000c, (**b**) GS000d, (**c**) GS030, and (**d**) GS031. This image was modified from the original version with permission from *DNA Research* (Takami et al. 2016). Triangle, photorespiration (M00532); diamond, ethylmalonyl pathway (M00373); square, C5-isoprenoid biosynthesis (M00095). The number of modules for which *Q*-value <0.5 in each WC is shown in each sample. Among those models with *Q*-value <0.5, the number of modules with that of nearly zero is shown in parentheses. MCR, module completion ratio

determine diversity in individual taxonomic rank (ITR) for completed modules and abundance of the modules for every ITR to clarify the difference in the functional potential of modules commonly completed across multiple samples. ITR is a second taxonomic rank defined in the KEGG Organisms database such as phylum, class, and order (http://www.genome.jp/kegg/catalog/org_list.html).

The total number of sequence reads assigned to each KO in constructing a module was divided by the average length of each KO group to normalize KO abundance. This normalized KO abundance is described in the lower right corner of each KO box when the mapping pattern of the query genes for the module are displayed (Fig. 7.12). The module abundance is calculated based on the normalized KO abundance. For example, the module M00529 (Figs. 7.10b–f and 7.12), defined as a denitrification reaction, is composed of four reaction steps. For each K number

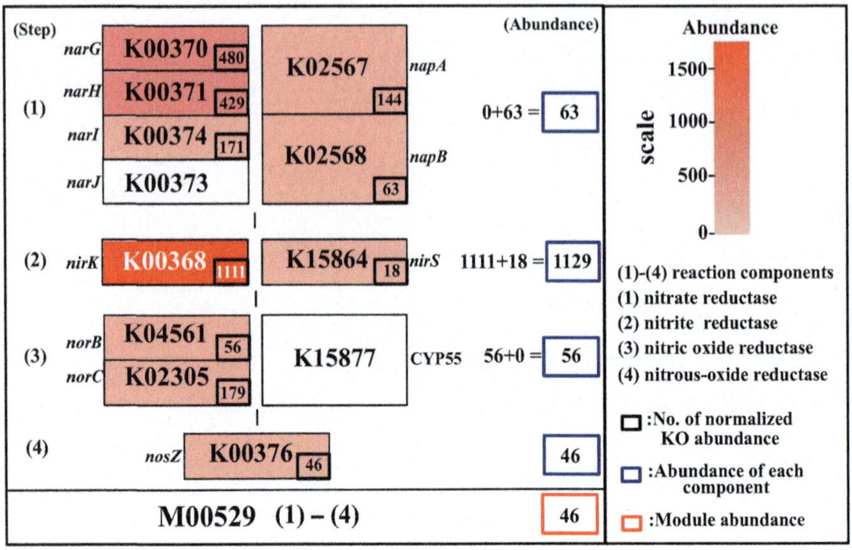

Fig. 7.12 Illustration of the module abundance concept. This diagram was slightly modified from the original version with permission from *DNA Research* (Takami et al. 2016). Module M00529 comprises four reaction components and is defined as a denitrification reaction. In each K number set, vertically arranged K numbers indicate a complex, whereas horizontally arranged K numbers indicate alternatives. Small numbers in the lower right of the boxes indicate the abundance of the Kyoto Encyclopedia of Genes and Genomes Orthology (KO)-assigned genes normalized by the mean number of genes categorized in each orthologous group (i.e., KO). In the case of a complex (1 and 3), the minimum number is defined as an abundance of a complex. When there are alternatives in the reaction component (i.e., 1, 2, and 3), the sum of both abundances is defined as the abundance of each reaction component. Finally, the minimum abundance of the four reaction components is defined as the module abundance

set, vertically arranged K numbers represent a complex, whereas horizontally arranged K numbers represent alternatives (Takami et al. 2012; Takami et al. 2016). Because the enzyme responsible for the first reaction (nitrate reductase) is composed of four (Fig. 7.12, left side) or two KO complexes (Fig. 7.12, right side), the abundance of the first reaction step becomes 0 unless all KOs vertically connected are filled with the KO-assigned genes. When all vertically connected KOs are filled, the minimal value of KO abundance in the vertically connected boxes becomes the abundance at the first step. Thus, the abundance of the first step in module M00529 is 63. As the second step, when two horizontally located KOs are filled with the KO-assigned genes, the abundance at the second step becomes 1129, which is the sum of both KOs. The abundance at the third step becomes 56 in a similar manner as the first step, and that of the last step is 46 (Fig. 7.12). Because the module abundance becomes the minimum value among all steps, the abundance of module M00529 is calculated to be 46. To facilitate normalization by ribosomal proteins, we defined a new virtual module (M91000) for all ribosomes (i.e., prokaryote + eukaryote ribosomes) comprising 130 ribosomal proteins (excluding accessory proteins) because the 31 KOs corresponding to the ribosomal proteins are shared by Bacteria and Archaea and 26 KOs are shared between Archaea and Eukaryote (Fig. 7.13a). Thus, we calculated the module abundance per ribosomal protein to enable the comparison among different environmental sites.

7.3.4 Microbial Community Structure Based on Ribosomal Proteins

In many metagenomic analyses, 16S rRNA gene sequences obtained by PCR amplification are used to compare microbial community structures among different environments. In recent studies, this PCR-based amplicon approach has been used to target the V4 region because different regions of the 16S rRNA gene yield varying degrees of accuracy in taxonomic assignments (Liu et al. 2007). However, prokaryotic species exhibit variation in copy number of the 16S rRNA gene even within the same species (https://rrndb.umms.med.umich.edu/) (Stoddard et al. 2015), and it is impossible to determine the copy numbers of individual unculturable and unknown microbes present in actual microbial communities. Thus, because taxonomic compositions based on amplicon sequences are strongly influenced by copy number in addition to basic PCR bias, this approach is not useful for the analysis of microbial community structure.

Ribosomal proteins are well conserved among all organisms and possess sequences specific to each individual organism; therefore, ribosomal proteins can be used for the identification of organisms. Accordingly, we examined how accurately KAAS, which is used in the MAPLE system, can annotate ribosomal proteins taxonomically. All prokaryotic ribosomal proteins available from the National Center for Biotechnology Information (NCBI) database (189,020 proteins) were annotated using MAPLE, and the results were then compared with the original taxonomic annotation. There was no significant difference between the original annotation at the phylum level and those inferred by KAAS, despite the fact that the MAPLE system uses the KEGG database, which contains only complete genome data (Fig. 7.13b). Accordingly, we concluded that MAPLE can be effectively used to identify organisms by constructing a metagenome based on ribosomal proteins (Takami et al. 2016). To apply MAPLE to taxonomic analysis, we calculated the proportions of Bacteria, Archaea, and Eukaryote in the metagenome based on the mapping pattern of the virtual module M91000 for all ribosomes (Fig. 7.13a) and the taxonomic annotation of each ribosomal protein, whereas eukaryotic taxonomic information is limited in the KEGG database. As mentioned above, because archaeal and eukaryotic ribosomes, which contain 58 and 77 ribosomal proteins, respectively, have 6 and 15 more proteins than the bacterial ribosome, we normalized the total number of archaeal ribosomal proteins to the number of bacterial ribosomal proteins by multiplying the archaeal and eukaryotic ribosome count data by 52/58 and 52/77, respectively. We summed the number of bacterial ribosomes and normalized archaeal and eukaryotic ribosomes and then used this sum as a denominator for calculating the proportions of Archaea, Eukaryote, and

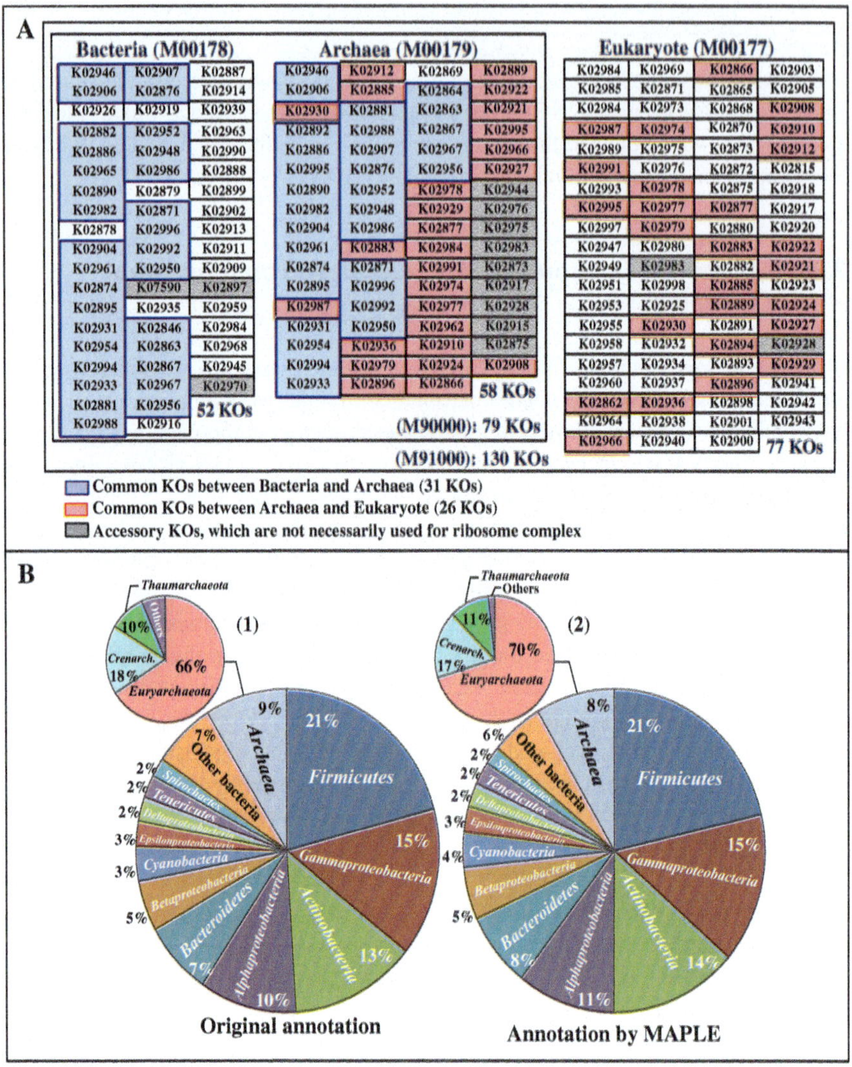

Fig. 7.13 Virtual modules for prokaryotic and all ribosome (M90000 and M91000) and taxonomical characterization of the ribosomal proteins by the MAPLE system. Panels A and B were modified from the original version with permission from *DNA Research* (Takami et al. 2016). (**a**) Organization of the module. The M90000 is composed of 79 (21 bacterial, 27 archaeal, and 31 common) and the M91000 is composed of 130 (M90000 + 51 eukaryotic) ribosomal proteins, respectively. Accessory KEGG Orthologies (KOs) are not used to define the module. Because most of the genes for broadly conserved ribosomal proteins are single copy within individual genomes and specific for each organism, the taxonomic information for each KO assigned to this module can be used for more precise analysis of the taxonomical composition of microbiomes from various environments. In contrast, 16S rRNA genes are of limited value because their copy number obviously varies among organisms. (**b**) Taxonomic breakdown of the pre-analyzed ribosomal proteins retrieved from public genomic databases. (1) Taxonomic breakdown at the phylum level based on the original taxonomic information. (2) Phylum-level taxonomic breakdown based on the results of the MAPLE system. A total of 189,020 ribosomal proteins retrieved from the National Center for Biotechnology Information (NCBI) database were subjected to the MAPLE system for a comparison of the original taxonomic annotation with those assigned by the KAAS

Bacteria. We can also calculate the proportion for each taxonomic level defined by KEGG, such as phylum and class, in the metagenome using the same method. When a metagenome is composed of only prokaryotic sequences, module M90000 is useful for analysis of prokaryotic community structure, but module M91000 for ribosomes of all organisms is also useful for community structure analyses when the microbial community also contains eukaryotic species together with prokaryotes.

On the other hand, a new method based on universal single-copy genes, which provides prokaryotic species boundaries at a higher resolution than possible using the 16S rRNA gene, has been used to estimate the relative abundances of known and unknown microbial community members with metagenomic data at a species-level resolution (Mende et al. 2013). However, community structure analysis at such a high resolution is not necessarily required in metagenomic analyses of natural environments, unlike that of the human gut microbiome, because many community members have not yet been cultivated or identified at the species level. Thus, community structure analyses at the phylum or class level based on ribosomal proteins using the latest version of the MAPLE system (version 2.3.1) are thought to be feasible for analyses of metagenomes from natural environments. MAPLE enables species-level taxonomic assignment to each ribosomal protein included in the human gut microbiome because many individual genomes of the community members composing the human gut microbiome, in contrast with environmental community microbes, have been registered in the KEGG database.

7.3.5 Homology Search Engine for Faster Analysis

MAPLE ver. 2.1.0 can accept 1 million amino acid sequences (<160 Mbytes) for each job; however, the 80-h running time to finish all calculation steps for 1 million sequences strongly motivated efforts to reduce the calculation time for many users. To resolve this urgent problem, we employed GHOSTX (Suzuki et al. 2014), a homology search program that is much faster than BLAST, which had been used in a previous version of MAPLE. We modified KAAS (Moriya et al. 2007) incorporated into MAPLE to use GHOSTX (http://www.genome.jp/tools/kaas/help.html). We then examined the difference in MAPLE results between the GHOSTX and BLAST implementations using GS030 data in a dataset from GOS (Venter et al. 2004).

In the current MAPLE version 2.3.1, approximately 18 h are required to complete jobs when users select GHOSTX instead of BLAST as a homology search program, depending on the size and content of the metagenomic sequences. For example, a job using the BBH method for a 4.4-Mb individual genome containing 4035 protein sequences can be completed within 45 min. In the case of metagenomic sequences containing 3 million sequence reads, a job will take about 12–18 h to complete when the SBH method is used. Fundamentally, the SBH method should be used for metagenomic analysis even when using contigs assembled from metagenomic sequences if contigs are a part of the genome of an individual organism (i.e., draft genome). An e-mail address can be specified for users to receive a message with a URL for the results from the system upon completion of the job.

7.3.6 Differences in MAPLE Results Between Using GHOSTX and BLAST

There was a slight difference in MCR results between MAPLE using GHOSTX and BLAST (Fig. 7.14). Among the 766 functional modules registered in the KEGG database (as of November 16, 2016), 171 were completed by GHOSTX and 173 by BLAST in the GS030 dataset from the GOS expedition (Venter et al. 2004) with 163 results in common (Arai et al. 2018). Almost all modules completed by neither homology search lacked KOs assigned to only one reaction step from among the multiple steps comprising the module. On the other hand, there were some differences in patterns of KO assignment between the GHOSTX and BLAST implementations of MAPLE, especially for short query sequences of fewer than 100 amino acid residues (Fig. 7.15). However, 96.4% of KO-assigned query sequences among 1 million input sequences were assigned to the same KOs by both homology search programs as shown in Figs. 7.15 and 7.16a (Arai et al. 2018). Additionally, there was no substantial difference in the breakdown of functional categories of Pfam domains included in the unique KOs assigned by either GHOSTX or BLAST, although one unique Pfam functional category was identified in each unique KO (Fig. 7.16B-2 and B-3). We examined distribution patterns of sequence length and average scores of KOs assigned by both programs and also specifically assigned by only GHOSTX or BLAST (Fig. 7.17). Though query sequences of fewer than 100 amino acid residues comprised more than half of the 1 million input sequences, the percentage of such short sequences, relative to all sequences receiving the same KO assignment from both programs, was only 15% (Fig. 7.17a). In contrast, there was not much difference in average score between sequences assigned to the same KO by both programs in each sequence length range. The distribution of lengths for sequences with KOs assigned by BLAST alone was similar to that of sequences with KOs assigned by both algorithms. In contrast with BLAST, the sequence length distribution of those sequences assigned to KOs by only GHOSTX totally differed from those assigned to the same KO by both algorithms, instead resembling the

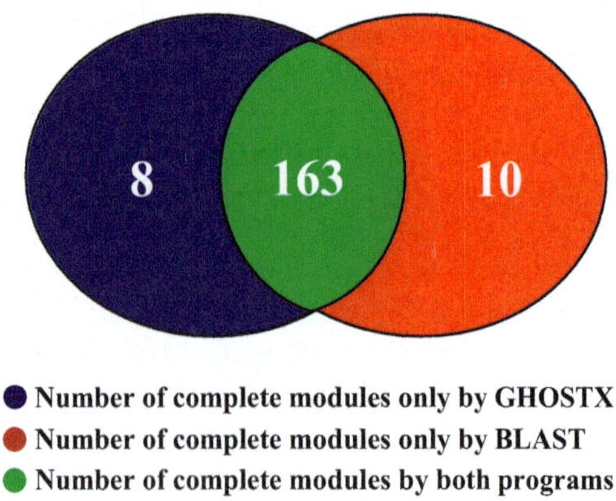

Fig. 7.14 Comparison of module completion ratios using GHOSTX with those using BLAST. Among 766 functional modules in 1 million metagenomic genes from the GS30 dataset, 163 modules were completed by both search methods

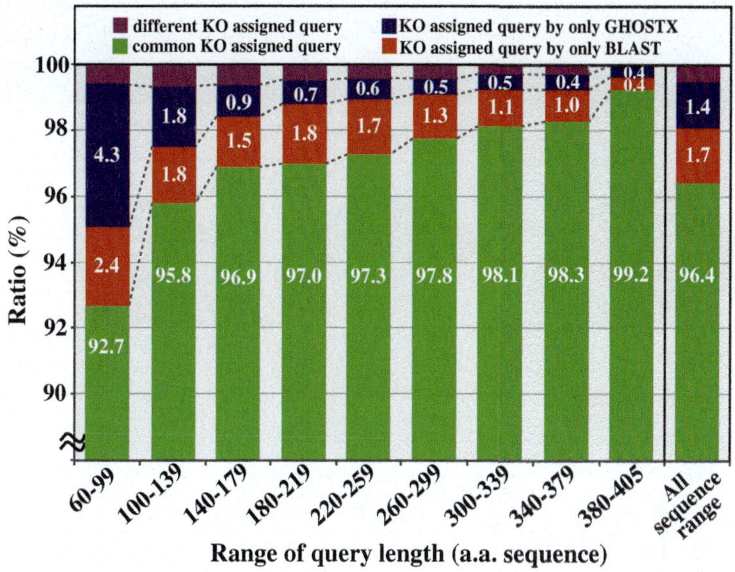

Fig. 7.15 Comparison of patterns of Kyoto Encyclopedia of Genes and Genomes Orthology (KO) assignment depending on query sequence length in GHOSTX with that in BLAST. Although 92.7% of KO-assigned query sequences were common between GHOSTX and BLAST in short sequences of fewer than 100 amino acid residues, 96.4% of KO-assigned query sequences were common between both homology search programs for all query length ranges (see Fig. 7.16)

distribution pattern of all query sequences (Fig. 7.17c, d) (Arai et al. 2018). Thus, GHOSTX provided more functional information than BLAST by assigning KOs to short query sequences.

7.4 User's Guide for MAPLE Version 2.3.1

7.4.1 Data Submission

Massive NGS datasets consist of raw data that cannot be directly imported into the MAPLE system as a standard FASTQ file because data submitted to MAPLE must be in multi-FASTA files of amino acid (a.a.) sequences with unique IDs in the comment lines and without tab delimiters (Fig. 7.6a). This compatibility issue can be an impediment to researchers who would like to utilize MAPLE but are unfamiliar with the relevant bioinformatic tools. Accordingly, we developed MAPLE Submission Data Maker (MSDM), an application that automatically converts paired-end FASTQ files into multi-FASTA files. This software can be installed on personal computers with MacOS X and Windows OS, as detailed in the installation manual. Users can obtain the software from the MAPLE software download site (https://maple.jamstec.go.jp/maple/maple-2.3.1/softdownload/) after

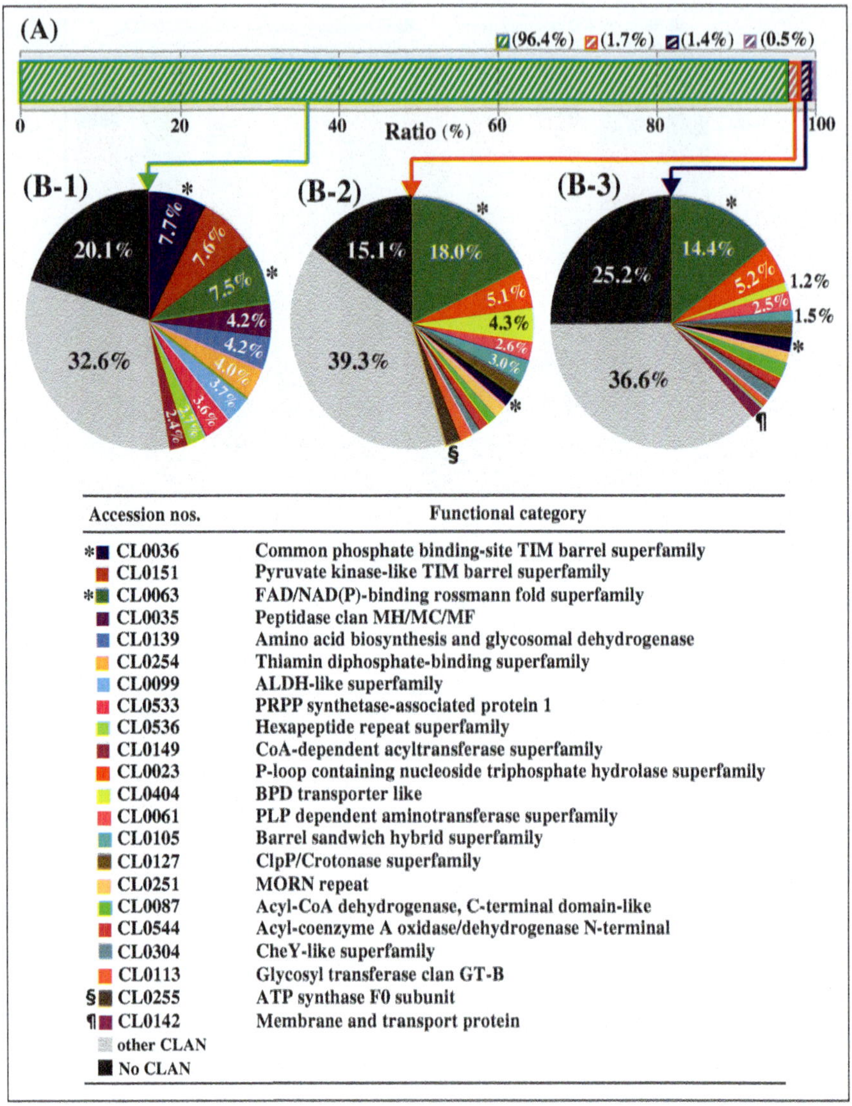

Fig. 7.16 Classification of functional categories of Pfam domains in Kyoto Encyclopedia of Genes and Genomes Orthology (KO)-assigned sequences based on BLAST and GHOSTX search. (**a**) Breakdown of KO-assigned sequences by BLAST and GHOSTX. Green hatching, common KO-assigned sequences by both search programs; red hatching, KO-assigned sequences only by BLAST; blue hatching, KO-assigned sequences only by GHOSTX; purple hatching, sequences for which different KOs were assigned depending on the engine used. (**b**) Breakdown of Pfam functional categories of KO-assigned sequences: (*B1*) green hatching, (*B2*) red hatching, (*B3*) blue hatching; *, commonly identified functions in B1 through B3; §, function identified only by BLAST; ¶, function identified only by GHOSTX

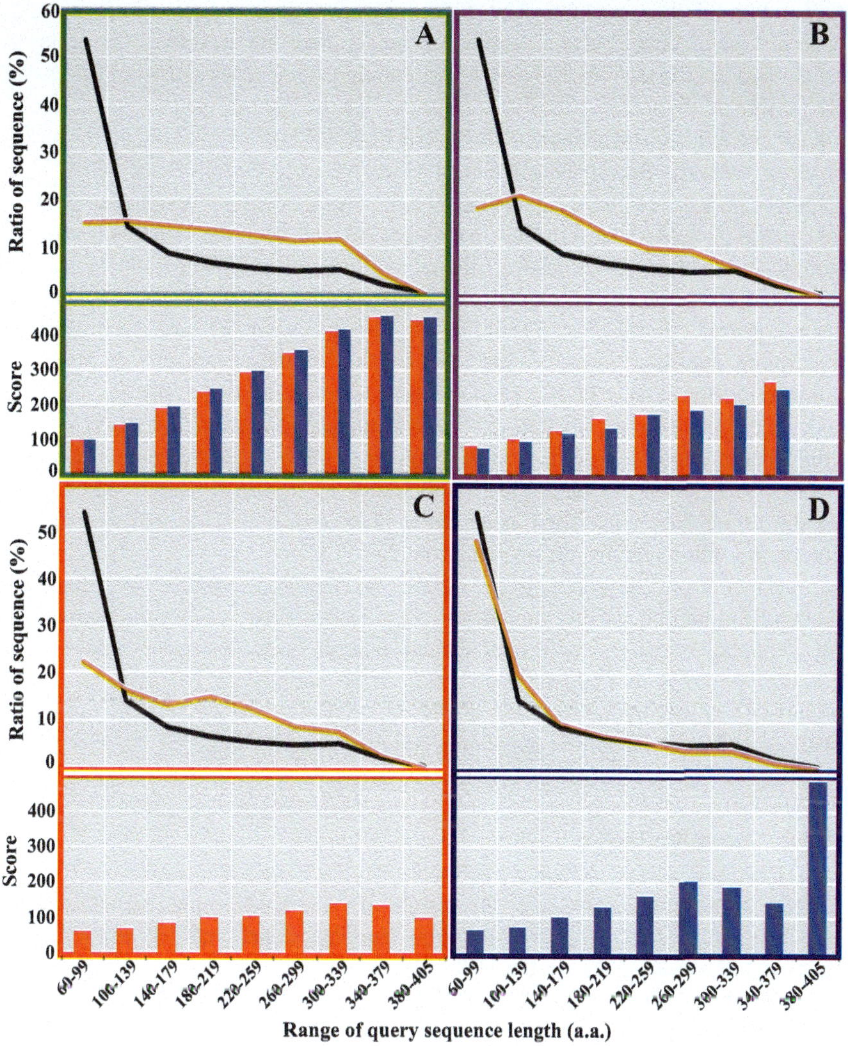

Fig. 7.17 Distribution patterns of sequence lengths and scores of the Kyoto Encyclopedia of Genes and Genomes Orthology (KO)-assigned queries. Line graphs and histograms indicate the distribution of sequence lengths and average scores of homology search, respectively. The frame colors for (**a**) through (**d**) correspond to the colors of the bar graphs in Figs. S2 and S3: (**a**) common KO-assigned queries by both homology search programs; (**b**) queries for which different KOs were assigned depending on the program; (**c**) KO-assigned queries only by BLAST; (**d**) KO-assigned queries only by GHOSTX; ■, all queries; ■, KO-assigned queries; ▬, average score by BLAST; ▬, average score by GHOSTX; a.a., amino acids

registration of their account. The query sequence does not necessarily need to be a complete gene, but a.a. sequences longer than 100 residues are generally recommended for accurate KO assignment, though some genes consist of fewer than 100 amino acids, such as ribosomal proteins. The number of query sequences may not exceed 3 million sequences owing to the limitations of computational resources; an error message is displayed when the file size exceeds this limit. When more sequence data are required for an analysis, the user can submit several sub-datasets consisting of fewer than 3 million sequences derived from the same metagenomic sample and then merge the results from all sub-datasets into one dataset by clicking the "Merge" button on the job list page. When two jobs with 3 million sequences are merged as one job (i.e., 6 million total sequences), it will take 10 h. Datasets containing 3 million sequences are ideal for the accurate evaluation of MCRs by considering the results of KO rarefaction curves to determine whether sufficient sequences have been included (https://maple.jamstec.go.jp/maple/maple-2.3.1/help.html), particularly for determining the abundance of completed modules. However, we can sufficiently elucidate the overall trends of the functionome indicated by metagenomic data even when using fewer than 3 million sequences. Rarefaction curves are automatically drawn when the MAPLE analysis is complete and can be accessed from the results page. Indeed, we have successfully determined the metabolic potential of the human gut microbiome from 13 healthy individuals by comparative analysis with total sequences from each consisting of fewer than 100,000 amino acids (Takami et al. 2012). After submission of a dataset, a URL address for accessing the results is displayed along with the job ID. The results are also displayed on the current page upon completion of the job.

7.4.2 Results Pages

To access the tables containing the results of completion ratios for all types of KEGG modules (i.e., pathway, structural complex, functional set, and signature modules), the user clicks on the URL address displayed on the submission page; alternatively, the user can be notified of job completion by e-mail and then click on the job ID in that e-mail. The user can view detailed information of the mapping results for each KEGG module by clicking on the module ID in each table. In addition, the user can access an overview of the MCR results by clicking on the "histogram (PNG)" button on the results page and accessing the mapping results of the KEGG module by placing the cursor on each module name (Fig. 7.7).

Additional information for each module, such as taxonomy, class, and definition, are also displayed on the results page (Fig. 7.7). Taxonomy is defined as the biological classification based on the MCR patterns of reference organisms with the determined genomic sequences. For example, if a module contains more than four prokaryotic species (i.e., Bacteria or Archaea) belonging different phyla, the module is represented by a prokaryotic taxonomy. Similarly, if a module contains only species belonging to *Proteobacteria*, the taxonomy of the module is *Proteobacteria*.

Class indicates the module type based on the MCR patterns of reference organisms as previously defined.

The distribution of MCRs among 3186 prokaryotic or 436 eukaryotic species (one genome per species) can be categorized into four patterns (i.e., universal, restricted, diversified, and nonprokaryotic/noneukaryotic) regardless of the module type (i.e., pathway, structural complex, signature, or functional set). A Boolean algebra-like equation is defined by KEGG for each module, and MCRs are calculated based on this equation. Since KOs are often assigned to two or more modules, all module IDs that share each KO composing a particular module are listed together with the pathway IDs containing the KO. For example, K01623, a member of M00167 (reductive pentose phosphate cycle), is shared by M00001 (glycolysis) and M00003 (gluconeogenesis) and is used in seven pathway maps (Fig. 7.7). In the case of metagenomic sequences, taxonomic information for KO-assigned genes is displayed, and the details of the phylum or class level for every KO, which facilitate the classification of organisms contributing to the completion of each functional module, are listed. For example, the module for virtual prokaryotic ribosomes (M90000), comprising 21 bacterial, 27 archaeal, and 31 common ribosomal proteins between Bacteria and Archaea (Fig. 7.13a), can be used to represent the taxonomic breakdown of prokaryotes in the metagenome instead of the 16S rRNA gene, whose copy number varies among prokaryotic species. Ribosomal proteins can be used effectively because most are encoded by single-copy genes in the genome, and there is only minor variation in length among orthologous groups. The results are removed from the server 14 days after the job is completed; however, the user can download the results by clicking the "MAPLE results" button (Fig. 7.7).

The user can browse all MAPLE data by re-uploading previously downloaded data from the first page. In addition, the user can also easily download the data in an Excel-readable format containing KO assignments by KAAS, MCRs, the abundances of KO-assigned genes and completed modules, analyses of module significance results, and taxonomic information for the KO-assigned genes mapped to each module from the results page (Fig. 7.8).

7.4.3 Comparison of Results

Users can compare results not only between their own jobs but also between job(s) and KEGG-annotated genomes by clicking the "MAPLE job comparison" button on the top of the page and then inputting an e-mail address to access the job list. To conduct a comparative analysis, the "Comparison" button can be selected after checking the job IDs to be compared on the job list page. When several IDs of jobs to be compared are checked in the job list and the "Comparison" button is clicked, the job arrangement page is displayed. The user can add pre-analyzed individual organisms from the organism list if necessary and change the display order. The comparison results of MCR values and mapping patterns for each KEGG module are displayed side by side in parallel (Fig. 7.9). Detailed information for each KEGG module and the taxonomy of the KO-assigned genes mapped to the module are displayed, and the MCR and taxonomy results of the KO-assigned genes can be downloaded, as previously mentioned. Comparisons between KEGG-annotated genomes, excluding the user's jobs, are also possible. The user can directly access the comparison page without submitting an e-mail address using the "MAPLE genome comparison" button at the top of the page.

7.4.4 Visualization of MAPLE Results Using MAPLE Graph Maker

MAPLE results, such as MCR, *Q*-value, and module abundance, can be easily downloaded as an Excel file. Drawing histograms using these files is laborious because there are 800 functional modules. To address this difficulty, we developed MAPLE Graph Maker (MGM) to automatically draw histograms of MAPLE results (Fig. 7.5). Users can easily create histograms by importing Excel files containing their MAPLE results. After a MAPLE job has been completed, the user can download an Excel file summarizing all the MAPLE results (MCR, *Q*-value, and module abundance) from the first results page (Fig. 7.7). When MGM is started, the initial menu is shown.

When the user selects an Excel file and clicks the "Read Excel" button, a new menu is displayed after 20 s, from which the user can select data (i.e., MCR%, *Q*-value, and module abundance) and drawing parameters (combination and order of the data and the elimination of the histogram showing that MCR% is zero in all samples used for comparison). When the user clicks the "Show Graph" button after making the appropriate selections, the histogram is automatically displayed. This histogram can be saved as a PDF file.

7.5 Biodiversity of Marine Samples Revealed by Functional Metagenomics

7.5.1 Functional Metagenomic Analysis of Ocean Sample

In addition to evaluating the presence or absence of module function based on MCR values, the MAPLE system enables the estimation of the abundance of complete functional modules and the biodiversity of the contributors to it. MAPLE can also indicate the working probability of the MCR results using statistical tests. In this section, we highlight the differences in microbial community structure and the functional potential among four metagenomic datasets from the GOS expedition (Venter et al. 2004) using these criteria. Among 83 samples, we selected four metagenomic samples including over 1 million protein-coding genes, two from the Sargasso Sea (GS000c and GS000d) and two from near the Galápagos Islands (GS030 and GS031). Approximately 1.21 million to 1.43 million amino acid sequences in the multi-FASTA format were submitted to MAPLE.

7.5.2 Microbial Community Structure of GOS Sites Based on Ribosome Proteins

To compare the microbial community structure among the four GOS sites, we analyzed the taxonomic composition at the phylum level based on mapping patterns of the metagenomic sequences to M91000, which corresponds to a virtual universal ribosome comprising 130 ribosomal proteins as mentioned above. Because filters with a pore size of 0.1–0.22 μm were used to recover the cells from seawater collected at the four GOS sites in the previous studies (Venter et al. 2004; Yutin et al. 2007), most of the recovered cells were expected to be prokaryotic after prefiltration with 0.8-μm pore filters. As anticipated, more than 99% of the microbial community was composed of prokaryotes for all GOS sites, particularly the GS030 site; Bacteria accounted for 99.9% of sampled cells at GS030, whereas Archaea accounted for about 2–6% at the other three sites (Fig. 7.18a). Although Archaea were more represented at site GS000c (5.9%) relative to other sites (GS000d, 2.3%; GS030, 0.07%; GS031, 3.5%), the archaeal proportion of the microbial community was generally very low (Takami et al. 2016).

The major bacterial taxon at all GOS sites was *Alphaproteobacteria*. The proportion of *Alphaproteobacteria* within the WC of microbes was approximately 60%, except for at site GS000c, where the proportion was only 36%. On the other hand, site GS000c exhibited the highest proportion of *Gammaproteobacteria* (26%), which was the second most common taxon across all sites (Fig. 7.18b).

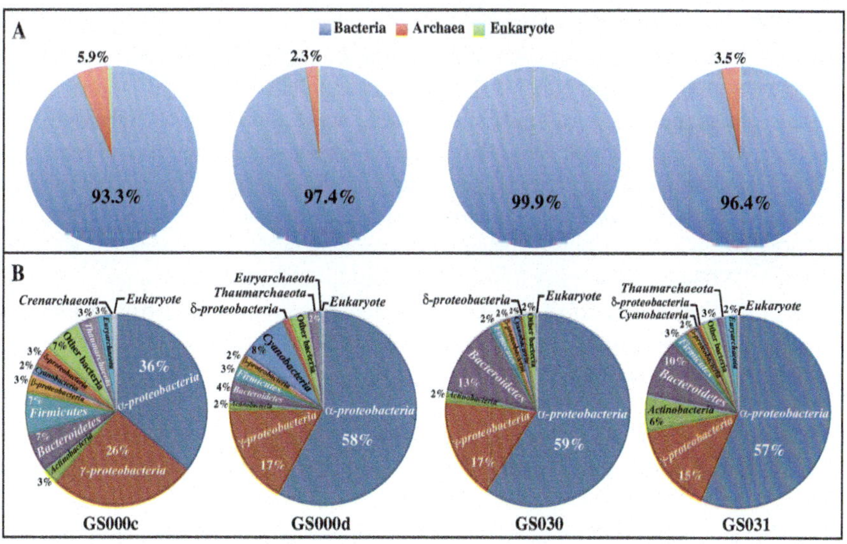

Fig. 7.18 Microbial community structure for the four Global Ocean Sampling expedition sites based on ribosomal proteins. (**a**) Microbial community structure at the kingdom level. (**b**) Prokaryotic community structure at the phylum (partially class) level. Reproduced with permission from *DNA Research* (Takami et al. 2016)

Another major difference between sites GS000c and GS000d was the proportion of *Firmicutes* and *Bacteroidetes*; the proportions of both organisms tended to be higher at site GS000c than at site GS000d. In contrast, the population of *Cyanobacteria* was four times higher in GS000d (8%) than in GS000c, despite the similarities in environmental and sampling conditions of these two sites. With respect to the other two sites near the Galápagos Islands, there were no substantial differences in bacterial proportions, with the exception that the proportion of *Actinobacteria* was three times higher at site GS031 (6%) than at site GS0030, whereas almost no Archaea were detected at site GS030 (Fig. 7.18b) (Takami et al. 2016).

7.5.3 Discrimination of GOS Sites by Differences in Module Abundance

Although MAPLE highlights the differences in the potential functions of organisms and environmental samples if the MCRs of various modules are different, it was not possible to clarify the differences in the functional potential of modules commonly completed in all samples. Out of 444 KEGG modules (excluding the nonprokaryotic modules), 155 modules were completed by the WC consisting of various taxonomic ranks, whereas 136 modules were completed by ITRs.

Among 155 complete modules (88 pathway, 55 complex, and 12 functional set modules), there were 42 pathway modules and 24 structural complex modules showing more than twofold differences in module abundance ratios among the four GOS sites; these data were normalized according to the average length of each KO and number of ribosomal proteins identified at each GOS site (Fig. 7.19). For example, module M00572, responsible for pimeloyl-ACP biosynthesis related to biotin metabolism, showed a more than tenfold difference in the module abundance ratio, although the module was completed by only one ITR (namely, *Gammaproteobacteria*-others, i.e., *Gammaproteobacteria* excluding the order *Enterobacteriales* [γ-others]) for the four GOS sites. This module, which included six reaction steps, contained two module-specific KOs, namely, K02169 and K02170, and there were substantial differences in the abundances of K02170 between GS031 and the other three sites. Indeed, most of the sequences assigned to K02170 were derived from γ-others, whereas some sequences from *Deltaproteobacteria*, having MCRs of 83.3%, also contributed to the increase in the total module abundance of WC (Takami et al. 2016).

7.5.4 Biodiversity of Contributors to Module Abundance

M00018 (threonine biosynthesis), which was categorized into the universal module (class A; for module class), was completed by various ITRs. However, there were no substantial differences in module abundance ratios among the four GOS sites. The population patterns of ITRs contributing to the module abundance were similar for the four GOS sites, with the exception of GS000c. *Alphaproteobacteria* accounted

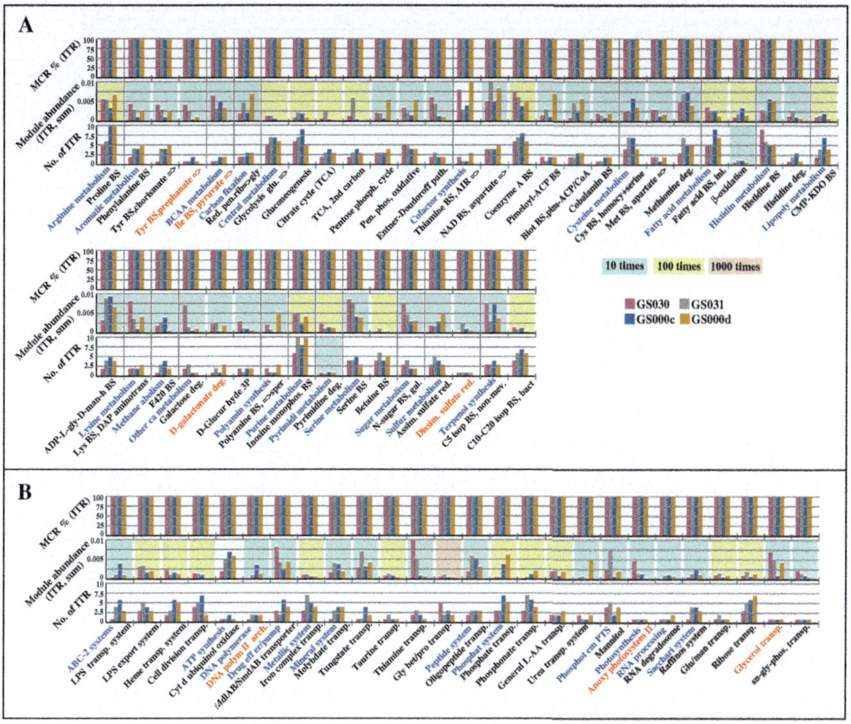

Fig. 7.19 Comparison of biodiversity contributing to the module completion ratio (MCR) and relative module abundance in the pathway and structural complex modules among the four Global Ocean Sampling (GOS) sites. The upper histogram shows the MCR patterns of the KEGG module. Red letters indicate rare modules. The middle histogram shows the module abundance. The lower histogram shows the number of individual taxonomic ranks that completed the functional module. The module abundance was calculated by dividing the abundance of each module by the minimum abundance among the four GOS sites. Background colors indicate the relative scale for each histogram

for 80% or more of the module abundance in GS000d, GS030, and GS031, and after adding the second and third major ITRs, namely, *Gammaproteobacteria* and *Bacteroidetes*, more than 98% of the modules were accounted for (Fig. 7.20). For GS000c, the ratio of *Alphaproteobacteria* was low (63%) and that of *Gammaproteobacteria* for the second major ITR was more than twice that of the other three sites. The remaining 15% nearly evenly comprised various ITRs, including *Bacteroidetes*, *Firmicutes–Clostridia*, *Deltaproteobacteria*, and *Thaumarchaeota*.

Modules M00338 and M00082, which were categorized as restricted (class B) and diversified (class C) modules, respectively, did not show significant differences in module abundance ratios, except for GS000d, which had a lower module abundance than the other sites (Fig. 7.20). As in the case of module M00018, the patterns of various ITRs, including those that completed module M00082 (fatty acid biosynthesis initiation), and the module abundance ratio of each ITR were similar

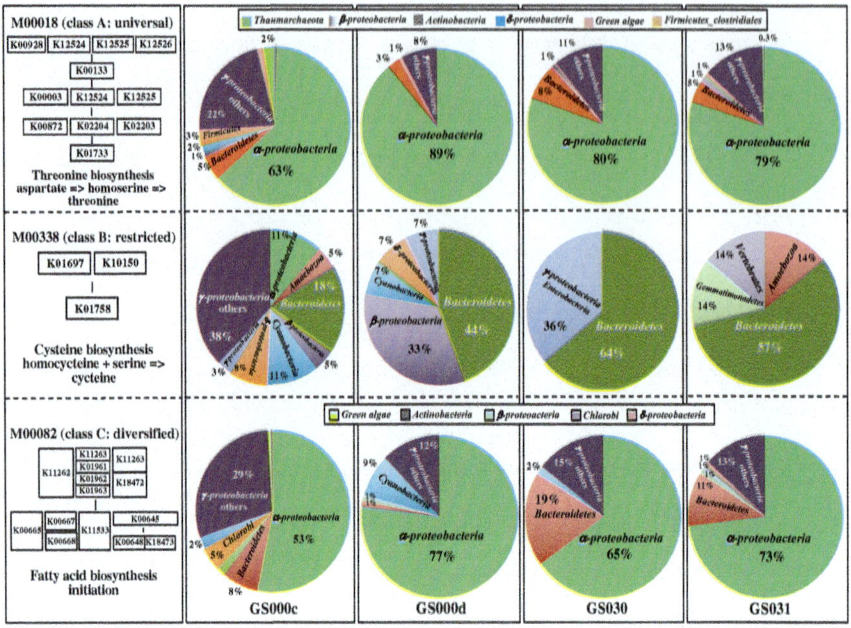

Fig. 7.20 Comparison of abundance ratios of the modules completed by individual taxonomic ranks (ITRs). There were no substantial differences in relative abundances of pathway modules among the four global ocean sampling sites selected for comparison of the contributions of ITR patterns to module abundance. The typical modules categorized into classes A, B, and C are shown. Reproduced with permission from *DNA Research* (Takami et al. 2016)

between GS030 and GS031, but obviously different for GS000c and GS000d, even though *Alphaproteobacteria* was the major ITR contributing to module abundance for both sites (Fig. 7.20). In contrast, with respect to M00338 (cysteine biosynthesis), various ITRs and the module abundance ratios were completely different from one another. However, *Bacteroidetes*, one of the major contributors to the module abundance, was shared among the four GOS sites. For example, this module was completed by eight types of ITR in GS000c but only by two types of ITR in GS030, and two eukaryotic ITRs (Vertebrata and Amoebozoa) accounted for 14% of the total module abundance (28% in total) in GS031. In GS000c, two ITRs (γ-others and *Alphaproteobacteria*) that never appeared at the other three sites accounted for 38% and 11% of the total module abundance, respectively.

7.6 Conclusion and Future Prospects

We developed a new method for evaluating the potential functionome based on calculating the completion ratio of four types of KEGG modules (Kanehisa et al. 2008), namely, pathways, molecular complexes, functional sets, and signatures, represented as the percentage of a module component filled with the input KO-assigned genes by KAAS (Takami et al. 2012). Based on this method, we also developed MAPLE, an automated system that can perform analyses used in the evaluation of the potential functionome corresponding to the genome and metagenome (Takami et al. 2016). MAPLE first assigns KOs to the query genes, maps the KO-assigned genes to the KEGG functional modules, and then calculates the MCR of each functional module to characterize the potential functionome corresponding to the genomic and metagenomic data.

The sequence lengths generated by Illumina HiSeq and MiSeq sequencers (up to 400–500 bp as an assembled contig of a paired-end sequence) are shorter than the old Sanger sequences. However, we have confirmed that there are no discernible differences in KO ID assignment and mapping ratios of KO-assigned coding sequences for the KEGG modules in the metagenomic sequences of the human gut microbiome (Takami et al. 2012) when comparing Sanger (Kurokawa et al. 2007) and Illumina reads (Qin et al. 2010). MAPLE was further improved, and two more useful functions have been added to this system to calculate module abundance and *Q*-value, which indicate the functional abundance and statistical significance of the MCR results, respectively. To increase user functionality, we also added another new function for automatically drawing rarefaction curves to facilitate determining if sufficient sequence data has been input, and we now supply MAPLE users with stand-alone automated histogram drawing program (MGM), which produce graphs based on MCR, *Q*-value, and module abundance outputs. In addition, we successfully reduced the calculation time by one fifteenth through the use of GHOSTX instead of BLAST in the current version of MAPLE. This MAPLE system has been used for metagenomic analysis targeting the human gut microbiome (Obregon-Tito et al. 2015) and environmental microbial communities (Pilgrim et al. 2017; Suzuki et al. 2017; Zeng et al. 2017) as well as for comparative functional analysis of individual microbes, even within candidate phyla (Hayatsu et al. 2017; Kuo et al. 2017; Shoemaker et al. 2015; Takami et al. 2015, 2017).

The MAPLE system is also well positioned to take advantage of recent technological advancements. For example, Oxford Nanopore Technologies has developed a nanopore DNA sequencer, MinION, which is a portable, real-time, long-read, low-cost sequencing device that has been designed to bring easy biological analyses to anyone in the scientific research community (https://nanoporetech.com/). Although there are still shortcomings in its sequencing accuracy, the performance of high-throughput MinION sequencing is approaching that of the Illumina MiSeq sequencer (https://www.illumina.com/), and the total length of sequence reads ranges from 5 to 30 Gb. Considering that functional metagenomic analysis using MAPLE requires numerous sequence reads longer than 400 bases, further improvement of nanopore DNA sequencers can be expected, together with enhancements of Illumina and PacBio (http://www.pacb.com/) DNA sequencers, and will accelerate functional metagenomics and metatranscriptomics research using the MAPLE system.

Acknowledgments The author thanks Professor S. Goto of the Database Center for Life Science and Professor K. Takemoto of Kyushu Institute of Technology for their great contribution to this work. The author also thanks W. Arai and T. Nakagawa of JAMSTEC, T. Taniguchi of Mitsubishi Research Institute, K. Yoshimura of NEC Ltd., and H. Uehara of Hewlett-Packard Japan Ltd., for their technical assistance. This work was supported in part by KAKENHI Grants-in-Aid for Scientific Research (Nos. 17H00793 and 15KT0039). This work was also supported in part by grants from the Collaborative Research Program of the Institute for Chemical Research, Kyoto University (Nos. #2013-23 and #2014-24), and by a grant from the Cross-ministerial Strategic Innovation Promotion Program.

References

Arai W, Taniguchi T, Goto S, Moriya Y, Uehara H et al (2018) MAPLE 2.3.0: an improved system for evaluating the functionomes of genomes and metagenomes. Biosci Biotechnol Biochem 82:1515–1517

Filippo CD, Ramazzotti M, Fontana R, Cavalieri D (2012) Bioinformatic approaches for functional annotation and pathway inference in metagenomic data. Brief Bioinform 13:696–710

Hayatsu M, Tago K, Uchiyama I, Toyoda A, Wang Y, Shimomura Y et al (2017) An acid-tolerant ammonia-oxidizing γ-proteobacterium from soil. ISME J 11:1130–1141

Huson DH, Mitra S, Ruscheweyh HJ, Weber N, Schuster SC (2011) Integrative analysis of environmental sequences using MEGAN4. Genome Res 21:1552–1560

Kanehisa M, Goto S (2000) KEGG: Kyoto encyclopedia of genes and genomes. Nucleic Acids Res 28:27–30

Kanehisa M, Araki M, Goto S, Hattori M, Hirakawa M, Itoh M et al (2008) KEGG for linking genomes to life and environment. Nucleic Acids Res 36:D480–D484

Kuo V, Shoemaker WR, Muscarella ME, Lennon JT (2017) Whole-genome sequence of the soil bacterium *Micrococcus* sp. KBS0714. Genome Announc 5:e00697–e00617

Kurokawa K, Itoh T, Kuwahara T, Oshima K, Toh H, Toyoda A et al (2007) Comparative metagenomics revealed commonly enriched gene sets in human gut microbiome. DNA Res 14:169–181

Liu Z, Lozupone C, Hamady M, Bushman FD, Knight R (2007) Short pyrosequencing reads suffice for accurate microbial community analysis. Nucleic Acids Res 35:e120

Mende DR, Sunagawa S, Aeller G, Bork P (2013) Accurate and universal delineation of prokaryotic species. Nat Methods 10:881–884

Meyer F, Paarmann D, D'Souza M, Olson R, Glass EM, Kubal M et al (2008) The metagenomics RAST server–a public resource for the automatic phylogenetic and functional analysis of metagenomes. BMC Bioinform 9:386

Moriya Y, Itoh M, Okuda S, Yoshizawa A, Kanehisa M (2007) KAAS: an automatic genome annotation and pathway reconstruction server. Nucleic Acids Res 35:W182–W185

Obregon-Tito AJ, Tito RY, Metcalf J, Sankaranarayanan K, Clemente JC, Ursell LK et al (2015) Subsistence strategies in traditional societies distinguish gut microbiomes. Nat Commun 6:6505

Overbeek R, Begley T, Butler RM, Choudhuri JV, Chuang HY, Cohoon M et al (2005) The subsystems approach to genome annotation and its use in the project to annotate 1000 genomes. Nucleic Acids Res 33:5691–5702

Pilgrim J, Ander M, Garros C, Baylis M, Hurst GDD, Siozios S (2017) Torix group *Rickettsia* are widespread in *Culicoides* biting midges (Diptera: Ceratopogonidae), reach high frequency and carry unique genomic features. Environ Microbiol 19:4238–4255

Qin J, Li R, Raes J, Arumugam M, Burgdorf KS, Manichanh C et al (2010) A human gut microbial gene catalogue established by metagenomic sequencing. Nature 464:59–65

Shoemaker WR, Muscarella ME, Lennon JT (2015) Genome sequence of the soil bacterium *Janthinobacterium* sp. KBS0711. Genome Announc 3:e00689–e00615

Stoddard SF, Smith BJ, Hein R, Roller BRK, Schmidt TM (2015) *rrn*DB: improved tools for interpreting rRNA gene abundance in bacteria and archaea and a new foundation for future development. Nucleic Acids Res 43:D593–D598

Suzuki S, Kakuta M, Ishida T, Akiyama Y (2014) GHOSTX: an improved sequence homology search algorithm using a query suffix array and a database suffix array. PLoS One 9:e103833

Suzuki S, Ishii S, Hoshino T, Rietze A, Tenney A, Morrill P et al (2017) Unusual metabolic diversity of hyperalkaliphilic microbial communities associated with subterranean serpentinization at The Cedars. ISME J 11:2584–2598

Takami H, Taniguchi T, Moriya Y, Kuwahara T, Kanehisa M, Goto S (2012) Evaluation method for the potential functionome harbored in the genome and metagenome. BMC Genomics 13:699

Takami H, Arai W, Takemoto K, Uchiyama I, Taniguchi T (2015) Functional classification of uncultured "*Candidatus* Caldiarchaeum subterraneum" using the MAPLE system. PLoS One 10:e0132994

Takami H, Taniguchi T, Arai W, Takemoto K, Moriya Y, Goto S (2016) An automated system for evaluation of the potential functionome: MAPLE version 2.1.0. DNA Res 23:467–475

Takami H, Toyoda A, Uchiyama I, Itoh T, Takaki Y, Arai W et al (2017) Complete genome sequence and expression profile of the commercial lytic enzyme producer *Lysobacter enzymogenes* M497-1. DNA Res 24:169–177

Venter JC, Remington K, Heidelberg JF, Halpern AL, Rusch D, Eisen JA et al (2004) Environmental genome shotgun sequencing of the Sargasso Sea. Science 304:66–74

Yutin N, Suzuki MT, Teeling H, Weber M, Venter JC, Rusch DB et al (2007) Assessing diversity and biogeography of aerobic anoxygenic phototrophic bacteria in surface waters of the Atlantic and Pacific Oceans using the Global Ocean sampling expedition metagenomes. Environ Microbiol 9:1464–1475

Zeng Q, Tian X, Wang L (2017) Genetic adaptation of microbial populations present in high-intensity catfish production systems with therapeutic oxytetracycline treatment. Sci Rep 7:17491

Part III

Applications in Ocean and Fisheries Sciences: Diversity and Function of Microbial Community

Comparison of Microscopic and PCR Amplicon and Shotgun Metagenomic Approaches Applied to Marine Diatom Communities

8

Tsuyoshi Watanabe and Tomoko Sakami

Abstract

Metagenomic analysis is recognized as an effective tool to monitor marine organisms. However, to date, the accuracy of the phytoplankton composition determined based on metagenomics has not been fully verified. Among phytoplankton, diatoms can be identified by silica cell wall morphology and enumerated using light microscopy (LM), although identification using either scanning electron microscopy (SEM) or DNA sequencing is required in some species. We examined the species composition of diatoms during the spring bloom in Sendai Bay using metagenomic analysis (whole metagenome, ribosomal small subunits (SSUs) and large subunits from shotgun metagenomic data, and PCR amplicons of the SSU V4 region) and compared the results with those determined by using morphological identification by microscopic observation. For all the metagenomic analysis methods, the abundance estimated by SSU sequences was closest to that revealed by LM. However, an adequate reference database and sufficient numbers of metagenome sequences (>10^7 reads) would improve estimation of species composition based on the SSU metagenomic data. Furthermore, seasonal dynamics based on microscopic observation and the SSU metagenomic approach were compared in the diatom communities at Sendai Bay from March 2012 to June 2014, revealing different tendencies in the annual changes both quantitatively and qualitatively. On the other hand,

T. Watanabe (✉)
Tohoku National Fisheries Research Institute, Japan Fisheries Research and Education Agency, Shiogama, Miyagi, Japan

JST, CREST, Saitama, Japan
e-mail: tsuyoshiw@affrc.go.jp

T. Sakami
National Research Institute of Aquaculture, Japan Fisheries Research and Education Agency, Minami-ise, Mie, Japan

T. Gojobori et al. (eds.), *Marine Metagenomics*,
https://doi.org/10.1007/978-981-13-8134-8_8

the metagenomic analysis showed changes in species composition in the genus *Skeletonema*, which could not be recognized by LM. These results indicate that the metagenomic approach can be a powerful tool for classification of species, but it is still ineffective for quantitation and precise identification of species. For phytoplankton monitoring, it is preferred to use morphological observation in combination with metagenomic analysis.

Keywords
Phytoplankton · Diatom · Phytoplankton monitoring methodology

8.1 The Necessity of Accurate Verification of the Metagenomic Analysis of Phytoplankton Composition

In recent years, metagenomic analysis of marine species by next-generation sequencing (NGS) has increased. Metagenomic studies of phytoplankton are also increasing. For example, the "Tara Oceans Project" generated an ocean microbial reference gene catalog including a large number of novel phytoplankton sequences from long-term global cruises (Bork et al. 2015; de Vargas et al. 2015), and Needham and Fuhrman (2016) revealed daily variation of phytoplankton, archaea, and bacteria during spring bloom using the NGS technique. Metagenomic approach based on direct cloning and shotgun sequencing of environmental DNA, which does not require PCR steps, is expected to be an alternative method of amplification (Not et al. 2009). Metagenomic analysis is thought to be a strong monitoring tool for marine organisms. However, the accuracy of the species composition determined by metagenomic analysis has not been verified in marine planktonic organisms. For example, in tintinnid ciliates, which can be identified based on morphological classification, the species composition determined by metagenomic analysis was roughly in agreement with that determined by morphological identification (Bachy et al. 2013). It was suggested that the difference is caused by misidentification of the morphologically very similar species in the direct observation and the variation in the copy number of rRNA genes among species in the metagenomic analysis. Accuracy of metagenomic analysis of species composition can be evaluated by comparison with the quantitative data. We suggest using diatoms for verification of phytoplankton composition, because (1) diatoms are characterized as major primary producers of high abundance and high diversity in coastal waters, (2) most diatom species can be identified by silica cell wall morphology and enumerated using light microscopy (LM), and (3) identification of some other species may require either scanning electron microscopy (SEM) or DNA sequencing.

Sendai Bay is the largest bay in Tohoku in northern Japan (Fig. 8.1). It is an area with a shallow continental shelf that opens into the Pacific Ocean, allowing frequent intrusion of the oceanic water. Freshwater from several rivers is also discharged into the bay. Diatom species composition of Sendai Bay is suitable for verification of metagenomic analysis because the species composition and

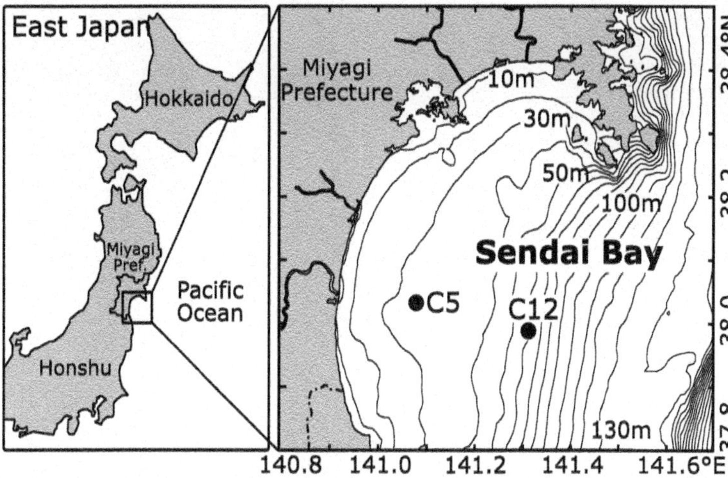

Fig. 8.1 Location of sampling stations C5 and C12 in Sendai Bay

seasonal dynamics of the phytoplankton community were sufficiently clarified using common methods (Taniuchi et al. 2017; Watanabe et al. 2017). Seasonal dynamics of the phytoplankton community in Sendai Bay are characterized as follows: diatoms are dominant from winter to spring, while pico-phytoplankton is dominant from summer to autumn (Taniuchi et al. 2017). The major species of the diatom community are *Chaetoceros* spp. in spring, *Skeletonema costatum* sensu lato at the end of spring bloom, *Leptocylindrus danicus* in summer, and *Thalassiosira* spp. from autumn to winter (Watanabe et al. 2017). We examined diatom species composition during the spring bloom using metagenomic analysis (whole shotgun metagenomic data, ribosomal small subunits (SSU) and large subunits (LSU) data from the shotgun metagenomic analysis, and PCR amplicons of the SSU V4 region) and compared the results with those obtained via microscopic identification. Furthermore, to verify the application of metagenomic analysis for phytoplankton monitoring, seasonal variations of species composition of the three dominant genera revealed by metagenomic analysis were compared with those determined by microscopic identification for about 2 years.

8.2 Metagenomic Approach for Analysis of the Phytoplankton Community

We collected samples on April 16, 2012, during the spring bloom. Water samples of about 9 L were obtained from the depth of the subsurface chlorophyll maximum (SCM) at stations C5 (18 m) and C12 (10 m) (Fig. 8.1). After filtering the water samples with 100 μm mesh, the samples were fractioned using 20 μm, 5 μm, 0.8 μm, and 0.2 μm pore-size nucleopore filters of 142 mm diameter (Fig. 8.2, see BioMarks

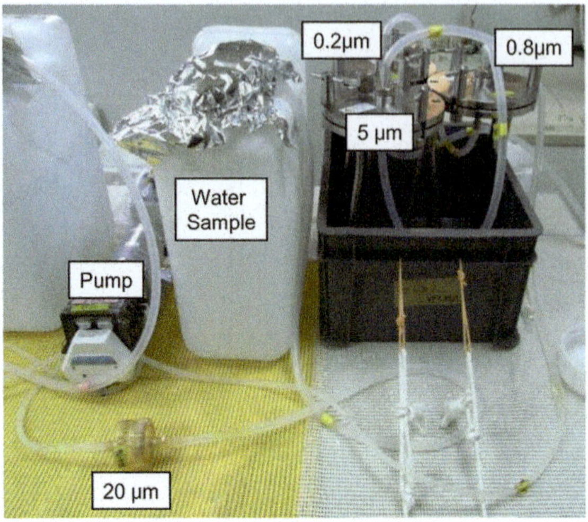

Fig. 8.2 The photograph of the filtration equipment for the DNA sampling in this study. Water sample (<100 μm) is pumped up by a peristaltic pump. First, the sample is filtered by 20 μm pore-size nylon mesh filter of 47 mm diameter. Subsequently, the water sample is filtered by 5 μm, 0.8 μm, and 0.2 μm pore-size polycarbonate filters of 142 mm diameter

Data Portal http://biomarks.eu/). In this study, 20 μm and 5 μm pore-size filters were used. DNA was extracted using a PowerWater DNA Isolation Kit (Mo Bio) in accordance with the manufacturer's instructions or the xanthogenate-SDS (XS) DNA extraction protocol (Leuko et al. 2008; Tillett and Neilan 2000). The extracted DNA was purified using a NucleoSpin gDNA Clean-up Kit (Macherey-Nagel) in accordance with the manufacturer's instructions. Sequencing was performed on the Illumina HiSeq or Miseq by 300 bp paired end run. Using the same DNA, we amplified the ribosomal SSU hypervariable V4 region (ca. 400 bp) by PCR using a primer set designed to amplify diatom genes (Zimmermann et al. 2011). The amplicons were also sequenced using the Illumina Miseq by 300 bp paired end run. We analyzed the metagenomic data via five methods: First, the PCR amplicons of the ribosomal small subunit (AMP) were analyzed. Next, the shotgun reads assigned as a ribosomal SSU were analyzed. Third, we analyzed those assigned as a ribosomal LSU. AMP, SSU, and LSU were identified with reference to the SILVA database of the Max Planck Institute (Quast et al. 2013). Fourth, the shotgun reads were identified by referring to a nucleotide database (nt) in the NCBI database (NT). Fifth, metagenomic analysis was performed by referring to a marine genome database, CAMERA (the Community Cyberinfrastructure for Advanced Microbial Ecology Research and Analysis) (CAM) in the UCSD database (CAMERA data can now be accessed from the iMicrobe Project database http://data.imicrobe.us). Sequences were annotated by a bit score of more than 100 with the homology search of BLAST using the above databases (Altschul et al. 1997). The sequences were classified by using the algorithms for distances to the lowest common ancestor (LCA) (Tarjan 1979). The dataset of species composition for verification were obtained by using LM. Phytoplankton species were based on Hasle et al. (1996). *Skeletonema* species were also identified and counted by SEM (Sarno et al. 2005).

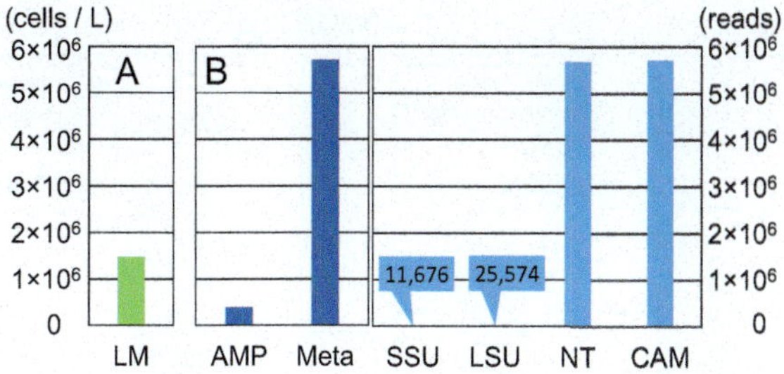

Fig. 8.3 Total cell density and read numbers (**a**) and annotated read numbers (**b**) at the surface of the station C5 in Sendai Bay in April 2012. AMP- and PCR-amplified ribosomal small subunit hypervariable V4 region identified by using the SILVA database; *CAM* whole shotgun sequences identified by using the CAMERA database, *LM* morphological identification by light microscopy, *LSU* ribosomal large subunit from shotgun identified by using the SILVA database; *Meta* whole sequences, *NT* whole shotgun sequences identified by using the nucleotide database, *SSU* ribosomal small subunit from shotgun identified by using the SILVA database

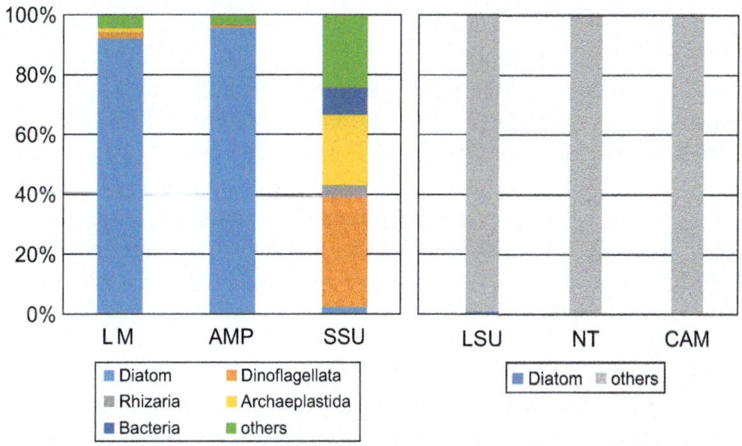

Fig. 8.4 Microbial composition in the sample from the surface of station C5 in Sendai Bay in April 2012. AMP, amplified ribosomal small subunit hypervariable V4 region identified by using the SILVA database; *CAM* whole shotgun sequences identified by using the CAMERA database, *LM* morphological identification by light microscopy, *LSU* ribosomal large subunit from shotgun identified by using the SILVA database, *NT* whole shotgun sequences identified by using the nucleotide database, *SSU* ribosomal small subunit from shotgun identified by using the SILVA database

Figure 8.3 shows the total cell density observed by using LM and the obtained read number of amplicon and shotgun sequences in each metagenomic data category. The NT and CAM of shotgun sequences obtained very large number of reads compared to the ribosomal gene reads. The composition of microorganisms was compared between the LM and SSU methods (Fig. 8.4a). In LM, diatoms (>90%) were dominant. However, in SSU, dinoflagellates were the dominant group,

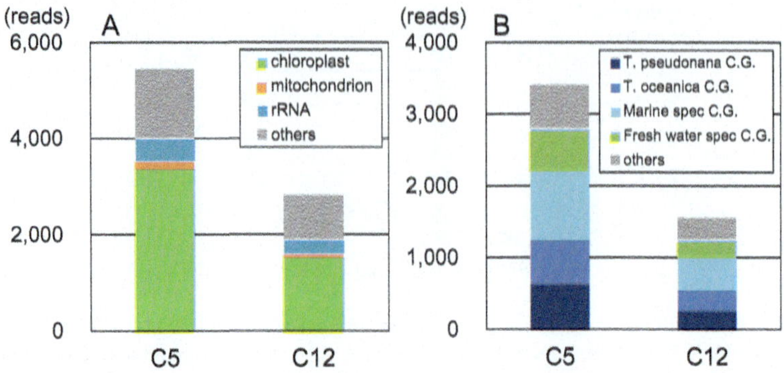

Fig. 8.5 Diatom gene composition in the nucleotide gene (NT) from shotgun sequences (**a**) and the chloroplast gene composition of diatoms in NT (**b**) in the sample from the surface of station C5 in Sendai Bay in April 2012. C.G., chloroplast complete genome

and diatom read number was very small (2.7%). Diatom reads were also very low for other metagenomic categories (LSU 2.0%, NT 0.1%, CAM 0.2%) (Fig. 8.4b). These results indicate that the phytoplankton composition obtained by shotgun metagenomic data was different from the ordinal microscopic observation. Most of the reads in AMP were diatoms because we used the primers specific to diatom genes.

Next, we analyzed the gene composition of shotgun sequences assigned to diatoms in NT (Fig. 8.5). Chloroplast genes showed the highest abundance (Fig. 8.5a). Within the chloroplast genes, those of *Thalassiosira pseudonana* and *T. oceanica*, for which whole-genome sequencing has been completed, were abundant (Fig. 8.5b). In addition, marine and fresh water species were also detected abundantly, and their chloroplast genomes have also been clarified completely. Thus, it seemed that a restricted species whose genome was sequenced completely was detected abundantly in this shotgun sequence analysis. Figure 8.6 shows the detected number of species and genera in LM and metagenomic analysis at station C5. CAM and NT detected much more species and genera than AMP and SSU methods. Those detected by LSU analysis were very low. Thus, there is a high possibility that NT and CAM overestimate the species and genus numbers.

Figure 8.7 shows the genus composition determined by the LM method and metagenomic analysis at station C5. In LM, *Skeletonema*, *Chaetoceros*, and *Thalassiosira* appeared at similar levels. In the metagenomic analysis, *Thalassiosira* was the most dominant genus in all categories. In particular, more than 90% of AMP reads belonged to the genus *Thalassiosira*. The second dominant genus was *Chaetoceros* in SSU analysis, but it was *Skeletonema* in LSU, NT, and CAM analyses. Thus, the genus composition was different between LM and metagenomics analysis. Figure 8.8 shows species composition of the dominant genus *Thalassiosira* at station C5. All of the species observed by LM were also detected by using metagenomic analysis. However, the species composition was different between

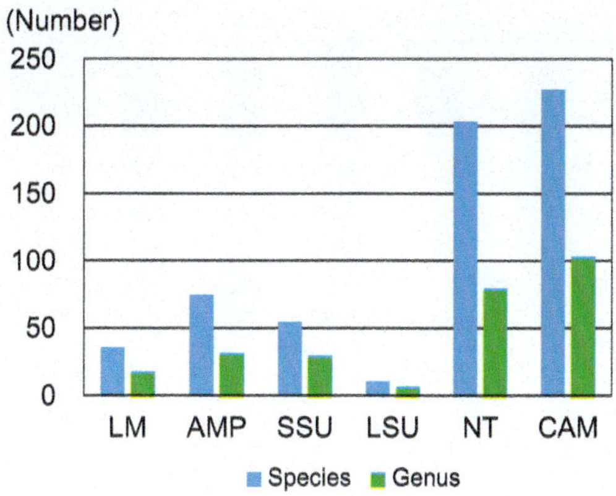

Fig. 8.6 Species and genus number of diatoms detected in the sample collected from the surface of station C5 in Sendai Bay, April 2012. AMP, amplified ribosomal small subunit hypervariable V4 region identified by using the SILVA database; *CAM* whole shotgun sequences identified by using the CAMERA database, *LM* morphological identification by light microscopy, *LSU* ribosomal large subunit from shotgun identified by using the SILVA database, *NT* whole shotgun sequences identified by using the nucleotide database, *SSU* ribosomal small subunit from shotgun identified by using the SILVA database

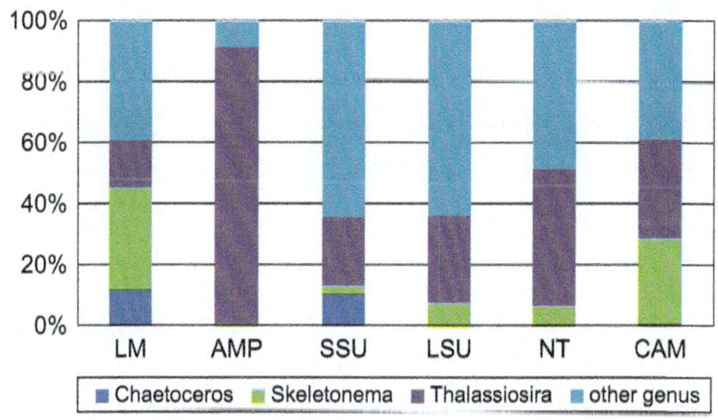

Fig. 8.7 Genus composition of diatoms in the sample collected from the surface of station C5 in Sendai Bay, April 2012. AMP, amplified ribosomal small subunit hypervariable V4 region identified by using the SILVA database; *CAM* whole shotgun sequences identified by using the CAMERA database, *LM* morphological identification by light microscopy, *LSU* ribosomal large subunit from shotgun identified by using the SILVA database, *NT* whole shotgun sequences identified by using the nucleotide database, *SSU* ribosomal small subunit from shotgun identified by using the SILVA database

LM and metagenomic analysis; *T. mala* and *T. nordenskioeldii* were dominant in LM. On the other hand, *T. auguste-lineata* and *T. curviseriata* were dominant in AMP and SSU analyses, and *T. oceanica* and *T. pseudonana* were dominant in LSU, NT, and CAM analyses. Even in the metagenomic analysis, the dominant species varied with the reference gene and/or database (AMP, SSU vs LSU, NT,

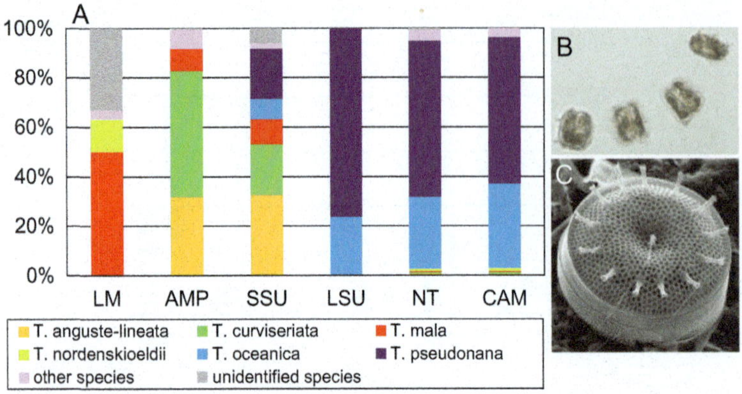

Fig. 8.8 Species composition in the genus *Thalassiosira* in the sample collected from the surface at station C5 in Sendai Bay, April 2012 (**a**). A colony of living cells (**b**) and a SEM photograph of silica cell wall (**c**) of *Thalassiosira nordenskioeldii*

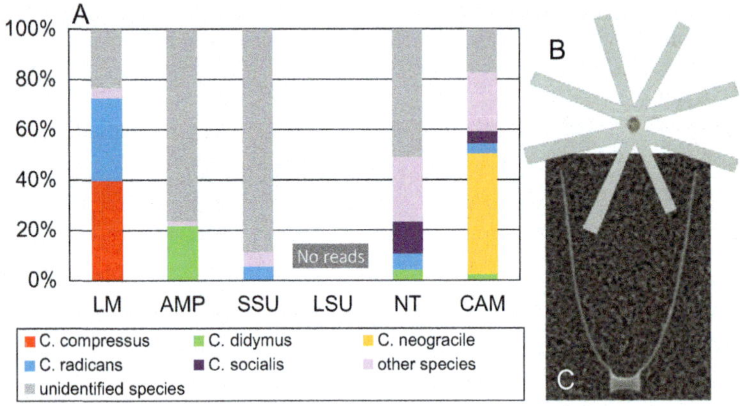

Fig. 8.9 Species composition of the genus *Chaetoceros* in the sample collected from the surface at station C5 in Sendai Bay, April 2012 (**a**). A living cell (**b**) and a SEM photograph of silica cell wall (**c**) of *Chaetoceros* sp.

CAM). In the genus *Chaetoceros*, the species observed under LM were not detected in the metagenomic analysis unlike the case of the genus *Thalassiosira* (Fig. 8.9). The dominant species differed by the analysis categories; they were *C. compressus* and *C. radicans* in LM, *C. didymus* in AMP, *C. radicans* in SSU, *C. socialis* in NT, and *C. neogracilis* in CAM. None of the *Chaetoceros* species was observed in LSU analysis. In addition, unidentified species were detected in AMP and SSU analyses at high percentages. Thus, the species composition detected using LM and metagenomic analysis in *Chaetoceros* was different, and the differences among the metagenomic analyses were dependent on the reference database. Because *Skeletonema* species cannot be identified using LM observation, SEM was used to

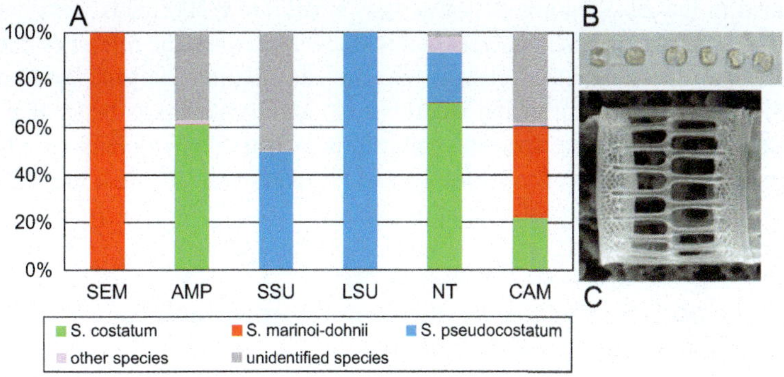

Fig. 8.10 Species composition of the genus *Skeletonema* in the sample collected from the surface at station C5 in Sendai Bay, April 2012 (**a**). A colony of living cells (**b**) and a SEM photograph of silica cell wall (**c**) of *Skeletonema marinoi-dohnii*

determine the species composition, and all of the *Skeletonema* cells were identified as *S. marinoi-domanii* (Fig. 8.10). On the other hand, *S. costatum* was the dominant species in AMP and NT. In SSU and LSU, *S. pseudocostatum* was the dominant species, while in CAM, *S. marinoi-domanii* and *S. costatum* were the dominant species. Thus, the species composition of *Skeletonema* was different between SEM observation and metagenomic analysis. On the other hand, the differences among the metagenomic analysis methods did not depend on the gene or the reference database, which was contrary to the other two genera.

8.3 Which Metagenomic Analysis Method Is Good for Understanding Phytoplankton Composition?

Our results showed that the species composition revealed by LM did not correspond with those obtained by metagenomic analysis. The characteristics of the metagenomic analysis methods were as follows: The PCR amplicon method (AMP) is effective to obtain diatom sequences, but the species composition is different from that observed by microscopic observation. This fact suggests that a PCR bias has considerable influence on the result (Hansen et al. 1998; Pinto and Raskin 2012). The whole-genome analysis (NT and CAM methods) can detect a large number of diatom sequences. However, the species composition is also different from that observed by LM, probably due to the effect of the species contained in the reference database, i.e., sequences containing highly conserved gene regions can be misidentified because of whole-genome data being available for other species, such as *T. pseudonana*. As a result, it can lead to overestimation of species number in NT and CAM analyses. The analyses using the ribosomal RNA genes (SSU and LSU methods) are free from the PCR bias. In addition, each sequence is registered for one species, in contrast to whole-genome analysis. However, the number of sequences contained in the metagenome data was very small. Only 300 reads were detected from 6×10^6 reads in SSU analysis in this study. Moreover, the reference database for LSU analysis seems insufficient for species composition analysis. Based on our results, we suggest the following solutions: (1) Metagenomics is better than PCR amplicons for determining species composition of marine phytoplankton, (2) SSU is the most practical reference currently, and (3) more than 10^7 reads are necessary to determine the diatom species composition.

8.4 Application of the Metagenomic Approach for Phytoplankton Monitoring

In order to verify the effectiveness of the metagenomic approach as an environmental monitoring tool, we compared the seasonal dynamics of diatoms determined by metagenomic analysis to those determined by microscopic observation. Abundances of the three dominant diatom genera (*Chaetoceros*, *Skeletonema*, *Thalassiosira*) were monitored from March 2012 to June 2014. Samples were taken from the surface (1 m) at station C5 in Sendai Bay (Fig. 8.1). The metagenomic data were obtained in the same manner as discussed earlier. SSU analysis was conducted using the 20 μm pore-size filters for metagenomic analysis. The species composition determined by SSU analysis is indicated as percentages of the reads of the species of the total shotgun reads. The dataset for verification of phytoplankton species composition was obtained using LM.

In the genus *Thalassiosira*, LM classified 13 species, 5 of which were unidentified. On the other hand, SSU analysis classified 23 species and 4 of them were unidentified. The mean abundance of *Thalassiosira* was 3.5×10^4 cells/L for LM and 2.6×10^{-4}% of the total shotgun reads for SSU. For seasonal dynamics, the cell density counted by LM increased from winter to spring (Fig. 8.11a). The percentage of SSUs increased in spring in 2012 and 2013 and in winter in 2013. For species composition, *Thalassiosira* sp. 2, *T.* cf. *mala*, and *T. nordenskioeldii* dominated in LM observation (Fig. 8.11b). *Thalassiosira* sp. 2 was remarkably dominant in spring, and *T.* cf. *mala* increased from autumn to spring. In SSU analysis, *Thalassiosira* S1, *T. mala*, and *T. curviseriata* were the dominant species (Fig. 8.11c). *Thalassiosira* S1 and the other species increased in spring. These results strongly suggest that *Thalassiosira* sp. 2 detected by using LM is the same as the *Thalassiosira* S1 detected by using SSU analysis. In SSU analysis, *T. mala* increased in autumn and was displaced by *T. curviseriata* in winter. These two species are very small and similar in morphology, and thus difficult to identify using LM observation. There is a possibility that *T.* cf. *mala* form observed by LM a complex of *T. mala* and *T. curviseriata*.

In the genus *Chaetoceros*, 24 species were classified by LM, and 5 of them were unidentified. In SSU, 18 species were classified and 12 species were unidentified. The species number classified by LM was more than that determined by SSU analysis. The mean abundance of the genus was 1.3×10^5 cells/L in LM and 2.5×10^{-3}% in SSU analysis. For seasonal dynamics, high cell densities were

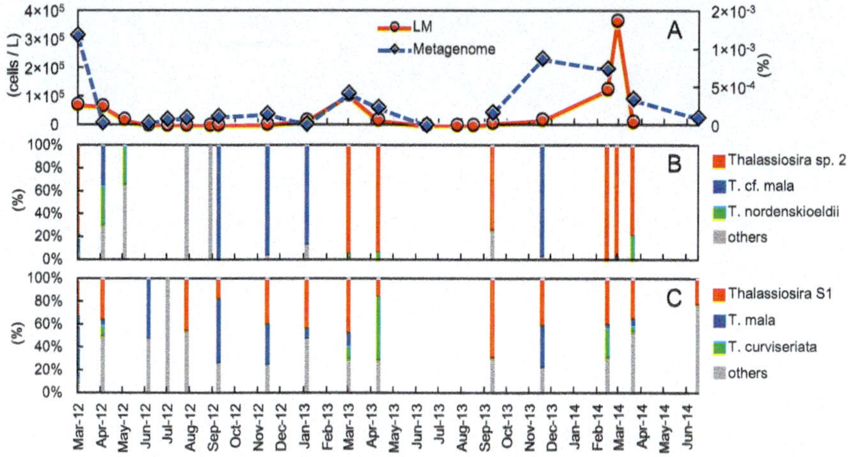

Fig. 8.11 Seasonal changes of the genus *Thalassiosira* in the surface water at station C5 in Sendai Bay from March 2012 to June 2014. Total cell densities determined by light microscopic observation (LM) and relative abundances of SSU reads (metagenome) (**a**), species compositions of *Thalassiosira* determined by light microscope (**b**), and those determined by SSU analysis (**c**). Metagenomic data are indicated as ratios of the read numbers assigned as SSU to the total shotgun reads

observed in LM in autumn and spring (Fig. 8.12a), though the percentage of SSUs showed three peaks in summer in 2012 and 2014 and spring in 2013. For species composition, *C. radicans*, *C. compressus*, *C. debilis*, and *C. socialis* dominated in LM observation (Fig. 8.12b). In LM, *C. debilis* and *C. socialis* increased from January to March, and *C. radicans* and *C. compressus* increased from April to May. In particular, *C. radicans* was remarkably dominant in April 2013. *Chaetoceros compressus* and other species increased in autumn. In SSU analysis, *C. radicans*, *Chaetoceros* S1, *Chaetoceros* S4, and *Chaetoceros* S5 dominated (Fig. 8.12c). *Chaetoceros radicans* reached their peak in April 2013, similar to that observed by LM. *Chaetoceros* S1 and *Chaetoceros* S4 increased in summer.

Species within the genus *Skeletonema* could not be identified by LM. On the other hand, ten species were classified by SSU analysis, and five of them were unidentified. The mean number of *Skeletonema* spp. per sample was 9.4×10^4 cells/L in LM and only $5.5 \times 10^{-5}\%$ in SSU analysis. For seasonal dynamics, the cell density in LM peaked in spring, and a small peak was observed in autumn in 2013 (Fig. 8.13a). These two peaks in spring and autumn in 2013 were also observed in the read numbers in SSU analysis. In addition, a relatively high percentage of SSUs were detected in summer in 2012 and winter in 2013. *Skeletonema costatum*, *Skeletonema* S1, and *Skeletonema* S5 dominated in SSU (Fig. 8.13b). *Skeletonema* S5 occurred in early spring, while *Skeletonema* S1 dominated at the peak in spring

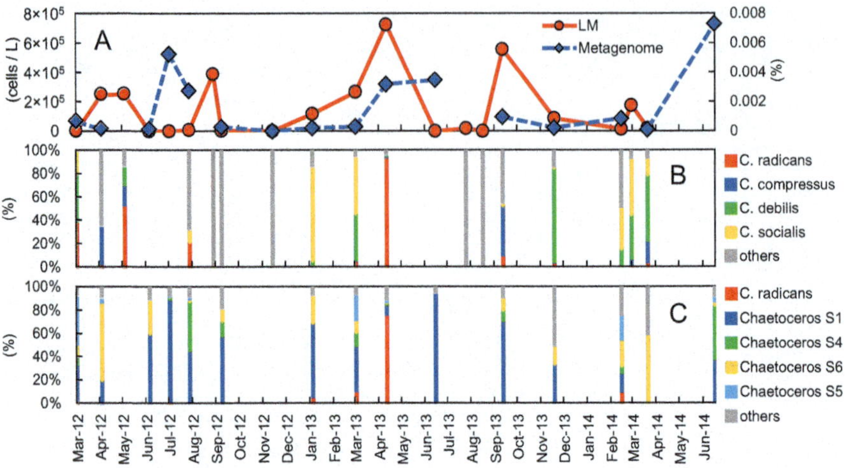

Fig. 8.12 Seasonal changes of the genus *Chaetoceros* in the surface water at station C5 in Sendai Bay from March 2012 to June 2014. Total cell densities determined by light microscopic observation (LM) and relative abundances of SSU reads (metagenome) (**a**), species compositions of the genus *Chaetoceros* determined by light microscopy (**b**), and those determined by SSU analysis (**c**). Metagenomic data are presented as ratios of the read numbers assigned as SSU to the total shotgun reads

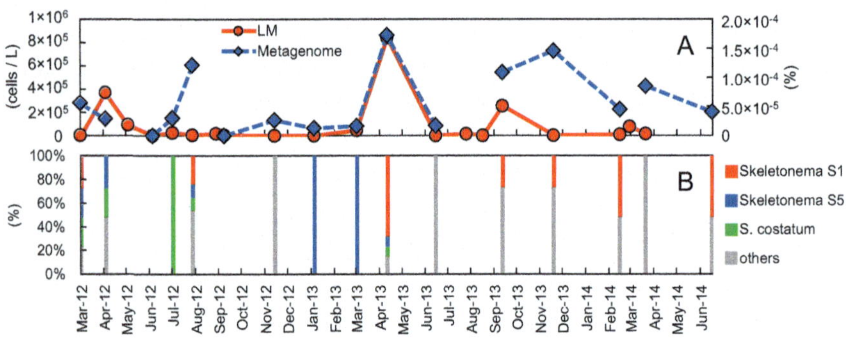

Fig. 8.13 Seasonal changes of the genus *Skeletonema* in the surface water at station C5 in Sendai Bay from March 2012 to June 2014. Total cell densities determined by light microscopic observation (LM) and relative abundance of SSU reads (metagenome) (**a**) and species compositions of the genus *Skeletonema* determined by SSU analysis (**b**). Metagenomic data are presented as a ratio of the read numbers assigned as SSU to the total shotgun reads

and some other species also occurred at the same time. These peaks occurring from autumn to winter were composed of several species having similar abundances. It seemed that (1) the major species changed during the spring peak, (2) the species composition was different between the bloom in spring and that in autumn, (3) the autumn bloom was a mixture of several species, and (4) the species composition of the bloom differed by year.

8.5 Possibility of Using the Metagenomic Approach to Monitor the Phytoplankton

It was shown that metagenomic analysis is effective for classifying the species that cannot be identified or detected by morphological observation in phytoplankton monitoring. A great improvement in the accuracy of determining species composition and seasonal variation of the community is expected via the use of metagenomes. However, this study showed that the metagenomic approach sometimes detected less species than the microscopic method. Therefore, at present, it is difficult to monitor the phytoplankton community by a metagenomic approach alone, and the best way is to consider combining a quantitative morphological observation with a metagenomic approach. Improvement of reference sequences and databases, together with morphological and genetic information of each species, is necessary for improving the accuracy of metagenomic analysis of the phytoplankton communities.

8.6 Conclusions

Since the metagenomic analysis of the phytoplankton community can obtain a precise species data, it is a suitable approach for the studies of phytoplankton species composition and/or community diversity. However, it is suggested that microscopic observation should be used together with metagenomic analysis to evaluate the accuracy and quantitative quality of the latter. The SSU method is most reliable because it is less affected by the bias of PCR or reference gene databases. As a condition, sufficient numbers of shotgun sequences ($>10^7$ reads) would be necessary to improve estimations. When the SSU analysis was applied for phytoplankton community monitoring at the coastal area in Sendai Bay, the dynamics detected by the metagenomic method were not in agreement with those determined by the ordinal microscopic observation. This result indicates that the metagenomic approach still has some problems, yet it can be a powerful tool for identification of species that cannot be classified by morphological observation. Thus, it is expected that more accurate phytoplankton monitoring can be performed by using a morphological observation and a metagenomic approach together.

Acknowledgments We thank Drs. Shigeho Kakehi, Yukiko Taniuchi, and Kazutoshi Yoshitake for their valuable collaboration on this research. We would like to thank Editage (www.editage.jp) for English language editing. This work was supported by CREST (Core Research for Evolutional Science and Technology) of Japan Science and Technology Corporation (JST).

References

Altschul SF, Madden TL, Schaffer AA, Zhang J, Zhang Z, Miller W, Lipman DJ (1997) Gapped BLAST and PSI-BLAST: a new generation of protein database search programs. Nucleic Acids Res 25(17):3389–3402

Bachy C, Dolan JR, López-García P, Deschamps P, Moreira D (2013) Accuracy of protist diversity assessments: morphology compared to cloning and direct pyrosequencing of 18S rRNA genes and ITS regions using the conspicuous tintinnid ciliates as a case study. ISME J 7(2):244–255

Bork P, Bowler C, de Vargas C, Gorsky G, Karsenti E, Wincker P (2015) Tara oceans studies plankton at planetary scale. Science 348(6237):873

de Vargas C, Audic S, Henry N, Decelle J, Mahé F, Logares R, Lara E, Berney C, Le Bescot N, Probert I, Carmichael M, Poulain J, Romac S, Colin S, Aury JM, Bittner L, Chaffron S, Dunthorn M, Engelen S, Flegontova O, Guidi L, Horák A, Jaillon O, Lima-Mendez G, Lukeš J, Malviya S, Morard R, Mulot M, Scalco E, Siano R, Vincent F, Zingone A, Dimier C, Picheral M, Searson S, Kandels-Lewis S, Coordinators TO, Acinas SG, Bork P, Bowler C, Gorsky G, Grimsley N, Hingamp P, Iudicone D, Not F, Ogata H, Pesant S, Raes J, Sieracki ME, Speich S, Stemmann L, Sunagawa S, Weissenbach J, Wincker P, Karsenti E, Carmichael M (2015) Eukaryotic plankton diversity in the sunlit ocean. Science 348(6237):1261605. https://doi.org/10.1126/science.1261605

Hansen MC, Tolker-Nielsen T, Givskov M, Molin S (1998) Biased 16S rDNA PCR amplification caused by interference from DNA flanking the template region. FEMS Microbiol Ecol 26(2):141–149. https://doi.org/10.1111/j.1574-6941.1998.tb00500.x

Hasle GR, Syvertsen EE, Steidinger KA, Tangen K, Tomas CR (1996) Identifying marine diatoms and dinoflagellates. Academic Press, San Diego

Leuko S, Goh F, Ibáñez-Peral R, Burns BP, Walter MR, Neilan BA (2008) Lysis efficiency of standard DNA extraction methods for *Halococcus* spp. in an organic rich environment. Extremophiles 12(2):301–308. https://doi.org/10.1007/s00792-007-0124-8

Needham DM, Fuhrman JA (2016) Pronounced daily succession of phytoplankton, archaea and bacteria following a spring bloom. Nat Microbiol 1:16005. https://doi.org/10.1038/nmicrobiol.2016.5

Not F, del Campo J, Balagué V, de Vargas C, Massana R (2009) New insights into the diversity of marine picoeukaryotes. PLoS One 4(9):e7143. https://doi.org/10.1371/journal.pone.0007143

Pinto AJ, Raskin L (2012) PCR biases distort bacterial and archaeal community structure in pyrosequencing datasets. PLoS One 7(8):e43093. https://doi.org/10.1371/journal.pone.0043093

Quast C, Pruesse E, Yilmaz P, Gerken J, Schweer T, Yarza P, Peplies J, Glöckner FO (2013) The SILVA ribosomal RNA gene database project: improved data processing and web-based tools. Nucleic Acids Res 41(D1):D590–D596. https://doi.org/10.1093/nar/gks1219

Sarno D, Kooistra WHCF, Medlin LK, Percopo I, Zingone A (2005) Diversity in the genus *Skeletonema* (Bacillariophyceae): II. An assessment of the taxonomy of *S. costatum*-like species, with the description of four new species. J Phycol 41:151–176. https://doi.org/10.1111/j.1529-8817.2005.04067.x

Taniuchi Y, Watanabe T, Kakehi S, Sakami T, Kuwata A (2017) Seasonal dynamics of the phytoplankton community in Sendai Bay, northern Japan. J Oceanogr 73(1):1–9. https://doi.org/10.1007/s10872-015-0334-0

Tarjan RE (1979) Applications of path compression on balanced trees. J ACM 26(4):690–715. https://doi.org/10.1145/322154.322161

Tillett D, Neilan BA (2000) Xanthogenate nucleic acid isolation from cultured and environmental cyanobacteria. J Phycol 36(1):251–258. https://doi.org/10.1046/j.1529-8817.2000.99079.x

Watanabe T, Taniuchi Y, Kakehi S, Sakami T, Kuwata A (2017) Seasonal succession in the diatom community of Sendai Bay, northern Japan, following the 2011 off the Pacific coast of Tohoku earthquake. J Oceanogr 73(1):133–144. https://doi.org/10.1007/s10872-016-0387-8

Zimmermann J, Jahn R, Gemeinholzer B (2011) Barcoding diatoms: evaluation of the V4 subregion on the 18S rRNA gene, including new primers and protocols. Organisms Diversity & Evolution 11(3):173–192. https://doi.org/10.1007/s13127-011-0050-6

Seasonal Dynamics of Bacterial Community Composition in Coastal Seawater at Sendai Bay, Japan

9

Tomoko Sakami, Tsuyoshi Watanabe, and Shigeho Kakehi

Abstract

Metagenome analysis could be a useful tool to monitor biological diversities in marine environments. However, environmental conditions fluctuate widely in coastal areas. We examined bacterial community compositions in coastal seawater from Sendai Bay for 2 years to clarify their temporal variation in relation to physical and biological environmental factors. Sequencing of 16S rRNA gene amplicons revealed obvious annual variation and the significant associations of temperature, salinity, and chlorophyll *a* concentration with the community composition change. The most dominant operational taxonomic unit (OTU) was assigned to the oceanic bacterium, SAR11. An OTU assigned to *Rhodobacteraceae* was also dominant during phytoplankton blooms. OTUs assigned to cyanobacteria or *Alteromonadaceae* were increased in the warm water period after phytoplankton blooming, from June to September. Of the top 20 OTUs, 12 OTUs had a significant correlation with water temperature, indicating seasonal fluctuation. Three OTUs significantly correlated with salinity, indicating that freshwater discharge influenced the bacterial community in the

T. Sakami (✉)
National Research Institute of Aquaculture, Japan Fisheries Research and Education Agency, Minami-ise, Mie, Japan
e-mail: sakami@affrc.go.jp

T. Watanabe (✉)
Tohoku National Fisheries Research Institute, Japan Fisheries Research and Education Agency, Shiogama, Miyagi, Japan

JST, CREST, Saitama, Japan
e-mail: tsuyoshiw@affrc.go.jp

S. Kakehi
Tohoku National Fisheries Research Institute, Japan Fisheries Research and Education Agency, Shiogama, Miyagi, Japan

T. Gojobori et al. (eds.), *Marine Metagenomics*,
https://doi.org/10.1007/978-981-13-8134-8_9

bay. Five OTUs that represented 38% of the total did not show any correlation with environmental factors. The dominant SAR11 OTU was included in this category, implying that hydrographical conditions are very important factors determining microbial community composition in coastal seawater.

Keywords
Microbial community · 16S rRNA gene · Illumina Miseq · Environmental factors

In the ocean, microorganisms have very important roles in organic matter cycling and geochemical material circulation. However, it is difficult to examine microbial communities in natural environments because most of the microorganisms are unculturable in ordinary culture media. Technologies to extract genetic information (metagenome) from whole microbial communities have been developed. These tools have revealed the nature and variation of the communities. The dynamics of microbial communities in marine environments have been explored using PCR amplicons of a marker gene, such as the 16S ribosomal RNA gene (Fuhrman et al. 2015).

Bacterial communities in seawater fluctuate regularly according to changes in temperature and/or salinity, with influences by a complicated network of environmental and biological factors (Fuhrman et al. 2006; Needham and Fuhrman 2016). Studies have indicated the strong association between bacterial communities and phytoplankton blooms (Rooney-Varga et al. 2005; Wemheuer et al. 2012). The composition of microbial communities in seawater also varies with physical water mass conditions, such as stratification and current movements (Morris et al. 2005). In coastal seawater, river water discharge from terrestrial areas and oceanic water from an offshore area gives complex effects on the microbial community (Fortunato et al. 2013).

The Sanriku-Joban coastal area is located at the Pacific coast of the northeast region of Japan. Seawater from the cold Oyashio Current and the warm Kuroshio Current mix offshore, which produces conditions that favor high biological productivity. Sendai Bay is located at the center of the Sanriku-Joban coastal line. The bay mouth is wide open to the Pacific Ocean, and seawater readily exchanges with the bay water, which receives the inflow of freshwater from the Abukuma River and Kitakami River. These hydrographical events result in marked fluctuations in the bay water environment (Kakehi et al. 2015).

We collected metagenomes from the seawater in Sendai Bay and analyzed the variation of microbial communities for 2 years. In this report, for clarifying the variation of the bacterial community, the composition of the microbial community was analyzed over time using the 16S rRNA gene amplicon. The relationship between major species and environmental factors of water temperature, salinity, and phytoplankton abundance was analyzed.

9.1 Materials and Methods

9.1.1 Sample Collection

From March 2012 to April 2014, seawater was collected at station C5 in Sendai Bay (Fig. 9.1) using Niskin and Vandorn samplers. Large particles like zooplankton were removed immediately after collection using 100 μm mesh-size plankton net. The seawater sample was filtered through 20 μm mesh-size plankton net and nucleopore membrane filters with sequentially smaller pore sizes of 5, 0.8, and 0.2 μm under positive pressure using a peristaltic pump. Each filter was immediately frozen and stored at −80 °C until DNA extraction was done. Environmental parameters were determined as described previously (Sakami et al. 2015). Water temperature, salinity, and chlorophyll concentration was measured in situ using an aqua quality sensor. Chlorophyll concentration was calibrated by fluorescence measurement. Nutrient salt was measured with an AutoAnalyzer.

9.1.2 DNA Extraction

DNA was extracted using a PowerWater DNA Isolation Kit (Mo Bio) in accordance with the manufacturer's instructions.

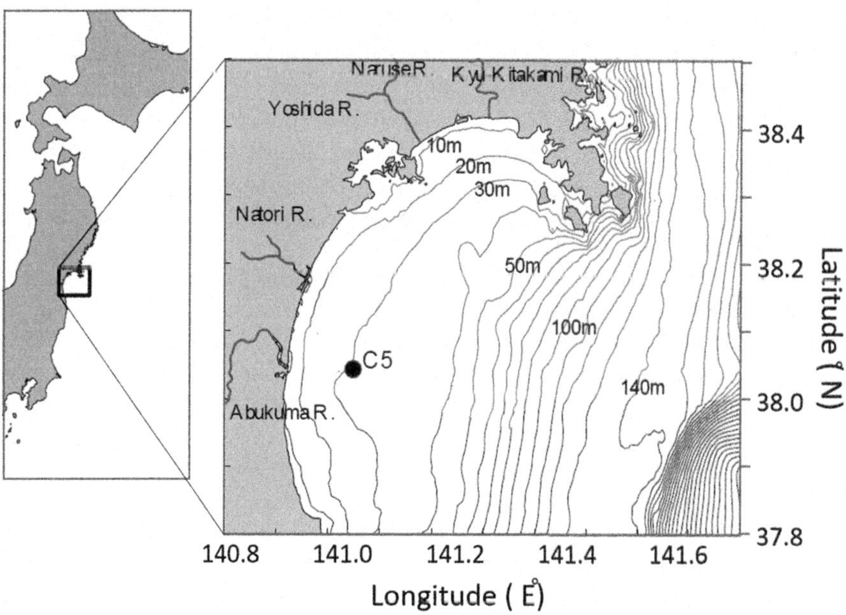

Fig. 9.1 Location of the observation station and bathymetry in Sendai Bay

9.1.3 Amplification and Sequencing of 16S rRNA Gene

PCR amplification, sequencing, and data analysis were conducted as described previously (Sakami et al. 2015). Briefly, the V1 and V2 regions of the 16S rRNA gene were amplified from the DNA extracted from the 0.2 μm filter. Sequencing was performed using the Miseq device (Illumina) by 300 bp paired-end run. The obtained sequence was divided into operational taxonomic units (OTUs) with 97% homology using Mothur software.

9.1.4 Statistical Analyses

The relationship between OTU composition and environmental factors was analyzed using statistical package R. The data were unified to the smallest output number using the rare fraction analysis commands, and nonmetric-multidimensional scaling (NMDS) analysis was performed. For OTUs having a high frequency of occurrence, relationships with each environmental factor were examined by Pearson correlation coefficients.

9.2 Results

Temperature and salinity profiles indicated water stratification from around April to August, with vertical mixing in September during the study period (Fig. 9.2). A remarkable decline in salinity was observed in July due to the heavy precipitation in the monsoon rainy season. Chlorophyll *a* concentration began to increase from February and was highest in April and May. It decreased in June and remained low until the next spring. The major members of the phytoplankton community were large diatoms during low temperature periods including the spring bloom, and pico-eukaryotes and/or cyanobacteria in warm temperature periods (Taniuchi et al. 2017; Watanabe et al. 2017).

The 24 samples yielded 207,287 sequences that were clustered in 7061 OTUs. The most dominant OTU was affiliated with SAR11 in alphaproteobacteria. Its average relative abundance in each sample was 31% (OTU1 in Fig. 9.3 and Table 9.1). The next dominant OTU was affiliated with *Rhodobacteraceae* in alphaproteobacteria, and its average relative abundance was 9% (OTU4). It appeared from January to April when the chlorophyll *a* concentrations were high. OTU7, which was also affiliated with *Rhodobacteraceae*, appeared similarly with OTU4. OTU2, which was affiliated with *Alteromonas* in gammaproteobacteria, and OTU3, which was affiliated with cyanobacteria, were observed in the other months and were especially abundant from July to September in 2012.

NMDS analysis revealed that temperature, salinity, and chlorophyll *a* concentration were the environmental factors that were significantly related to the bacterial community variation in Sendai Bay (all $p < 0.01$; Fig. 9.4). The bacterial communities in winter and spring tended to plot at the positive direction of both

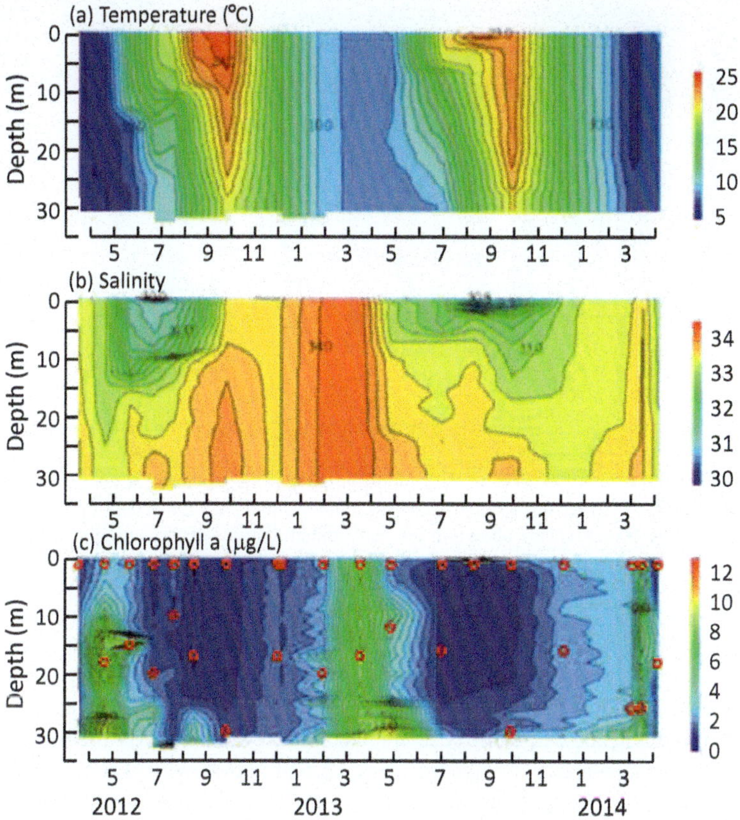

Fig. 9.2 Contour plots of (**a**) temperature, (**b**) salinity, and (**c**) chlorophyll *a* at station C5 from March 2012 to March 2014 in Sendai Bay. Circles in (**c**) indicate date and depths from which water samples for bacterial community analysis were taken

axes, with those in summer and autumn plotting at the opposite side. They were aligned along with the vectors of chlorophyll *a* concentration and temperature, indicating that these two environmental factors affected the seasonal variation of the bacterial community. Moreover, salinity also influenced the aforementioned seasonal differences in plots (Fig. 9.4).

Among the 20 most dominant OTUs, 15 displayed significant correlations with the environmental factors (Table 9.1). Four OTUs negatively and positively correlated with temperature and chlorophyll *a* concentration, respectively, indicating their increase during the spring bloom (Group I). They were assigned to *Rhodobacteraceae*, *Flavobacteriaceae*, and unclassified gammaproteobacteria. They comprised 9–35% of the total OTUs when they were at their maximum and decreased to around 0.1% or were undetectable level when they were at their lowest (Fig. 9.5). Conversely, the other four OTUs had a significant positive correlation with water

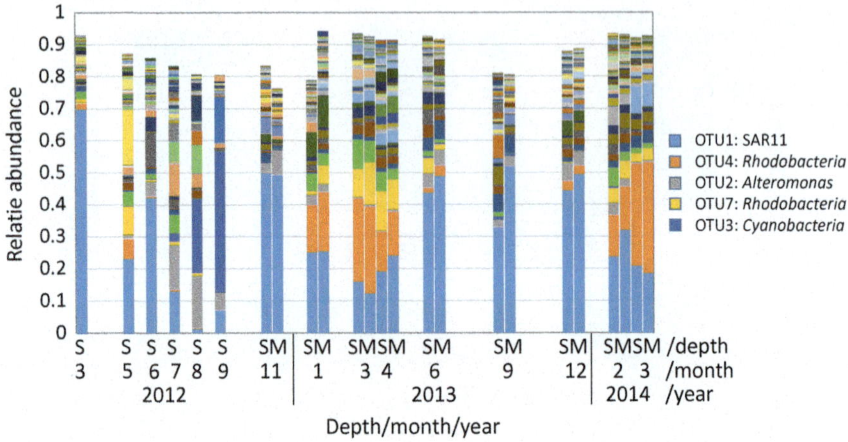

Fig. 9.3 Temporal variation of bacterial composition in seawater collected at surface (from March to September in 2012) or surface and subsurface chlorophyll maximum depth at station C5 in Sendai Bay. Top 100 OTUs are shown in the bars, and affiliations of the 5 dominant OTUs were indicated

temperature and a negative correlation with chlorophyll *a* concentration; that is, they appeared from summer to autumn (Group II). They belong to cyanobacteria, *Alteromonas*, and *Rhodobacteria*. They occupied 7–44% of the total when they were at their maximum and decreased to an undetectable level, except for the *Alteromonas* OTU. Four OTUs had a significant negative correlation with water temperature. They belonged to betaproteobacteria, *Methylophilaceae*, and unclassified alpha- or gammaproteobacteria (Group III). Compared to Group I, the fluctuations of Group III members were small, and most were observed throughout the year (Fig. 9.5). Three OTUs were related with salinity. They belong to *Microbacteriaceae* in *Actinobacteria* and *Burkholderiaceae* in *Betaproteobacteria* (Group IV). They appeared more abundantly in 2012 than in 2013. Five OTUs did not have any significant relationships with the examined environmental factors (Group V). The most abundant OTU of SAR11 was in this group.

9.3 Discussion

This study indicates that the bacterial community in seawater fluctuates according to changes in water temperature, salinity, and chlorophyll concentration in Sendai Bay. Bacterial communities vary periodically in marine environments (Fuhrman et al. 2015). In the coastal seawater of Sendai Bay, the bacterial community also varied seasonally, even with the marked fluctuation of the environmental conditions due to precipitation and/or sporadic intrusion of ocean water. Among the top 20 OTUs, 15 had significant relationships with environmental factors (Table 9.1). Their relative abundance fluctuated by two or three orders of magnitude, indicating that

Table 9.1 Top 20 OTUs taxonomy, abundance ratio, and correlation coefficients with environmental factors

Group	OTU code	Taxonomy		Relative abundance (%)			Correlation coefficient[b]			
				Max	Min	Average	Temp	Sal	NO_3	Chl *a*
I	OTU04	Alphaproteobacteria	*Rhodobacteraceae*	34.7	ND[a]	9.4	**−0.81**	0.44		**0.66**
	OTU07	Alphaproteobacteria	*Rhodobacteraceae*	13.2	ND	3.6	**−0.66**			**0.51**
	OTU15	Flavobacteria	*Flavobacteriaceae*	8.5	ND	1.3	**−0.71**			**0.69**
	OTU10	Gammaproteobacteria	Unclassified	10.1	ND	1.5		0.44		0.49
II	OTU02	Gammaproteobacteria	*Alteromonadaceae*	16.3	0.18	3.8	**0.69**			**−0.71**
	OTU03	Cyanobacteria	Family II	44.2	ND	3.1	**0.79**	−0.45		**−0.74**
	OTU12	Cyanobacteria	Family II	14.3	ND	0.7	**0.63**			−0.48
	OTU18	Alphaproteobacteria	*Rhodobacteraceae*	7.0	ND	0.6	**0.69**			
III	OTU23	Betaproteobacteria	*Methylophilaceae*	5.7	ND	1.0	**−0.72**			
	OTU06	Gammaproteobacteria	Unclassified	9.2	ND	2.7	−0.47			
	OTU09	Gammaproteobacteria	Unclassified	5.6	0.04	1.7	−0.47			
	OTU11	Alphaproteobacteria	Unclassified	4.5	0.18	1.5	−0.42			
IV	OTU46	Actinobacteria	*Microbacteriaceae*	17.5	ND	0.8		**−0.56**	−0.46	
	OTU16	Actinobacteria	*Microbacteriaceae*	8.8	ND	0.7	0.49	**−0.64**		−0.41
	OTU14	Betaproteobacteria	*Burkholderiaceae*	10.0	ND	1.1		−0.42		
V	OTU01	Alphaproteobacteria	SAR11	70.1	1.48	31.2				
	OTU05	Alphaproteobacteria	SAR11	6.1	0.11	2.3				
	OTU13	Alphaproteobacteria	*Rhodobacteraceae*	4.7	0.03	1.8				
	OTU17	Alphaproteobacteria	Unclassified	4.6	0.02	0.8				
	OTU08	Unclassified	Unclassified	10.7	0.23	1.7				

[a]ND not detected
[b]Bold type, $p < 0.01$; normal type, $p < 0.05$, $N = 24$

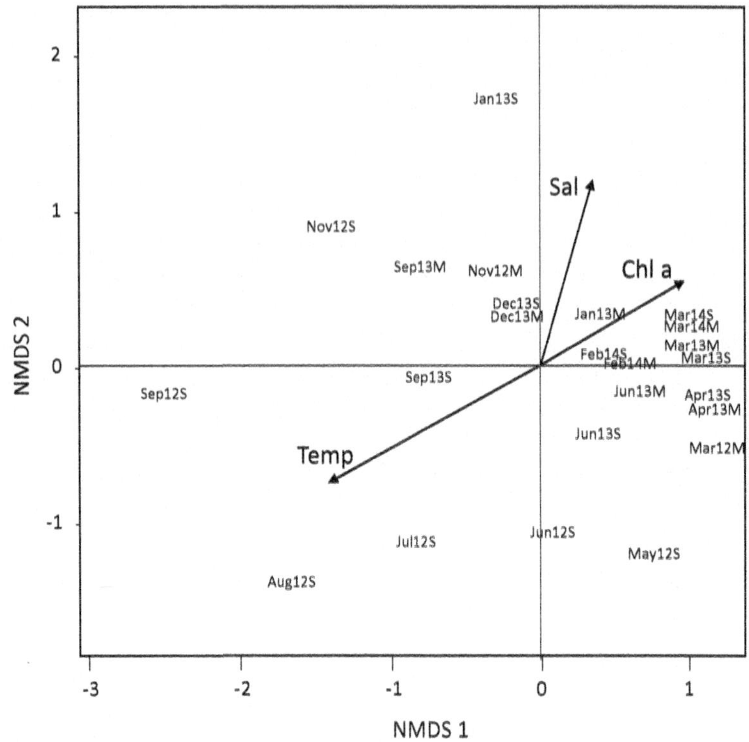

Fig. 9.4 NMDS analysis of 24 bacterial communities defined using 16S rDNA gene diversity at station C5 in Sendai Bay from March 2012 to March 2014. Sample codes indicate the sampling month, year, and depth (*S* surface, *M* subsurface chlorophyll maximum). Vectors indicate environmental variables significant at $p < 0.05$

bacteria in the coastal seawater sensitively respond to environmental changes. The bacterial growth rate in seawater differs between species or phylogenetic groups and by season (temperature) or sites (salinity) in the estuarine area (Yokokawa et al. 2004; Lankiewicz et al. 2016). The bacterial OTUs in group IV (*Betaproteobacteria* or *Actinobacteria*) displayed a negative correlation with salinity. Bacteria affiliated with these phylogenetic groups have been shown to be abundant or to actively grow in low-salinity estuarine water (Yokokawa et al. 2004; Fortunato et al. 2013). River water discharge into the bay probably promoted the growth of these bacteria during the rainy season.

Among the environmental factors, low water temperature and high chlorophyll *a* concentration, which indicated the spring phytoplankton bloom, were very influential on the bacterial community composition (Fig. 9.4). The bacterial OTUs in Group I were related to these variables. OTU4 and OTU7 in Group I are affiliated with *Rhodobacteraceae*. OTU15, which also displayed a strong relationship, is affiliated with *Flavobacteriaceae*. These two bacterial groups are active and/or increase in

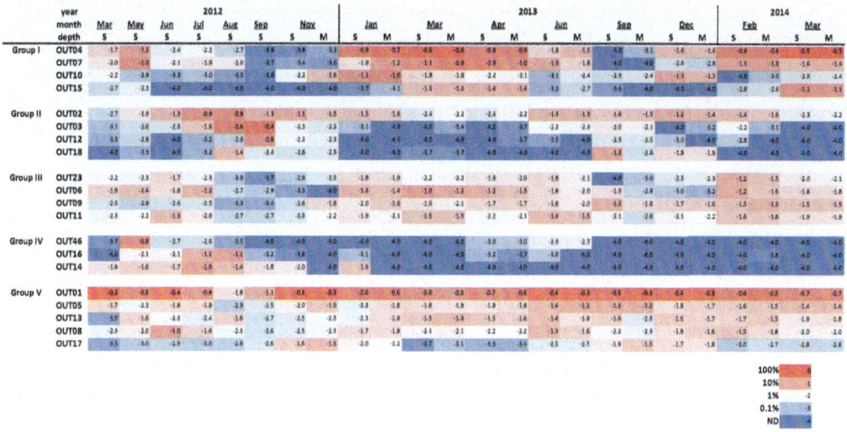

Fig. 9.5 Heat map of the top 20 bacterial OTUs in seawater collected at a depth of surface (S) or subsurface chlorophyll maximum (M) from March 2012 to March 2014 in Sendai Bay. Numbers indicate $\log_{10}$ values of the ratio to the total abundance

abundance during phytoplankton blooms (González and Moran 1997; Alonso-Sáez et al. 2007; Teeling et al. 2012; Wemheuer et al. 2014; Sakami et al. 2015). The most dominant OTU during the spring bloom was OTU4, comprising over 40% of the total at its peak. OTUs that became abundant during a phytoplankton bloom might have a copiotrophic character and increase their abundance in the bloom water by utilizing organic matter excreted from the phytoplankton. Lankiewicz et al. (2016) estimated the in situ growth rate of oligotrophic and copiotrophic bacterial taxa using the ratio of ribosomal RNA and DNA (rRNA/rDNA) and described that the copiotrophic bacteria grew as slowly as the oligotrophic bacteria in estuarine seawater. The fact that OTU4 remained in the bay for the whole year, decreasing to a minimum of <0.1% (Fig. 9.5), suggests that the bacterial species is indigenous in the coastal seawater.

The most dominant OTU, OTU1, was observed throughout the year. It comprised 30% of the total bacteria on average. This OTU belongs to the SAR11 group, which is an oceanic bacterial taxon. SAR11 is abundant in the seawater that intrudes into the bay from the outside oceanic area (Sakami et al. 2015), suggesting that the bacteria are supplied from out of the bay by occasional seawater exchange. Relative abundance of OTU1 varied widely from 1.4 to 70% at station C5, which is located in the central part of the bay. Seawater eddies are often formed near the bay entrance and external seawater intermittently enters the bay. Thus, the ratio of SAR11 might have exhibited a large fluctuation irrespective of environmental factors, such as water temperature. On the other hand, SAR11 activity (ratio of rRNA- to DNA-derived sequences) is decreased in estuarine waters (Wemheuer et al. 2014). The SAR11 OTUs seemed to decrease while it stayed within the bay. In August 2012, when the OTU1 was lowest, OTUs of cyanobacteria and *Alteromonas* were dominant. Copiotrophs, such as *Alteromonas*, potentially have a growth rate that is more than an order of magnitude higher than oligotrophic SAR11 (Lankiewicz et al. 2016). In eutrophic coastal seawater as compared with oceanic seawater, copiotrophs may grow relatively quickly, and the SAR11 decreases where the water mass is stable and seawater exchange is small. Incidentally, a virus infection has also been suggested to participate in the decrease of the SAR11 population (Zhao et al. 2013; Vage et al. 2013). In the shotgun metagenomic analysis conducted simultaneous with this research, we detected sequences that could presumptively be assigned to SAR11 phage abundantly in Sendai Bay compared to oceanic areas (unpublished data). SAR11 might decrease due to phage infection because its physiological condition had changed after entering the bay.

In conclusion, the bacterial community in Sendai Bay has a clear annual variation depending on environmental conditions, such as water temperature and chlorophyll *a* concentration. Together with the fact that its spatial variation is also closely related to environmental changes (Sakami et al. 2015), bacterial community composition is considered to be a good biological indicator for coastal environmental changes.

References

Alonso-Sáez L, Balagué V, Sà EL, Sánchez O, González JM, Pinhassi J, Massana R, Pernthaler J, Pedrós-Alió C, Gasol JM (2007) Seasonality in bacterial diversity in North-West Mediterranean coastal waters: assessment through clone libraries, fingerprinting and FISH. FEMS Microbiol Ecol 60:98–112

Fortunato CS, Eiler A, Herfort L, Needoba JA, Peterson TD, Crump BC (2013) Determining indicator taxa across spatial and seasonal gradients in the Columbia River coastal margin. ISME J 7:1899–1911

Fuhrman JA, Hewson I, Schwalbach MS, Steele JA, Brown MV, Naeem S (2006) Annually reoccurring bacterial communities are predictable from ocean conditions. PNAS 103(35):13104–13109

Fuhrman JA, Cram JA, Needham DM (2015) Marine microbial community dynamics and their ecological interpretation. Nature Rev Microbiol 13:133–146

González JM, Moran MA (1997) Numerical dominance of a group of marine bacteria in the alpha-subclass of the class Proteobacteria in coastal seawater. Appl Environ Microbiol 63:4237–4242

Kakehi S, Ito S, Kuwata A, Saito H, Tadokoro K (2015) Phytoplankton distribution during the winter convective season in Sendai Bay, Japan. Continent Shelf Res 97:43–53

Lankiewicz TS, Cottrell MT, Kirchman DL (2016) Growth rates and rRNA content of four marine bacteria in pure cultures and in the Delaware estuary. ISME J 10:823–832

Morris RM, Vergin KL, Cho J-C, Rappe MS, Carlson CA, Giovannoni SJ (2005) Temporal and spatial response of bacterioplankton lineages to annual convective overturn at the Bermuda Atlantic time-series study site. Limnol Oceanogr 50(5):1687–1696

Needham DM, Fuhrman JA (2016) Pronounced daily succession of phytoplankton, archaea and bacteria following a spring bloom. Nat Microbiol. https://doi.org/10.1038/NMICROBIOL.2016.5

Rooney-Varga JN, Giewat MW, Savin MC, Sood S, LeGresley M, Martin JL (2005) Links between phytoplankton and bacterial community dynamics in a coastal marine environment. Microb Ecol 49(1):163–175

Sakami T, Watanabe T, Kakehi S, Taniuchi Y, Kuwata A (2015) Spatial variation of bacterial community composition at the expiry of spring phytoplankton bloom in Sendai Bay, Japan. Gene 576:610–617

Taniuchi Y, Watanabe T, Kakehi S, Sakami T, Kuwata A (2017) Seasonal dynamics of the phytoplankton community in Sendai Bay, northern Japan. J Oceanogr 73(1):1–9. https://doi.org/10.1007/s10872-015-0334-0

Teeling H, Fuchs BM, Becher D, Klockow C, Gardebrecht A, Bennke CM, Kassabgy M, Huang S, Mann AJ, Waldmann J, Weber M, Klindworth A, Otto A, Lange J, Bernhardt J, Reinsch C, Hecker M, Peplies J, Bockelmann FD, Callies U, Gerdts G, Wichels A, Wiltshire KH, Glöckner FO, Schweder T, Amann R (2012) Substrate-controlled succession of marine bacterioplankton populations induced by a phytoplankton bloom. Science 4:608–611

Vage S, Storesund JE, Thingstad TF (2013) SAR11 viruses and defensive host strains. Nature 499:E3

Watanabe T, Taniuchi Y, Kakehi S, Sakami T, Kuwata A (2017) Seasonal succession in the diatom community of Sendai Bay, northern Japan, following the 2011 off the Pacific coast of Tohoku earthquake. J Oceanogr 73(1):133–144. https://doi.org/10.1007/s10872-016-0387-8

Wemheuer B, Wemheuer F, Daniel R (2012) RNA-based assessment of diversity and composition of active Archaeal communities in the German bight. Archaea, ID 695826:8. https://doi.org/10.1155/2012/695826

Wemheuer B, Güllert S, Billerbeck S, Giebel HA, Voget S, Simon M, Daniel R (2014) Impact of a phytoplankton bloom on the diversity of the active bacterial community in the southern North Sea as revealed by metatranscriptomic approaches. FEMS Microbiol Ecol 87:378–389

Yokokawa T, Nagata T, Cottrell MT, Kirchman DL (2004) Growth rate of the major phylogenetic bacterial groups in the Delaware estuary. Limnol Oceanogr 49(5):1620–1629

Zhao Y, Temperton B, Thrash JC, Schwalbach MS, Vergin KL, Landry ZC, Ellisman M, Deerinck T, Sullivan MB, Giovannoni SJ (2013) Abundant SAR11 viruses in the ocean. Nature 494:357. https://doi.org/10.1038/nature11921

Shotgun Metagenome Analyses: Seasonality Monitoring in Sendai Bay and Search for Red Tide Marker Sequences

10

Kaoru Matsumoto, Norikazu Kitamura, and Kazuho Ikeo

Abstract

Along with the advancement of sequencing technology, shotgun metagenomics has been applied for microbial community study. Shotgun metagenomics enables to examine not only taxonomic structure but also functional profile based on the gene repertory. It also allows broad comparative analysis including unknown sequences without annotation. Here we first mention about overview of shotgun metagenomics for microbial community study and then introduce two examples of study that applied different approaches. One is seasonality monitoring based on taxonomic and functional gene composition. This is monthly-bimonthly monitoring of surface water in Sendai Bay, Japan, for about a year. We observed typical seasonality in the taxonomic composition and also found seasonality in the overall functional gene composition. The other is search for red tide marker sequences that applied broad comparative analysis. This search is for sequences that showed different abundance between red tide samples and control samples in Buzen Sea, Japan, using the assembled contigs as the reference. As the candidate of the red tide marker sequences, we obtained 1220 contigs including those without taxonomic annotation. As in these examples, shotgun metagenomic studies provide insights to help understanding marine microbial community.

Keywords

Coastal ocean · Bacterioplankton · Red tide · Seasonality · Shotgun metagenomics

---

K. Matsumoto · N. Kitamura · K. Ikeo (✉)
National Institute of Genetics, Shizuoka, Japan
e-mail: kikeo@nig.ac.jp

T. Gojobori et al. (eds.), *Marine Metagenomics*,
https://doi.org/10.1007/978-981-13-8134-8_10

Shotgun metagenome contains sequences of various genomic locations for variety of organisms present in the environmental sample, while amplicon metagenome targets a specific marker gene of target taxa (Sharpton 2014). Although amplicon sequencing of 16S rRNA gene has revealed broad phylogenetic diversity of microbes in nature (Rappé and Giovannoni 2003), there are some limitations; it provides only diversity of marker gene, does not provide biological function of the microbes, and is applicable only to taxa with marker gene amplifiable by selected primers (Edwards and Rohwer 2005; Klindworth et al. 2013). Therefore, shotgun sequencing has been developed as a way to explore uncultured microbes more comprehensively and used, for example, to reconstruct genomes or metabolic pathways or to explore novel biomolecules (Simon and Daniel 2011; Sharpton 2014).

In microbial community studies, shotgun sequencing is mainly conducted to obtain functional profile and taxonomic profile of environmental samples, based on sequence similarity of the genes or proteins in reference database (Sharpton 2014). Functional profile, which describes what kind of functional gene sequences are included and their relative frequency in metagenome, is used to infer metabolic capacity of the community. By comparing such functional profiles, researchers have investigated the characteristics of ecological functions of microbial communities in various marine environments (Burke et al. 2011; Raes et al. 2011; Ganesh et al. 2014; Haggerty and Dinsdale 2017). In obtaining taxonomic profile from shotgun metagenome, we may extract and use only marker gene sequence and/or may use the whole predicted gene or nucleotide sequences (Smith et al. 2013l; Ruvindy et al. 2016). The former facilitate clear taxonomic annotation and comparison with amplicon studies, while the latter analyze much sequence data and include taxa without marker gene.

Although metagenomic analysis is generally based on reference databases, some studies have applied methods which do not rely on databases (Segata et al. 2013). Such methods have been proposed because many of metagenome sequences may remain unidentified due to lack of reference sequence, and comparative analysis excluding such unknown sequences may overlook trends of the community (Sunagawa et al. 2013). One of the methods is to use assembled contigs of the metagenome as a reference, in which case they map the metagenomic sequences to the contigs and obtain contig profile that describes relative abundance of each contig in each sample (contig coverage, Coutinho et al. 2017). While metagenome assembly has some difficulties (Sharpton 2014), such contig profile enables analysis including unknown sequence, as the reference contigs includes those cannot be annotated based on reference databases.

Here we introduce two shotgun metagenomic studies that investigate surface marine microbial community in Japanese coastal area: seasonality monitoring in Sendai Bay and search for red tide marker sequences. In both studies, we classified shotgun sequences by category based on sequence similarity, calculated relative abundances, and compared the compositions among samples. However, in the former study, as we aimed to associate the taxonomic/functional variation with the environmental change, we used taxonomic/functional profiles of predicted peptides

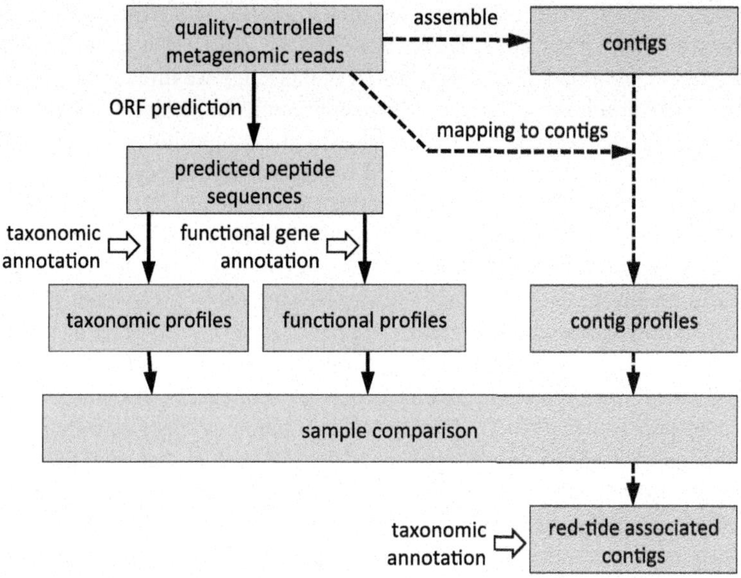

Fig. 10.1 Simple flowchart of our analysis method. Solid arrow indicates the process in "seasonality monitoring in Sendai Bay." Dashed arrow indicates the process in "search for red tide marker sequence." Outlined arrow indicates annotation process based on reference databases

based on reference database and excluded unannotated sequences (solid arrow in Fig. 10.1). In the latter study, on the other hand, we aimed to extract nucleotide sequences related to red tide, and we used contig profile by mapping sequences to assembled contigs including those without annotation (dashed arrow in Fig. 10.1).

10.1 Seasonality Monitoring in Sendai Bay

Our metagenomic sequencing mainly targets bacterial community. Bacteria plays key role in marine microbial food web, including assimilating phytoplankton-derived dissolved organic matter, and the community composition quickly changes in response to the nutrient and other environmental condition (Buchan et al. 2014; Fuhrman et al. 2015). Therefore, with checking the metagenomic data, we can expect to capture dynamics of the bacterial community relating to phytoplankton activity and other environmental variation and to detect change or unusual in the coastal environment (Paerl et al. 2003).

Sendai Bay locates at northeast part of Japan and widely opens onto the Pacific Ocean. This place was affected by the 2011 Japan earthquake and tsunami (Urabe et al. 2013). Therefore one purpose of this monitoring is to check and understand the present condition of the microbial community here. In addition, we aimed to investigate seasonal change of functional gene profile. While seasonal cycle in phylogenetic diversity has been well studied using 16S rRNA gene sequence (Giovannoni and Vergin 2012; Bunse and Pinhassi 2017), that in functional gene profile is rarely reported (Ward et al. 2017). In this study we investigated taxonomic and functional profiles of surface microbial community in Sendai Bay using shotgun metagenome. Here we introduce the overview of the compositions and show that the compositional differences were correlated with seasonal difference.

We used 22 samples collected from surface seawater at 2 sites (C12, mouth side; C5, coastal side) for about a year. DNA was extracted from 0.8–5.0 μ fraction and was sequenced with Ion Torrent technology. After quality control and duplication removal, the preprocessed sequences were 0.6–1.8 M reads per sample. Longest coding region was predicted on each read, and 0.3–1.0 M predicted peptide sequences (> 30 amino acids), which was from 48 to 62% of the preprocessed reads, were obtained. We used these predicted peptide sequences to obtain taxonomic/-functional profiles (Fig. 10.1). Taxonomic annotation of these peptide sequences was obtained by blastp search against NCBI-NR and lowest common ancestor (LCA) algorithm with annotation software MEGAN (Huson et al. 2016) using the default value. We obtained taxonomic profiles in various taxonomic levels, by summing the read number by taxonomic group and normalizing by total number of predicted peptides that were annotated in domain level. Functional annotation of the peptide sequences was based on KEGG Orthology (KO), conducted by Web service GhostKOALA (Kanehisa et al. 2016). To obtain functional profile, we collapsed KEGG Orthology into KEGG pathways and KEGG categories. Then read number was summed by KEGG pathways or KEGG categories and normalized by total number of predicted peptides assigned to KO.

The taxonomic profiles were shown in Fig. 10.2. Among the predicted peptide sequences, 32–53% were annotated with domain level taxonomy (Fig. 10.2a). Although eukaryotic sequences were also abundantly observed especially in November and January, most of the annotated sequences were bacterial ones. Figure 10.2b shows phylum level composition in the whole peptides annotated at the domain level. Here the taxonomic composition seemed different mainly by sampling season, while between sites difference seemed relatively small. In both sites, generally, *Bacteroidetes* and *Proteobacteria* dominated in April and May 2012, *Cyanobacteria* and *Verrucomicrobia* increased from June to November, *Chlorophyta* increased in November and January, and then the composition in March 2013 became similar to those in spring 2012. The compositions in April and June 2013 were similar to those in the same month in 2012. To test if the taxonomic compositions were different by sampling season, we calculated Bray-Curtis dissimilarity that quantifies compositional difference between two samples (here arcsine square root transformation was applied on relative abundances to stabilize variance relative to the mean). Then we tested correlation between the Bray-Curtis dissimilarity and seasonal differences of sampling date (interval of sampling date, |365 – interval| was used if interval > 182 days), using samples from both sites. As a result, Bray-Curtis dissimilarity was positively correlated with seasonal difference at each level from phylum to species ($R > 0.63$, $p < 0.001$). This indicates that the compositional difference increased as the sampling interval became longer until half

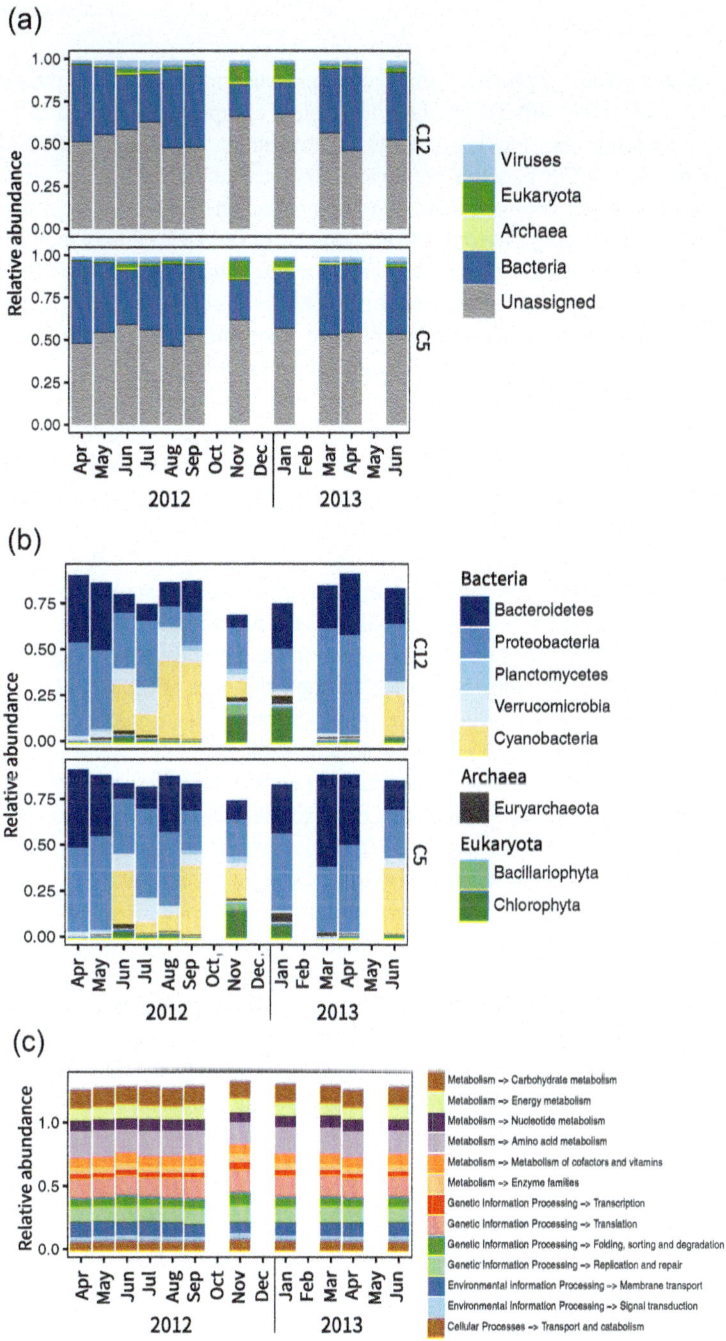

Fig. 10.2 Taxonomic and functional composition of predicted peptides in seasonal metagenome from Sendai Bay. Months with blank have no data. (**a**) Domain level composition (% of total number of predicted peptides) at C12 (mouth side) and C5 (coastal side). (**b**) Phylum level composition (% of total predicted peptides assigned in domain level). Only taxa with max relative abundance >3% were shown. (**c**) Composition of KEGG categories (% of total predicted peptides assigned to any KO) at C5. The composition at C12 was as similar. Only categories with max relative abundance >5% were shown. As some genes are assigned to more than one category, the total is greater than 100%

a year, but it decreased again after that and increased again after 1 year passed. The content of the seasonal difference of the taxonomic composition seemed reasonable in general, considering previous reports (Giovannoni and Vergin 2012; Bunse and Pinhassi 2017). For example, relative abundance of *Bacteroidetes*, which was composed mainly of *Flavobacteriaceae* (data not shown), increased from March to May (Fig. 10.2b). Spring phytoplankton bloom is observed in this place (Taniuchi et al. 2017), and members of *Flavobacteriaceae* are generally known to efficiently assimilate complex organic matters and to increase at phytoplankton bloom (Buchan et al. 2014). In addition, increase of relative abundance of *Cyanobacteria* and *Verrucomicrobia* in summer and *Euryarchaeota* in winter (Fig. 10.2b) is also consistent with the previous studies in other temperate ocean (Bunse and Pinhassi 2017). As for *Cyanobacteria*, there seemed to be two peeks in relative abundance in the year (Fig. 10.2b). Checking lower taxonomic annotation (data not shown), this was considered to be due to exchange of *Synechococcus* ecotypes likely adapted to different temperature and nutrient condition (Sohm et al. 2016).

In functional annotation, 20–33% of the predicted peptides were assigned to KO identifiers. Majority (50–84%) of those were peptides assigned to bacteria in taxonomic annotation, while 13–24% of those were peptides that could not be assigned in domain level. Comparing to taxonomic composition, functional gene composition was much more consistent through the year (Fig. 10.2c). However, Bray-Curtis dissimilarity of functional composition was also positively correlated with seasonal difference (KEGG pathway, $R = 0.48$, $p < 0.001$; KEGG category, $R = 0.45$, $p < 0.001$). This result suggests that although the microbial community kept majority of functions through the year, seasonal community change altered their metabolic potential. The seasonal difference may reflect difference of organic matter source utilized by bacterial community (Ward et al. 2017).

Thus, we recorded surface microbial communities in Sendai Bay and showed that the seasonal variation of taxonomic composition seemed reasonable. Further, we showed that their functional potential was also seasonally different. We are going to investigate what kind of functional genes were different and how those relative abundances were different by season. Although we need several years of observation to confirm seasonality, accumulation of such survey would promote understanding of the dynamics of microbial community and would enable to detect change of microbial community and environment.

10.2 Search for Red Tide Marker Sequences: A Case of Microbial DNA Marker Associated with *Chattonella* Red Tide

Chattonella marina var. *antiqua* is a microalga belonging to raphidophytes and has sometimes formed red tide bloom in closed or semi-closed coastal environment, which has given rise to large economic damage for farmed fishery (Imai et al. 2006). In general, phytoplankton bloom produces organic matters in water through exudate and the biomass, resulting in proliferation and increase of productivity of bacteria (Cole et al. 1988; Smith et al. 1995; Tada et al. 2011). During the bloom, bacterial composition also changes dynamically (Riemann et al. 2000; Fandino et al. 2001; West et al. 2008). In addition to the impact of phytoplankton bloom on bacterial composition, bacteria have various and complex relationships with phytoplankton (Croft et al. 2005; Amin et al. 2012). Thus, investigating change of microbial assemblage caused by red tide could lead to better understanding of the red tide phenomena from perspective of microbial system. In this part, we present a study of search for microbial DNA marker that is associated with *C. marina* var. *antiqua* red tide, by using metagenomics approach. In order to conduct comprehensive survey, we used a reference-free method that was based on mapping of metagenomic reads onto contigs produced by assembly (Fig. 10.1). Then, sequences differentially abundant in red tide or control samples were extracted.

Water samples, including both six red tide and seven control samples, were collected from coastal surface within Buzen Sea in Japan. In addition, 36 samples were obtained from neighboring areas. DNA extraction was conducted from microbes trapped on 0.2 μm filter membrane (1–0.2 μm fraction), resulting in metagenome mainly consisting of free-living bacteria. Metagenomic sequencing was performed with Roche 454 platform, and obtained sequences were preprocessed through removal of adapters, quality filtering, and elimination of duplicated sequences. The number of the read sequences for a sample was 532,829 ± 160,299 (mean ± SD). Then, all the read sequences were pooled and assembled by using Newbler 2.9 (Margulies et al. 2005). Among contigs produced, 500 bp or longer contigs were used for the following analyses. Approximately 484 K contigs were obtained, with an N50 length of 2051 bp.

Relative abundance of the contigs was estimated by mapping of all the reads from Buzen Sea samples onto all contigs, using blastn (Altschul et al. 1990) at an e-value cutoff of 1.0e-30. Alignments, of which over 90% of mapped read length could be aligned to a contig with more than 90% identity, were used for analysis. As a result, the number of mapped reads meeting the alignment condition was 384,735 ± 69,469 (mean ± SD), with mapping rate of 66.6 ± 5.5% (mean ± SD). For each contig, the number of match sites in alignments was summed up by the sample; then the values were normalized by a sum of the length of the mapped reads in the sample. These values were regarded as relative abundance of the contig in the sample. Then, a matrix of relative abundance of contig was obtained for all combinations of contigs and samples. Contigs without aligned reads were removed from analysis.

To extract differentially abundant contigs in red tide or control samples, Wilcoxon rank sum test (also referred to as Mann-Whitney U-test) was performed. Using the obtained p-values, false discovery rate (FDR) was estimated according to Benjamini and Hochberg (1995). Since the estimated FDR was relatively high (minimum was 0.405) due to the small sample size ($n = 6$ for red tide and $n = 7$ for control) and a great number of contigs (335,646 sequences), extraction were conducted based on not FDR for multiple testing but just lower p-value ($p < 0.0041$). In total, 1220 contigs were extracted as marker candidates. Of these contigs, 1046 and 174 were abundant in red tide and control samples, respectively.

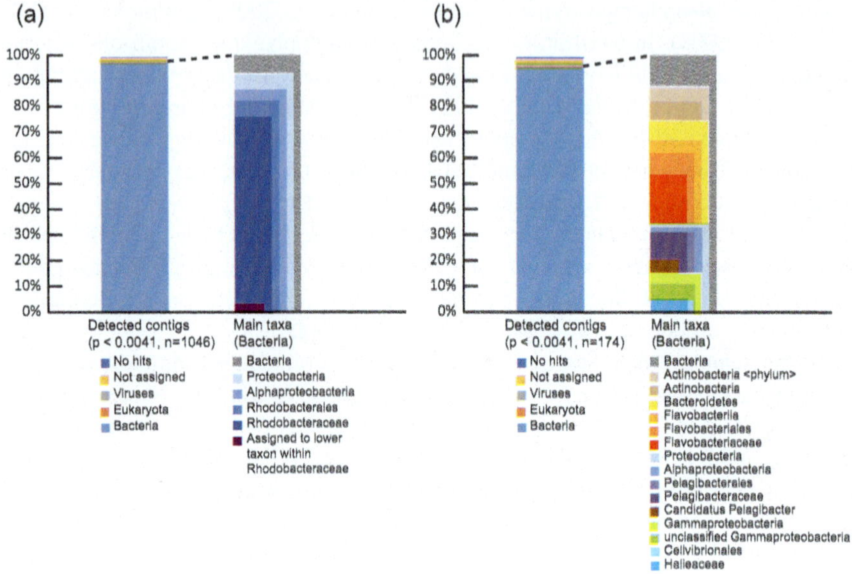

Fig. 10.3 Taxonomic assignment of extracted marker candidates. (**a**) Main taxa abundant in red tide samples. (**b**) Main taxa abundant in control samples

To infer the taxonomy of the extracted contig sequences, we performed blast search, blastx against NCBI-NR database. Then, to determine taxonomic assignment of each sequence, the blast result was analyzed according to the lowest common ancestor (LCA) algorithm using MEGAN software (Huson et al. 2016). Taxonomic assignment of those contigs is shown in Fig. 10.3. While family *Rhodobacteraceae* (*Alphaproteobacteria*, *Rhodobacterales*) was dominant in the contigs differentially abundant in red tide (Fig. 10.3a), various taxa were identified in the contigs differentially abundant in control (Fig. 10.3b). Rhodobacteraceae includes the *Roseobacter* group, members of which are known to be associated with phytoplankton (Geng and Belas 2010; Buchan et al. 2014). Given that some marine phytoplankton were reported to interact with members of *Roseobacter* group via signaling and nutrient (Amin et al. 2015), whether or not significant relationship exists between *C. marina* var. *antiqua* and the bacteria could be the question to be investigated.

Lastly, since we adopted the reference-free strategy, comprehensive survey could be performed. Actually, a substantial proportion of contigs used for the test could not be assigned to specific taxa, especially at lower taxonomic level; at family and genus-or-lower taxa levels, assignment rate was, respectively, 33.1–41.7% and 24.2–33.3% when the range of minimum threshold of blast bit score is 50–200

Table 10.1 Taxonomic assignment rate of contigs at each taxonomic level, determined by MEGAN software

Minimum threshold for the bit score of blast hits	Domain-or-virus	Phylum	Class	Order	Family	Genus-or-lower taxa
50	89.9%	71.5%	59.8%	48.3%	41.7%	33.3%
200	64.9%	55.8%	47.7%	38.2%	33.1%	24.2%

Note: assignment conditions other than minimum threshold of bit score were consistent among above all cases; minimum e-value of 1.0e-5, naive LCA within top 10% of bit score

(Table 10.1). However, in the present study, such unassigned sequences could be also analyzed as subjects of testing. As a result, we could extract marker candidate sequences, which might include those from taxa absent in reference database (Fig.10.3). In general, extracted marker candidates are provided for further selection to determine an optimal marker set, and validation is performed to assess the ability of the marker set using other new data set (e.g., Qin et al. 2012). Thus, although the detected sequences in the present study are still marker candidates, their ability can be evaluated in future studies in which new dataset is available for validation. Establishment of markers and further studies using the markers will help to circumvent issues caused by the red tide.

References

Altschul SF, Gish W, Miller W et al (1990) Basic local alighnment search tool. J Mol Biol 215(3):403–410

Amin SA, Parker MS, Armbrust EV (2012) Interaction between diatoms and bacteria. Microbiol Mol Biol R 76(3):667–684

Amin SA, Hmelo LR, van Tol HM et al (2015) Interaction and signaling between a cosmopolitan phytoplankton and associated bacteria. Nature 522(4):98–101

Benjamini Y, Hochberg Y (1995) Controlling the false discovery rate: a practical and powerful approach to multiple testing. J Roy Stat Soc B Met 57(1):289–300

Buchan A, LeCleir GR, Gulvik CA et al (2014) Master recyclers: features and functions of bacteria associated with phytoplankton blooms. Nat Rev Microbiol 12(10):686–698

Bunse C, Pinhassi J (2017) Marine bacterioplankton seasonal succession dynamics. Trends Microbiol 25(6):494–505

Burke C, Steinberg P, Rusch D et al (2011) Bacterial community assembly based on functional genes rather than species. PNAS 108(34):14288–14293

Cole JJ, Findlay S, Pace ML (1988) Bacterial production in fresh and saltwater ecosystems: a cross-system overview. Mar Ecol Prog Ser 43:1–10

Coutinho FH, Silveira CB, Gregoracci GB et al (2017) Marine viruses discovered via metagenomics shed light on viral strategies throughout the oceans. Nat Commun 8:15955

Croft MT, Lawrence AD, Raux-Deery E et al (2005) Algae acquire vitamin B12 through a symbiotic relationship with bacteria. Nature 438:90–93

Edwards RA, Rohwer F (2005) Viral metagenomics. Nat Rev Microbiol 3(6):504–510

Fandino LB, Riemann L, Steward GF et al (2001) Variations in bacterial community structure during a dinoflagellate bloom analyzed by DGGE and 16S rDNA sequencing. Aqua Microb Ecol 23:119–130

Fuhrman JA, Cram JA, Needham DM (2015) Marine microbial community dynamics and their ecological interpretation. Nat Rev Microbiol 13(3):133–146
Ganesh S, Parris DJ, DeLong EF et al (2014) Metagenomic analysis of size-fractionated picoplankton in a marine oxygen minimum zone. ISME J 8(1):187–211
Geng H, Belas R (2010) Molecular mechanisms underlying roseobacter-phytoplankton symbioses. Curr Opin Biotech 21:332–338
Giovannoni SJ, Vergin KL (2012) Seasonality in ocean microbial communities. Science 335(6069):671–676
Haggerty JM, Dinsdale EA (2017) Distinct biogeographical patterns of marine bacterial taxonomy and functional genes. Glob Ecol Biogeogr 26(2):177–190
Huson DH, Beier S, Flade I et al (2016) MEGAN community edition-interactive exploration and analysis of large-scale microbiome sequencing data. PLoS Comput Biol 12(6):e1004957
Imai I, Yamaguchi M, Hori Y (2006) Eutrophication and occurrences of harmful algal blooms in the Seto Inland Sea, Japan. Plankton Benthos Res 1(2):71–84
Kanehisa M, Sato Y, Morishima K (2016) BlastKOALA and GhostKOALA: KEGG tools for functional characterization of genome and metagenome sequences. J Mol Biol 428(4):726–731
Klindworth A, Pruesse E, Schweer T et al (2013) Evaluation of general 16S ribosomal RNA gene PCR primers for classical and next-generation sequencing-based diversity studies. Nucleic Acids Res 41(1):e1
Margulies M, Egholm M, Altman WE et al (2005) Genome sequencing in microfabricated high-density picolitre reactors. Nature 437:376–380
Paerl HW, Dyble J, Moisander PH et al (2003) Microbial indicators of aquatic ecosystem change: current applications to eutrophication studies. FEMS Microbiol Ecol 46(3):233–246
Qin J, Li Y, Cai Z et al (2012) A metagenome-wide association study of gut microbiota in type 2 diabetes. Nature 490:55–60
Raes J, Letunic I, Yamada T et al (2011) Toward molecular trait-based ecology through integration of biogeochemical, geographical and metagenomic data. Mol Syst Biol 7:473
Rappé MS, Giovannoni SJ (2003) The uncultured microbial majority. Annu Rev Microbiol 57:369–394
Riemann L, Steward GF, Azam F (2000) Dynamics of bacterial community composition and activity during a mesocosm diatom bloom. Appl Environ Microbiol 66(2):578–587
Ruvindy R, White RA III, Neilan BA et al (2016) Unravelling core microbial metabolisms in the hypersaline microbial mats of Shark Bay using high-throughput metagenomics. ISME J 10(1):183–196
Segata N, Boernigen D, Tickle TL et al (2013) Computational meta'omics for microbial community studies. Mol Syst Biol 9:666
Sharpton TJ (2014) An introduction to the analysis of shotgun metagenomic data. Front Plant Sci 5:209
Simon C, Daniel R (2011) Metagenomic analyses: past and future trends. Appl Environ Microbiol 77(4):1153–1161
Smith DC, Steward GF, Long RA et al (1995) Bacterial mediation of carbon fluxes during a diatom bloom in a mesocosm. Deep-Sea Res Pt II 42(1):75–97
Smith MW, Allen LZ, Allen AE et al (2013) Contrasting genomic properties of free-living and particle-attached microbial assemblages within a coastal ecosystem. Front Microbiol 4:120
Sohm JA, Ahlgren NA, Thomson ZJ et al (2016) Co-occurring Synechococcus ecotypes occupy four major oceanic regimes defined by temperature, macronutrients and iron. ISME J 10(2):333–345
Sunagawa S, Mende DR, Zeller G et al (2013) Metagenomic species profiling using universal phylogenetic marker genes. Nat Methods 10(12):1196–1199
Tada Y, Taniguchi A, Nagao I et al (2011) Differing growth responses of major phylogenetic groups of marine bacteria to natural phytoplankton blooms in the western North Pacific Ocean. Appl Environ Microbiol 77(12):4055–4065
Taniuchi Y, Watanabe T, Kakehi S et al (2017) Seasonal dynamics of the phytoplankton community in Sendai Bay, northern Japan. J Oceanogr 73(1):1–9

Urabe J, Suzuki T, Nishita T et al (2013) Immediate ecological impacts of the 2011 Tohoku earthquake tsunami on intertidal flat communities. PLoS One 8(5):e62779

Ward CS, Yung CM, Davis KM et al (2017) Annual community patterns are driven by seasonal switching between closely related marine bacteria. ISME J 11(6):1412–1422

West NJ, Obernosterer I, Zemb O et al (2008) Major differences of bacterial diversity and activity inside and outside of a natural iron-fertilized phytoplankton bloom in the southern ocean. Environ Microbiol 10(3):738–756

Distribution and Community Composition of Ammonia-Oxidizing Archaea and Bacteria in Coastal Sediments in Response to Sediment Material Gradients at Sendai Bay, Japan

11

Tomoko Sakami and Shigeho Kakehi

Abstract

To examine how ammonia-oxidizing organisms in coastal sediments are affected by environmental changes, the distributions of ammonia-oxidizing archaea (AOA) and ammonia-oxidizing bacteria (AOB) were determined along an environmental gradient from the coastal mud to the offshore coarse sand at Sendai Bay, Japan. Sediment samples were collected in December 2011 and July 2012. The abundance of AOA ammonia monooxygenase alpha subunit gene (*amoA*) was high in the coastal muddy areas and low in the offshore sandy areas during both months. There was a strong positive correlation between AOA-*amoA* abundance and ammonia content in the sediment. AOB-*amoA* abundance was remarkably low in the muddy sediments in December. However, the distribution of AOB-*amoA* was similar to that of AOA-*amoA* in July. Clone library analysis indicated that the community composition for both types of organisms differed in sandy and muddy sediments and that the diversity was considerably lower in the muddy sediments during both months. These results suggest that the abundance of ammonia-oxidizing organisms was controlled by the ammonia levels in the sediment. However, there are some inhibitive conditions for AOB: presumably, the low organic matter supply to the surface oxic layer during autumn in the muddy sediment in Sendai Bay.

T. Sakami (✉)
Tohoku National Fisheries Research Institute, Japan Fisheries Research and Education Agency, Shiogama, Miyagi, Japan

National Research Institute of Aquaculture, Japan Fisheries Research and Education Agency, Minami-ise, Mie, Japan
e-mail: sakami@affrc.go.jp

S. Kakehi
Tohoku National Fisheries Research Institute, Japan Fisheries Research and Education Agency, Shiogama, Miyagi, Japan

T. Gojobori et al. (eds.), *Marine Metagenomics*,
https://doi.org/10.1007/978-981-13-8134-8_11

Keywords
Ammonia monooxygenase alpha subunit gene (*amoA*) · Ammonia-oxidizer · Coastal sediment · Sendai Bay

11.1 Introduction

Oxidation of ammonia is one of the key processes in geochemical nitrogen cycling; it is particularly important in the bottom sediments in coastal shallow water where organic matter is actively degraded to generate ammonia (Henriksen and Kemp 1988; Thandrup and Dalsgaard 2008). Bottom sediments are supplied with organic matter primarily from phytoplankton sedimentation, but the temporal fluctuation in phytoplankton abundance is great, especially in semi-arctic coastal areas (Jensen et al. 1990). Moreover, a decrease in dissolved oxygen (DO) occurs easily in such areas, especially in eutrophicated areas with high artificial nutrient inputs (Rabalais et al. 2009). The DO concentration typically affects the ammonia oxidation process because it is primarily an oxidative biological reaction (Thandrup and Dalsgaard 2008). When the nitrification process is suppressed, biological production declines because mineralization of organic matter is inhibited and a high concentration of ammonia is toxic to benthic organisms. In addition to oxygen depletion, other physical and chemical conditions show considerable fluctuations at the bottom of shallow coastal areas (Henriksen and Kemp 1988). Thus, it is important to the health of coastal ecosystems to understand how the ammonia oxidation process is affected by environmental changes.

Recent studies have enabled an estimation of the abundance and diversity of ammonia-oxidizing microorganisms in the natural environment by examining the abundance of ammonia monooxygenase alpha subunit (*amoA*), a gene related to the ammonia-oxidizing process (Francis et al. 2005; Rotthauwe et al. 1997). Two types of ammonia-oxidizing microorganisms, ammonia-oxidizing archaea (AOA) and ammonia-oxidizing bacteria (AOB), are involved in this process. Measurement of the abundance of *amoA* has been applied to marine environment monitoring (Preston et al. 2011; Qin et al. 2014), although any primers can generate some bias in PCR amplification (Baptista et al. 2014; Meinhardt et al. 2015; Tolar et al. 2013). The *amoA*-1F/*amoA*-2R primer set, which cannot detect AOB species of the gamma subclass of proteobacteria (Rotthauwe et al. 1997), is used in many studies. Moreover, some studies have failed to detect gammaproteobacterial AOB from a marine environment using qPCR (Christman et al. 2011; Erguder et al. 2009; Rotthauwe et al. 1997; Wuchter et al. 2006), although these organisms are widely distributed in seawater (Ward and O'Mullan 2002). Therefore, the AOB referred to in this paper are betaproteobacterial AOB. In general, AOA are predominantly present in oligotrophic environments, while AOB are relatively eutrophic (Erguder et al. 2009; Wuchter et al. 2006), as indicated by the substrate affinity for ammonia uptake, which is much greater in AOA than in AOB (Martens-Habbena et al. 2009). AOA are thought to be the major contributors to ammonia oxidization in marine environments (Bouskill et al. 2012); however, some reports suggest that AOB also contribute to nitrification to some extent, and in coastal areas, even more than

AOA do (Li et al. 2015; Magalhães et al. 2009; Tait et al. 2014). In addition, Some AOA are mixotrophs and are relatively tolerant to a low DO concentration (French et al. 2012; Park et al. 2010; Qin et al. 2014). In contrast, AOB are obligate aerobic chemoautotrophs that are suffocated by an oxygen reduction (French et al. 2012; Park et al. 2010). Therefore, these two organisms are differentially affected by oxygen depletion in bottom sediments (Abell et al. 2011). There are many reports of the distributions and diversities of *amoA* genes in coastal sediments. They revealed that the abundance and community composition vary under the influence of freshwater discharge and eutrophication level of the studied areas and could be explained by environmental factors such as salinity, temperature, DO concentration, ammonia concentration, and sulfide concentration in pore water (Abell et al. 2011; Bouskill et al. 2012; Dang et al. 2008, 2010; Santoro et al. 2008). However, some reports are contradictory, which suggests that the combination of environmental factors may complicate the dynamics of AOA and AOB due to differences in their physiological requirements (Tait et al. 2014).

Nitrification activity is typically higher in sandy sediment than in muddy sediment owing to the high oxygen penetration of sand (Hansen et al. 1981; Henriksen and Kemp 1988). Penetration of sediments by DO depends on the sediment materials, and therefore, the ammonia-oxidizing microbial community is affected by sediment type (Beman et al. 2012; Tait et al. 2014). Tait et al. (2014) found that sediment particle size was related to AOB-*amoA* abundance, but not with AOA-*amoA* abundance in coastal areas. A positive correlation was also observed between AOB-*amoA* abundance and sediment sand content in a very eutrophic area (Dang et al. 2010). Sediment material type is often associated with ammonia concentration and other environmental factors that influence ammonia-oxidizing microorganisms, but the precise relationship between sediment material type and AOA or AOB abundance remains to be clarified. Moreover, the diversity of ammonia-oxidizing organisms is influenced by the sediment type (Cao et al. 2011; Dang et al. 2008, 2010; Tait et al. 2014; Wankel et al. 2011). Smith et al. indicated that different AOA ecotypes differentially contribute to nitrification in coastal water (Smith et al. 2014). Therefore, we expect that the diversity of ammonia oxidizers may vary with environmental change and that it may be indicative of nitrification potential at a given site.

Sendai Bay is located in the northeast region of Japan; it is shaped as a simple semicircle with a wide opening to the Pacific Ocean. Several rivers discharge here and oceanic water flows into the bay, both contributing to a high nutrient concentration in the bay, which has a considerable effect on primary productivity (Kakehi et al. 2015). In general, there are fine-grained sandy areas along the beach and very fine-grained areas of silt offshore. From the end of the 1980s, a hypoxic water mass has been observed frequently at around 25 m depth in autumn in the bay (Iwai 2004). The Tohoku earthquake off the Pacific coast and the subsequent major tsunami on 11 March 2011 affected much on the environments in Sendai Bay. Huge amounts of terrestrial matter entered the bay, and bottom sediments were disturbed, especially in the coastal area (Pilarczyk et al. 2012; Siswantoa and Hashima 2012). To evaluate the environmental impact on nitrogen cycling in Sendai

Bay, it is important to elucidate how different sediment types influence ammonia-oxidizing microorganisms.

The purpose of this study was to clarify how the distribution of ammonia-oxidizing microorganisms varies in coastal sediments in Sendai Bay and to ascertain the sediment characteristics affecting this distribution. We determined the abundance and community composition of both types of ammonia-oxidizing organisms in different types of sediments collected in Sendai Bay in regions with varying sediment quality parameters.

11.2 Materials and Methods

11.2.1 Sample Collection and Sediment Quality Analysis

In Sendai Bay, the muddy line of the coastal area is about 12–14 meters in depth. The central part of the bay has regions of coarse- to very coarse-grained sand and granules (Kan-no 1966). Sediment samples were collected at five sites on the across-shoreline (C-line), at two sites on the along-shoreline at approximately 30 m depth (P-line), and at two additional sites in the northern part of Sendai Bay with the R.V. *Wakataka-Maru*, in December 2011 and July 2012 (Fig. 11.1). Stations C2, P3, and E4 were located in muddy regions (mud content was greater than 40%), and C5, C6, C9, and C12 were located in sandy regions (mud content was 0% at both months). Stations P7 and E1 were located in an intermittent area. The color

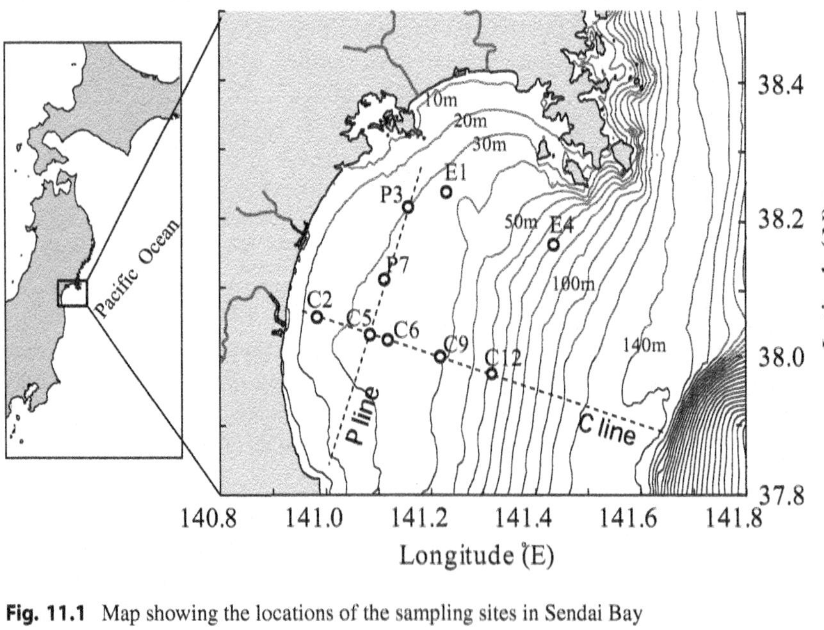

Fig. 11.1 Map showing the locations of the sampling sites in Sendai Bay

of the sediment was brown or light gray, indicating a highly oxidized environment. A Smith McIntyre Grab sediment sampler was used to collect sediment samples. A surface sediment sample (top 1 cm) was collected using a plastic core of 20 mm diameter and frozen immediately in a sterilized plastic bag. Hydrographic observations were conducted along the two transect lines. Temperature, salinity, chlorophyll *a*, and DO were measured using an AAQ aqua quality sensor (AAQ-1186; JFE Advantech, Kobe, Japan) from September 2011 to September 2012.

The following variables were analyzed: (i) particle size was determined using a laser diffraction particle size analyzer (SALD-3100; Shimadzu, Kyoto, Japan), (ii) ignition loss was evaluated by igniting the sediment samples at 500 °C for 4 h, (iii) total phosphorus levels were estimated by extracting orthophosphate from the ignited sediment samples by boiling for 15 min in 1 N HCl, followed by colorimetric measurement (Anderson 1975), (iv) ammonium levels were determined colorimetrically after extraction from the sediment samples in 1 M KCl (Kandeler and Gerber 1988), and (v) β-D-glucosidase activity in the sediment samples was estimated as an indicator of microbial organic matter degradation using a fluorophore-labeled analog substrate, 4-methylumbelliferyl (MUF)-β-D-glucoside (Hoppe 1983). This analog substrate was added to a sediment sample suspended in phosphate buffer (pH 8.0) at a final concentration of 100 μM and incubated for 20–80 min at 37 °C. Periodically, subsamples were taken and centrifuged in bicarbonate buffer (pH 10). The fluorescence intensity of the supernatant was measured using a spectrophotometer (RF-5300PC; Shimadzu).

11.2.2 DNA Extraction and Quantitative PCR

DNA was extracted from 0.4 to 0.6 g sediment sample using the Fast DNA SPIN Kit for soil (MP Biomedicals LLC., OH) according to the manufacturer's instructions. The extracted DNA was dissolved in 100 μL TE buffer and stored at −30 °C until further analysis. To determine *amoA* copy numbers, quantitative PCR (qPCR) was performed using the ABI StepOne™ System (Applied Biosystems, Foster City, CA) with SYBR Premix Ex Taq (Takara Bio, Otsu, Japan) and the primers crenAMO F/crenAMO R for crenarchaeal *amoA* (Ando et al. 2009) and primers *amoA*-1F/*amoA*-2R for betaproteobacterial *amoA* (Rotthauwe et al. 1997). Each PCR reaction mixture (10 μL) contained 1 μL template DNA solution, 0.2 μg μL^{-1} bovine serum albumin, 0.2 μM each primer, and 5 μL SYBR Premix Ex Taq. PCR was performed with an initial denaturation at 95 °C for 30 s, followed by 40 cycles of 95 °C for 15 s, 57 °C for 15 s, and 72 °C for 30 s (Ando et al. 2009). The fluorescence intensity was measured at 72 °C. Standards consisted of cloned AOA- or AOB-*amoA* fragments derived from biofilter materials that contained the region of each primer set; the GenBank accession numbers are AB571246 for AOA and AB571283 for AOB (Sakami et al. 2012).

11.2.3 Cloning of amoA

The same primer sets were used to amplify *amoA*s from the sediment DNAs. PCR was performed using Taq EX HS polymerase (Takara Bio) with an initial denaturation at 94 °C for 2 min, followed by 30 cycles of denaturation at 94 °C for 45 s, annealing at 58 °C for 30 s, and extension at 72 °C for 45 s; the final elongation was performed at 72 °C for 7 min. The amplified *amoA*s were cloned using the TA cloning kit (DynaExpress DNA Ligation Kit ver. 2; BioDynamics Laboratory Inc., Tokyo, Japan). The M13F/M13R PCR products were sequenced from both sides in a cycle sequencing reaction using a sequencing kit (BigDye Terminator Cycle version 3.1; Applied Biosystems) on a capillary DNA sequencer (ABI 3130; Applied Biosystems).

The *amoA* sequences were aligned and clustered in operational taxonomic units (OTUs) using MOTHUR (Schloss et al. 2009), which were defined as groups of nucleotide sequences that differed by 5% or less. The inverse of the Simpson index was also calculated using MOTHUR. Representative sequences of each OTU were compared with those available from the DDBJ/EMBL/GenBank databases by using nucleotide–nucleotide BLAST software. Phylogenetic analyses by the neighbor-joining method were implemented using MEGA version 4 software (Tamura et al. 2007).

11.2.4 Nucleotide Sequence Accession Numbers

The *amoA* sequences deposited at the DDBJ/EMBL/GenBank have been assigned accession numbers AB984797–AB985220.

11.3 Results

11.3.1 Abundance of amoA Genes and Relationship with Sediment Quality Parameters

Bottom water temperature was the highest in September and the lowest in March at stations P3 and C6, which represent the muddy and sandy regions, respectively (Fig. 11.2). DO concentrations varied with the water temperature but remained greater than 80 μM. The DO at station P3 was slightly lower than that at station C6 in September, but the DO at station P3 was higher than that at station C6 from April to May. Water column-integrated chlorophyll *a* levels were high in February and May and low from July to December at both stations, although the peak values were higher at station C6 than at station P3.

Bottom water temperatures at the sediment sample collection sites were uniform among the sampling stations (Table 11.1). Salinity was similar in both months at all the sampling points (ca. 33.5), indicating that freshwater discharge had little influence at the bottom. DO concentration was around 60% saturation at all stations in December but decreased to around 40% saturation at stations P3, P7, and C5 in

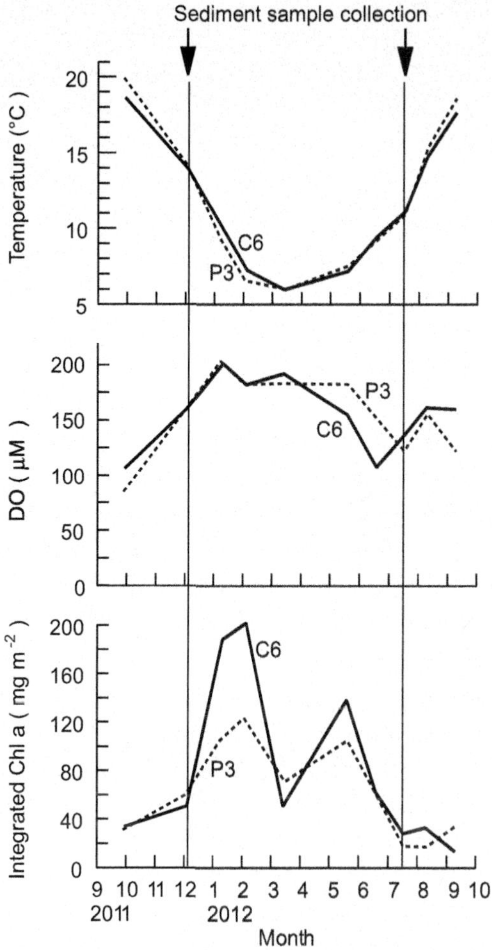

Fig. 11.2 Seasonal patterns of temperature, dissolved oxygen concentration in the bottom water, and chlorophyll *a* amount integrated over the water column of sediment at stations P3 (muddy area) and C6 (sandy area) in Sendai Bay

July. A high chlorophyll *a* concentration was observed at stations C2 and C5 in July, suggesting a high organic matter load in the sediment at these stations. The bottom material was mud at stations E4, P3, and C2, fine sand at stations E1 and P7, and coarse sand at stations C5 to C12. Ignition loss and total phosphorus content were high in the muddy sediments. Ammonia content and β-D-glucosidase activity were also high in the muddy sediments. Bottom water temperature, DO concentration, and ammonia content in the sediments in December were significantly different from these values in July (paired *t* test, $P < 0.01$).

The abundance of AOA-*amoA* was greater at the stations where the sediment material was mud or fine sand (E1, E4, P3, P7, and C2) than at stations with coarse sand (C5–C12) in both December and July (Fig. 11.3). The differences among the stations were greater in July than in December; the AOA-*amoA* abundance at site P3 was 67 times greater than that at site C6 in July, but the abundance at

Table 11.1 Overview of the environmental parameters of the different sites in Sendai Bay

Sampling time	*2011 Dec*								
Sites	E1	E4	P3	P7	C2	C5	C6	C9	C12
Bottom water									
Depth (m)	37	79	30	31	23	30	33	41	57
Temp (°C)	15.0	13.6	14.3	13.9	13.7	13.9	14.0	14.3	14.1
Salinity	33.5	33.8	33.5	33.6	33.5	33.5	33.5	33.6	33.8
DO (μM)	178	181	181	181	184	184	184	181	181
DO (% of saturation)	62	61	62	62	62	63	63	62	62
Chl. *a* (μg/L)	1.6	0.6	1.9	1.9	1.2	1.6	1.5	2.7	1.2
Sediment									
Median grain size (μm)	466	23	61	390	76	1221	981	776	698
Mud content (%)	0	80	52	0	43	0	0	0	0
IL (%)	1.4	7.4	5.7	1.2	7.7	0.8	0.8	1.0	0.6
TP (μmol/dwg)	3.6	16	15	6.4	21	3.2	3.7	5.5	3.4
NH_4^+ (μmol/dwg)	0.10	0.64	0.59	0.06	0.78	0.04	0.08	0.18	0.03
Glc (relative activity)	4.6	16	50	6.4	99	1.4	4.8	9.2	2.5
Sampling time	*2012 Jul*								
Sites	E1	E4	P3	P7	C2	C5	C6	C9	C12
Bottom water									
Depth (m)	30	76	32	34	26	33	35	43	59
Temp (°C)	11.1	9.5	10.9	10.9	12.4	10.8	11.1	10.2	9.9
Salinity	33.5	33.8	33.5	33.6	33.3	33.5	33.5	33.6	33.7
DO (μM)	147	178	138	125	184	119	156	169	178
DO (% of saturation)	47	55	44	40	61	38	50	53	56
Chl. *a* (μg/L)	1.8	0.2	1.1	0.7	12.3	6.9	1.8	0.5	0.4
Sediment									
Median grain size (μm)	185	60	60	353	75	831	1048	705	841
Mud content (%)	26	53	53	6	42	0	0	0	0
IL (%)	5.0	7.0	7.0	2.0	7.9	0.8	0.8	0.9	0.7
TP (μmol/dwg)	11	17	17	6.9	22	2.9	2.7	3.7	2.9
NH_4^+ (μmol/dwg)	0.92	1.6	1.6	0.57	0.94	0.29	0.12	0.24	0.41
Glc (relative activity)	7.4	30	30	16	41	2.9	2.7	5.3	1.4

Chl chlorophyll, DO dissolved oxygen, Glc β-D-Glucosidase activity, IL ignition loss, TP total phosphorus

site P3 was only 4.3 times greater than that at site C6 in December. The AOA-*amoA* abundance correlated negatively with median grain size and positively with total phosphorus and ammonia content, as well as β-D-glucosidase activity in the

Fig. 11.3 Abundance of archaeal and bacterial *amoA* copy numbers in the surface sediments of Sendai Bay. Error bars represent standard deviations of triplicate qPCR measurements

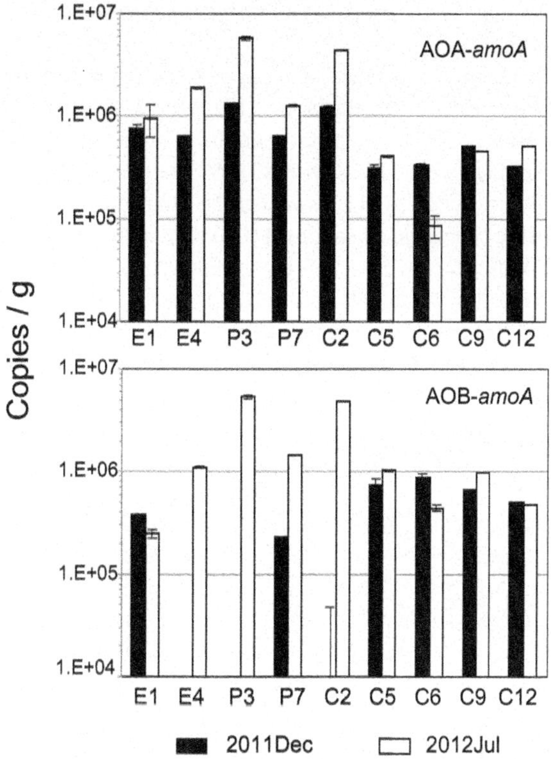

Table 11.2 Pearson's correlation coefficients of the environmental parameters with *amoA* abundance

		Median grain size	Mud content	IL	TP	NH_4^+	Glc
AOA	2011 Dec	−0.80*	0.58	0.74*	0.81*	0.79*	0.84*
	2012 Jul	−0.72*	0.65	0.61	0.78*	0.95*	0.78*
AOB	2011 Dec	0.96*	−0.78*	−0.84*	−0.84*	−0.80*	−0.64
	2012 Jul	−0.57	0.50	0.46	0.67*	0.86*	0.74*

Asterisk (*) means significant at $P < 0.05$
Glc β-D-Glucosidase activity, IL ignition loss, TP total phosphorus

sediment (Table 11.2). The highest correlation coefficient was observed between AOA-*amoA* abundance and ammonia content in July ($r = 0.95$).

AOB-*amoA* abundance was very low, around the detectable level (ca. 10^2 copies g^{-1}), in the muddy sediments at stations E4, P3, and C2 in December. Therefore, the correlations with the environmental parameters contrasted those of AOA-*amoA*. AOB-*amoA* abundance was higher in July than in December, but the distribution pattern was similar to that of AOA-*amoA*. The relationships among the sedimentary parameters for AOB-*amoA* abundance were similar to those of AOA-*amoA*, with the exception of median grain size.

11.3.2 Composition of the *amoA* Genes

Sediment samples collected at stations P3 and C6 were chosen to construct *amoA* clone libraries, as these sites are representative of the muddy (P3) and the course sandy (C6) areas and have similar bottom water depths. The AOA-*amoA* libraries included 9–19 OTUs with 82–95% coverage (Table 11.3). Both of the community diversity indices (the inverse of the Simpson index and the Chao index) were greater in the coarse sandy sediment (C6) than in the muddy sediment (P3). The AOB-*amoA* libraries included 1–5 OTUs with 96–100% coverage, indicating that the community diversity was smaller for AOB-*amoA* than for AOA-*amoA*. As with AOA-*amoA*, the community diversity indices for AOB-*amoA* were greater in the coarse sandy sediments. The community compositions were similar for both months, suggesting that the temporal variation was small compared to the influences of the sediment materials.

Eight OTUs accounted for more than 3% of the total in the AOA-*amoA* clone libraries (Fig. 11.4). Among them, three OTUs (OTUa1–a3) were abundant in the muddy sediments. They were phylogenetically close to *Nitrosopumilus maritimus*. Another five OTUs (OTUa4–a8) were abundant in the coarse sandy sediments. OTUa5 and OTUa7, as well as most of the other minor OTUs, fell into the same cluster as OTUa1–a3, indicating that most of the AOA clones were phylogenetically close to *N. maritimus* in the Sendai Bay sediment. In contrast, OTUa4 and OTUa6 fell into another cluster, which contained sequences detected in marine sediments. OTUa8, which was detected only in the coarse sandy sediments, was phylogenetically different from the other major OTUs and close to the sequence derived from macroalgal epiphytes.

Most of the AOB-*amoA* clones belonged to the dominant OTUb1 group in the muddy sediments (Fig. 11.5). However, OTUb2 and OTUb3 were abundant in the coarse sandy sediments. These OTUs were phylogenetically related to the "unculturable *Nitrosomonas*" group. The sequence closest to OTUb1 was isolated from an anoxic marine sediment. OTUb2 and OTUb3 were from the sediments at eutrophic coastal areas. Only one clone, OTUb4, was isolated from the muddy sediment in December, and it belonged to *Nitrosomonas*, which was close to the Nm143 lineage originally isolated from a marine aquaculture biofilm.

11.4 Discussion

This study was conducted after the big tsunami occurred in Sendai Bay following the offshore Pacific coast Tohoku earthquake in March 2011. The obtained results may well be affected by environmental disturbances caused by the tsunami. However, we conducted a preliminary investigation focusing on *amoA* gene abundance in the sediments in the summer before the tsunami (2008 and 2009) in Sendai Bay that produced results equivalent to those from July 2012 (data not shown). Therefore, we regard the current results as generalizable. Moreover, we cannot avoid some

Table 11.3 Diversity and richness indices of the *amoA* gene sequences from the clone libraries constructed from the muddy (P3) or the sandy (C6) sediment collected in Senday Bay

Type of *amoA*	*Archaea*				*Betaproteobacteria*			
Sites	P3		C6		P3		C6	
Sampling time	Dec	Jul	Dec	Jul	Dec	Jul	Dec	Jul
No. of clones	50	52	51	55	54	56	52	54
No. of OTUs	9	9	19	13	2	1	5	3
% coverage	82	82	84	95	98	100	96	100
Inv. Simpson	4.8	4.9	15.5	8.9	1.1	1.0	3.1	3.1
(95% CI)	(3.9–6.4)	(3.9–6.4)	(11–25)	(6.5–15)	(1.0–1.1)	(1.0–1.0)	(1.7–3.1)	(1.9–3.1)
Chao1 score	15	15	26	14	2	1	6	3

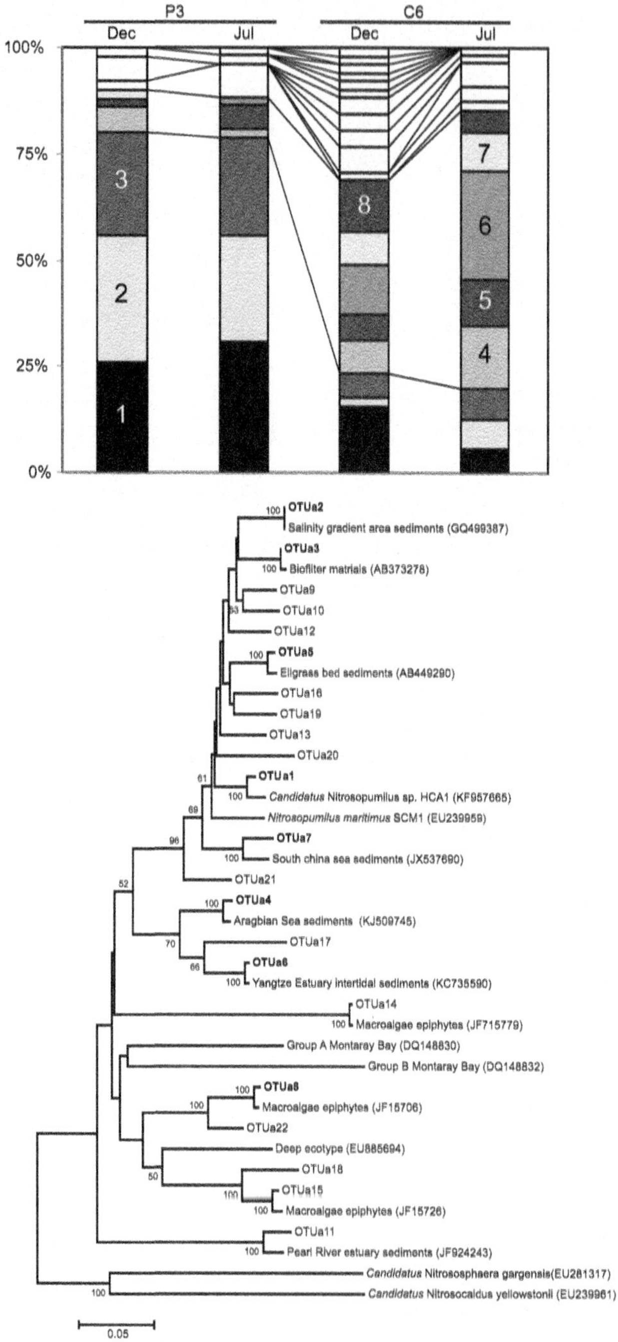

Fig. 11.4 Composition and phylogenetic relationships of archaeal *amoA* sequences (588 bp) recovered from muddy (P3) and sandy (C6) sediment samples in Sendai Bay. OTUs were defined by 97% nucleotide similarity. Numbers in the column indicate the top eight OTUs (OTUa1–OTUa8 in the text) which accounted for more than 3% of the total in the AOA-*amoA* clone libraries. The tree was estimated by neighbor joining and rooted with *Can. Nitrososphaera gargensis* and *Can. Nitrosocaldus yellowstonii*. The scale bar represents an estimated sequence divergence of 5%. Numbers at nodes indicate bootstrap values (> 50%)

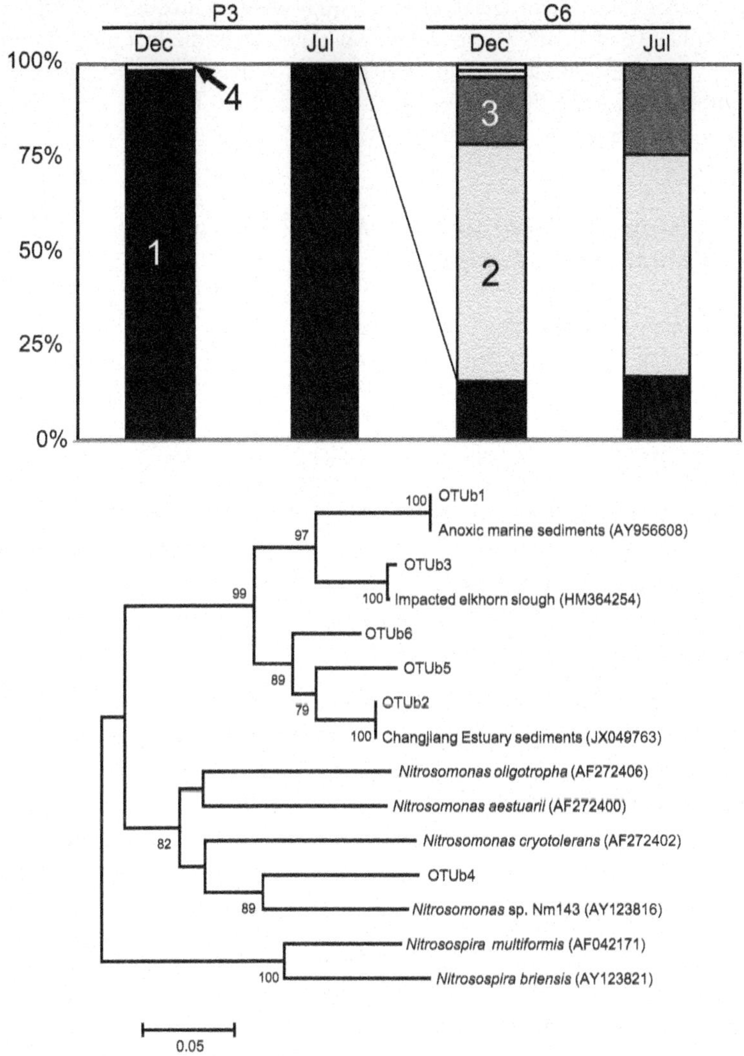

Fig. 11.5 Composition and phylogenetic relationships of bacterial *amoA* sequences (404 bp) recovered from muddy (P3) and sandy (C6) sediment samples in Sendai Bay. OTUs were defined by 97% nucleotide similarity. Numbers in the column indicate the code of each OTU (OTUb1–OTUb4 in the text). The tree was estimated by neighbor joining and rooted with *Nitrosospira multiformis* and *Nitrosospira briensis*. The scale bar represents an estimated sequence divergence of 5%. Numbers at nodes indicate bootstrap values (> 50%)

technical issues such as DNA extraction efficiency or primer specific bias in PCR when analyzing microbial communities in environmental samples. Meinhardt et al. (Meinhardt et al. 2015) indicated that the copy number of an AOA-*amoA* clone varied by more than one order of magnitude, depending on the specific qPCR primers. Though the primer sets used in this study are orthodox ones (Könneke et al. 2005; Rotthauwe et al. 1997), we should pay attention to the fact that the reported gene copy number is not an absolute value, but a relative one, which is comparable within this study. Nevertheless, we found a relationship between the distribution pattern of *amoA* and sediment quality in Sendai Bay.

11.4.1 Distribution of AOA-*amoA* Was Related to the Ammonia Content in the Sediment

The abundance of AOA-*amoA* was strongly correlated with ammonia content in the sediment, consistent with previous studies conducted in coastal marine sediments (Ando et al. 2009; Beman and Francis 2006; Sakami 2011; Schloss et al. 2009). In fact, the ammonia concentration in pore water must be lower than the values reported here, because in this study, ammonia was extracted under low pH conditions from frozen sediment, which causes an overestimation in ammonia concentration (García-Robledo et al. 2010). Therefore, the value is regarded as a relative one that indicates the maximum potential ammonia concentration. Moreover, the ammonia content in the sediment is not equal to the amount of ammonia available for ammonia oxidizers. Other biological processes such as nitrite oxidization, ammonification, assimilation by benthic microorganisms, and anammox also participate in ammonia dynamics; ammonia flux is multifaceted. As for generation of ammonia, the β-D-glucosidase activity in the sediment also shows a positive correlation with AOA-*amoA* abundance in this study, implying that the increase in organic matter degradation, presumably relating to the increase in ammonification, promotes ammonia oxidization in the bottom surface sediment.

Nevertheless, the pore water ammonia concentration could be useful for inferring physiological characteristics of ammonia oxidizers in the environment. In general, the ammonia concentration in pore water is reported to be around 1 μM (low organic matter sandy sediment (Santoro et al. 2008)) to 300 μM (eutrophicated muddy sediment (Urakawa et al. 2006)) in surface sediments. Although high concentrations of ammonia (1–3 mM) or organic matter may inhibit the growth of cultured AOA (Könneke et al. 2005), the ammonia concentration of pore water might not be high enough to suppress the AOA community, even in the muddy sediments. The half saturation constant (K_m) for ammonia uptake was reported as 60–100 μM NH_4^+ for NH_3 oxidation in slurries of an estuarine sediment, and pore water concentration in the surface sediment ranged from 20 to 200 μM NH_4^+ (Henriksen and Kemp 1988). If AOA contribute considerably to ammonia oxidization in the sediment, we can suppose that there are some AOA species that have adapted to the relatively high pore water ammonia concentration in the muddy sediments. In fact, the K_m for ammonia uptake was reported to be 0.1–0.6 μM from the AOA strains isolated from seawater where ammonia concentration is much lower than that in sediment pore water (Martens-Habbena et al. 2009). Other studies have suggested that a high ammonia concentration promotes AOA activity; Treusch et al. (2005) found considerably higher levels of archaeal *amoA* transcripts in samples that had been supplemented with 10 mM ammonia. Park and Noguera (2007) also indicated that when marine sediment was cultured in medium amended with 1 mM ammonia and 0.1–1 mM sodium thiosulfate, AOA was enriched in the medium, but AOB disappeared. Together, these results imply that an ammonia concentration in muddy sediment up to several hundred micromoles per liter may promote AOA, at least when the increase is sporadic.

In addition to ammonia availability, the AOA community in the muddy sediment may differ physiologically from the AOA community in the sandy area. Some OTUs (OTUa1–a3) were dominant in the muddy sediment. The OTUa1 was closest to the sequence of *Can didatus Nitrosopumilus* sp. strain HCA1. This organism was isolated from coastal seawater and was shown to be an obligate mixotroph (Qin et al. 2014). Moreover, the cell yield per oxidized ammonia was greater when alpha-ketoglutaric acid was added to the culture medium, suggesting that mixotrophic growth may be the preferred lifestyle of marine AOA (Qin et al. 2014). Organic matter content was much greater in the muddy sediments than in the sandy sediments. It is likely that some AOA utilize organic matter together with ammonia to flourish in muddy sediment. Some reports also indicate that the continual supply of ammonia through the mineralization of organic matter is associated with AOA growth, although inorganic ammonia input does not have an effect in soil environments (Offre et al. 2009; Zhalnina et al. 2012). In addition, *Can didatus Nitrosopumilus* sp. strain HCA1, which was closest to the sequence of the OTUa1, can grow at a relatively low temperature (~10 °C) (Qin et al. 2014). The optimal pH and temperature fall within the typical pH and temperature range in Sendai Bay, suggesting that OTUa1 and HCA1 have similar physiological properties.

The community diversity was high in the sandy sediment, whereas the AOA abundance was low. On the other hand, Wankel et al. (2011) reported that estuarine sediment that had relatively high nitrification activity had a high diversity of AOA-*amoA*, which contrasts our results. There seems to be enough ammonia to maintain an AOA community in the sandy sediment because the lowest AOA abundance level in this study was 8.7×10^4 copies g^{-1}, which is similar to the values determined for sandy sediments in previous studies (Ando et al. 2009; Beman et al. 2012). It seems that some kinds of AOA that are physiologically adapted to the environment probably flourish in the muddy sediment, resulting in a simple community composition. Nevertheless, not all of the AOA or AOB in the sediment can be detected by the PCR method used in this study. Some AOA phylotypes could be difficult to detect using the primers from this study. Further quantitative studies, such as 16S rRNA gene or FISH (Nunoura et al. 2015; Tolar et al. 2013), are required in order to elucidate how differences in *amoA* diversities relate to the in situ nitrification process as a whole.

AOA-*amoA* abundance is known to vary with salinity, and terrestrial and marine types of AOA, and that AOA-*amoA* composition are influenced by river water discharge (Erguder et al. 2009; Robidart et al. 2012; Sahan and Muyzer 2008; Santoro et al. 2008). However, the offshore seawater flowing into Sendai Bay through the bottom layer was estimated to be 27–29 times greater than the river water discharge into the bay (Kakehi et al. 2012), suggesting that river water discharge is not very influential at the bottom. In fact, we found that there was little influence of fresh water discharge on gene distribution and that all of the AOA-*amoA* clones detected in this study were similar in sequence to sequences derived from marine environments in Sendai Bay.

11.4.2 AOB-*amoA* Abundance Fluctuated Widely in the Muddy Sediment

The distribution of AOB-*amoA* was similar to that of AOA in July, and a strong correlation was observed between gene abundance and ammonia concentration in the sediment. The K_m of ammonia uptake by AOB was 10–100 μM in *Nitrosospira,* which often distribute in oligotrophic natural environments, and around 1000 μM in *Nitrosomonas* (Martens-Habbena et al. 2009). Park et al. (2010) also reported that, in an enriched marine sediment, AOB have a K_m of ammonia uptake of 60–1300 μM. These values were similar to or higher than the range of pore water ammonia concentrations, suggesting that a change in ammonia concentration might directly affect AOB communities in July.

On the other hand, the abundance of AOB-*amoA* was remarkably low in the muddy sediments in December. The particular low abundance seemed peculiar, and we observed a low abundance of AOB in the muddy area from September to March, although the duration was station-dependent (data not shown). In previous studies, AOB-*amoA* abundance was high in winter and spring, and it decreased from summer to autumn in coastal sediments (Ando et al. 2009; Tait et al. 2014). Ando et al. (2009) also indicated that the seasonal fluctuation was related to ammonia content in the sediment. Phytoplankton abundance in Sendai Bay had been low for several months before the sampling time in December, suggesting that organic matter levels in the bottom sediment were lower in December than in July (Fig. 11.2). In addition, microbial degradation of organic matter was probably high when the water temperature was high. These facts imply that organic matter might have been exhausted in the thin surface layer of the bottom sediment in December, although the total ammonia content of the top 1 cm remained high. Besides ammonia, DO is another important environmental factor affecting the distribution of ammonia-oxidizing microorganisms. DO concentration becomes zero at several mm (sandy sediment) or less than 1 mm (muddy sediment) below the surface of marine bottom sediments (Beman and Francis 2006; Brotas et al. 1990). The K_m of oxygen uptake by AOB was 16–200 μM (Park et al. 2010), indicating that AOB are very sensitive to a change in DO concentration. Moreover, AOB is also sensitive to sulfide (Abell et al. 2011; Erguder et al. 2009; Park et al. 2010). Considering the diffusion of sulfide from the anoxic layer, it seems that AOB can exist in a limited surface layer, especially in muddy sediments. Tait et al. (2014) reported a significant relationship between AOB-*amoA* abundance and sediment particle size, but not ammonia concentration, suggesting that oxygen penetration is related to the difference. Together with oxygen and ammonia concentrations, low

AOB abundance is likely influenced by a decrease in the ammonia concentration at the surface layer where DO could penetrate the muddy sediment in December. On the other hand, Beman et al. (2006) indicated that AOB-*amoA* was constantly present at depths up to 10 cm in both sandy and muddy sediments. We do not have enough data for further discussion of this hypothesis. Future evaluations of fine vertical profiles for both ammonia-oxidizing microorganisms, together with related environmental parameters, are needed to fully understand the mechanism controlling their distributions in natural environments.

AOA is considered relatively tolerant to low DO concentrations (Erguder et al. 2009). The K_m of oxygen uptake by enriched AOA culture from marine sediment was 2–4 μM (Park et al. 2010), and AOA enriched from a freshwater environment maintained a constant growth rate with DO concentrations of 6–22 μM (French et al. 2012). Because DO concentrations in the bottom water were not depleted at the sampling time, the dissolved oxygen concentration seemed to not strongly influence the AOA communities in this study.

In another study, AOB-*amoA* diversity was low in nitrogen- and phosphorus-rich sediments (Lage et al. 2010), which is consistent with the results of this study. The sequence of OTUb1, which was the only component of the AOB community in the muddy sediment in December, was closest to the sequence derived from anoxic sediment. It seemed that the AOB community population was reduced in December, and thus, only the AOB that was tolerant to low oxygen remained (Park and Noguera 2007). Only one clone, OTUb4, which was found in the muddy sediment in July, was assigned to *Nitrosomonas* sp. and was similar in sequence to the Nm143-lineage, which is often detected in coastal sediments (Cao et al. 2011; Dang et al. 2010; Freitag et al. 2006). However, *Nitrosospira* spp. were detected more often than *Nitrosomonas* spp. in natural systems that are low in ammonium (Kowalchuk et al. 1997; McCaig et al. 1999). The occurrence of *Nitrosomonas* sp. also suggests high organic matter load in the muddy sediment.

11.5 Conclusion

In the bottom sediments of Sendai Bay, the AOA-*amoA* abundance was dependent on the sediment quality and correlated with ammonia content in the sediment. The dominant OTU in the muddy sediment was close to a mixotrophic AOA strain, suggesting that organic matter content influences the AOA distribution. In contrast, the AOB-*amoA* abundance was remarkably low in the muddy sediments in December, but it produced a distribution pattern similar to that of AOA in July. Together with the low community diversity, the instability in abundance suggests that some factors suppress AOB growth in the coastal muddy area in Sendai Bay. In addition, we did not consider the quality of organic matter in the sediment in this study, which may influence the ammonia-oxidizer community in coastal areas. Further studies considering fine vertical profiles of the AO community or organic matter quality are necessary to clarify the mechanisms causing the fluctuations in

AOA and AOB abundance. The study of *amoA* dynamics can be applied to coastal environment monitoring as a biological indicator of nitrogen regeneration status.

Acknowledgments The authors thank the crew of the R.V. *Wakataka-Maru* for their support during the cruise and Dr. Hajime Saito of Fisheries Research Agency for the sediment material analysis. This study was funded by the Fisheries Agency of Japan, the Ministry of Agriculture, Forestry and Fisheries of Japan.

Conflict of Interest The authors declare no conflict of interest.

References

Abell GCJ, Banks J, Ross DJ, Keane JP, Robert SS, Revill AT, Volkman JK (2011) Effects of estuarine sediment hypoxia on nitrogen fluxes and ammonia oxidizer gene transcription. FEMS Microbiol Ecol 75:111–122

Anderson JM (1975) An ignition method for determination of total phosphorus in lake sediments. Water Res 10:329–331

Ando Y, Nakagawa T, Takahashi R, Yoshihara K, Tokuyama T (2009) Seasonal changes in abundance of ammonia-oxidizing archaea and ammonia oxidizing bacteria and their nitrification in sand of an eelgrass zone. Microbes Environ 24:21–27

Baptista JDC, Lunn M, Davenport RJ, Swan DL, Read LF, Brown MR, Morais C, Curtis TP (2014) Agreement between *amoA* gene-specific quantitative PCR and fluorescence in situ hybridization in the measurement of ammonia-oxidizing bacteria in activated sludge. Appl Environ Microbiol 80:5901–5910

Beman JM, Francis JA (2006) Diversity of ammonia-oxidizing archaea and bacteria in the sediments of a hypernutrified subtropical estuary. Appl Environ Microbiol 72:7767–7777

Beman JM, Bertics VJ, Braunschweiler T, Wilson JM (2012) Quantification of ammonia oxidation rates and the distribution of ammonia-oxidizing archaea and bacteria in marine sediment depth profiles from Catalina Island, California. Front Microbiol 3:263. https://doi.org/10.3389/fmicb.2012.00263

Bouskill NJ, Eveillard D, Chien D, Jayakumar A, Ward BB (2012) Environmental factors determining ammonia-oxidizing organism distribution and diversity in marine environments. Environ Microbiol 14:714–729

Brotas V, Amorim-Ferreira A, Vale C, Catarino F (1990) Oxygen profiles in intertidal sediments of Ria Formosa (S. Portugal). Hydrobiologia 207:123–129

Cao H, Hong Y, Li M, Gu JD (2011) Diversity and abundance of ammonia-oxidizing prokaryotes in sediments from the coastal Pearl River estuary to the South China Sea. Antonie Van Leeuwenhoek 100:545–556

Christman GD, Cottrell MT, Popp BN, Gier E, Kirchman DL (2011) Abundance, diversity, and activity of ammonia-oxidizing prokaryotes in the coastal Arctic Ocean in summer and winter. App Environ Microbiol 77:2026–2034

Dang H, Zhang X, Sun J, Li T, Zhang Z, Yang G (2008) Diversity and spatial distribution of sediment ammonia-oxidizing crenarchaeota in response to estuarine and environmental gradients in the Changjiang Estuary and East China Sea. Microbiol 154:2084–2095

Dang H, Li J, Chen R, Wang L, Guo L, Zhang Z, Klotz MG (2010) Diversity, abundance, and spatial distribution of sediment ammonia-oxidizing Betaproteobacteria in response to environmental gradients and coastal eutrophication in Jiaozhou Bay. China Appl Environ Microbiol 76:4691–4702

Erguder TH, Boon N, Wittebolle L, Marzorati M, Verstraete W (2009) Environmental factors shaping the ecological niches of ammonia-oxidizing archaea. FEMS Microbiol Rev 33:855–869

Francis CA, Roberts KJ, Beman JM, Santoro AE, Oakley BB (2005) Ubiquity and diversity of ammonia-oxidizing archaea in water columns and sediments of the ocean. PNAS 102:14683–14688

Freitag TE, Chang L, Prosser JI (2006) Changes in the community structure and activity of betaproteobacterial ammonia-oxidizing sediment bacteria along a freshwater–marine gradient. Environ Microbiol 8:684–696

French E, Kozlowski JA, Mukherjee M, Bullerjahn G, Bollmanna A (2012) Ecophysiological characterization of ammonia-oxidizing archaea and bacteria from freshwater. App Environ Microbiol 78:5773–5780

García-Robledo E, Corzo A, Papaspyrou S, Jiménez-Arias JL, Villahermosa D (2010) Freeze-lysable inorganic nutrients in intertidal sediments: dependence on microphytobenthos abundance. Mar Ecol Prog Ser 403:155–163

Hansen II, Henriksen K, Blackburn TH (1981) Seasonal distribution of nitrifying bacteria and rates of nitrification in coastal marine sediments. Microb Ecol 7:297–304

Henriksen K, Kemp WM (1988) Nitrification in estuarine and coastal marine sediment. In: Blackburn TH, Sorensen J (eds) Nitrogen Cycling in Coastal Marine Environments. Wiley, New York, pp 207–249

Hoppe HG (1983) Significance of exoenzymatic activities in the ecology of brackish water: measurements by means of methylumbelliferyl-substrates. Mar Ecol Prog Ser 11:299–308

Iwai T (2004) The recent occurrence and examination for occurred causes of oxygen depleted water in Sendai Bay. Miyagi Pref Rep Fish Sci 4:1–12. (in Japanese)

Jensen MH, Lomstein E, Sorensen J (1990) Benthic NH_4^+ and NO_3^- flux following sedimentation of a spring phytoplankton bloom in Aarhus Bight, Denmark. Mar Ecol Prog Ser 61:87–96

Kakehi S, Ito S, Yagi H, Wagawa T (2012) Estimation of the residence time of fresh and brackish water in Sendai Bay. J JSCE Div B, Hydr Coast Environ Engineer 68:951–955. (in Japanese with English abstract)

Kakehi S, Ito S, Kuwata A, Saito H, Tadokoro K (2015) Phytoplankton distribution during the winter convective season in Sendai Bay, Japan. Cont Shelf Res 97:43–53

Kandeler E, Gerber H (1988) Short-term assay of soil urease activity using colorimetric determination of ammonium. Biol Fertil Soils 6:68–72

Kan-no H (1966) Bottom environments of the ark shell, *Scapharca broughtonii* (Schrenck), in Sendai Bay. Bull Tohoku Nat Fish Res Inst 26:55–75. (in Japanese with English abstract)

Könneke M, Bernhard AE, de la Torre JR, Walker CB, Waterbury JB, Stahl DA (2005) Isolation of an autotrophic ammonia-oxidizing marine archaeon. Nature 437:543–546

Kowalchuk GA, Stephen JR, de Boer W, Prosser JI, Embley TM, Woldendorp JW (1997) Analysis of ammonia-oxidizing bacteria of the β subdivision of the class Proteobacteria in coastal sand dunes by denaturing gradient gel electrophoresis and sequencing of PCR-amplified 16S ribosomal DNA fragments. Appl Environ Microbiol 63:1489–1497

Lage MD, Reed HE, Weihe C, Crain CM, Martiny JBH (2010) Nitrogen and phosphorus enrichment alter the composition of ammonia-oxidizing bacteria in salt marsh sediments. ISME J 4:933–944

Li J, Nedwell D, Beddow J, Dumbrell AJ, McKew BA, Thorpe EL, Whitby C (2015) *amoA* gene abundances and nitrification potential rates suggest that benthic ammonia-oxidizing bacteria (AOB) not archaea (AOA) dominate N cycling in the Colne estuary, UK. Appl Environ Microbiol 81:159–165

Magalhães CM, Machado A, Bordalo AA (2009) Temporal variability in the abundance of ammonia oxidizing bacteria vs. archaea in sandy sediments of the Douro River estuary, Portugal. Aquat Microb Ecol 56:13–23

Martens-Habbena W, Berube PM, Urakawa H, de la Torre JR, Stahl DA (2009) Ammonia oxidation kinetics determine niche separation of nitrifying Archaea and Bacteria. Nature 461:976–981

McCaig AE, Phillips CJ, Stephen JR, Kowalchuk GA, Harvey SM, Herbert RA, Embley TM, Prosser JI (1999) Nitrogen cycling and community structure of proteobacterial β-subgroup ammonia-oxidizing bacteria within polluted marine fish farm sediments. Appl Environ Microbiol 65:213–220

Meinhardt KA, Bertagnolli A, Pannu MW, Strand SE, Brown SL, Stahl DA (2015) Evaluation of revised polymerase chain reaction primers for more inclusive quantification of ammonia-oxidizing archaea and bacteria. Environ Microbiol Rep 7:354–363

Nunoura T, Oida H, Nakaseama M, Kosaka A, Ohkubo SB, Kikuchi T, Kazama H, Hosoi-Tanabe S, Nakamura K, Kinoshita M, Hirayama H, Inagaki F, Tsunogai U, Ishibashi J, Takai K (2015) Archaeal diversity and distribution along thermal and geochemical gradients in hydrothermal sediments at the Yonaguni Knoll IV hydrothermal field in the southern Okinawa Trough. Appl Environ Microbiol 76:1198–1211

Offre P, Prosser JI, Nicol GW (2009) Growth of ammonia oxidizing archaea in soil microcosms is inhibited by acetylene. FEMS Microbiol Ecol 70:99–108

Park H–D, Noguera DR (2007) Characterization of two ammonia-oxidizing bacteria isolated from reactors operated with low dissolved oxygen concentrations. J Appl Microbiol 102:1401–1417

Park B–J, Park S–J, Yoon D–N, Schouten S, Sinninghe Damste JS, Rhee S–K (2010) Cultivation of autotrophic ammonia-oxidizing archaea from marine sediments in coculture with sulfur-oxidizing bacteria. App Environ Microbiol 76:7575–7587

Pilarczyk JE, Horton BP, Witter RC, Vane CH, Chagué-Goff C, Goff J (2012) Sedimentary and foraminiferal evidence of the 2011 Tōhoku-oki tsunami on the Sendai coastal plain, Japan. Sed Geo 282:78–89

Preston CM, Harris A, Ryan JP, Roman B, Marin R, Jensen S (2011) Underwater application of quantitative PCR on an ocean mooring. PLoS One 6:e22522

Qin W, Amin SA, Martens-Habbena W, Walker CB, Urakawa H, Devol AH, Ingalls AE, Moffett JW, Armbrust EV, Stahl DA (2014) Marine ammonia-oxidizing archaeal isolates display obligate mixotrophy and wide ecotypic variation. Proc Natl Acad Sci U S A 111:12504–12509

Rabalais N, Turner RE, Díaz RJ, Justić D (2009) Global change and eutrophication of coastal waters. ICES J Mar Sci 66:1528–1537

Robidart JC, Preston CM, Paerl RW, Turk KA, Mosier AC, Francis CA, Scholin CA, Zehr JP (2012) Ecophysiology of an ammonia-oxidizing archaeon adapted to low-salinity habitats. Microb Ecol 64:955–963

Rotthauwe JH, Witzel KP, Liesack W (1997) The ammonia monooxygenase structural gene *amoA* as a functional marker: molecular fine-scale analysis of natural ammonia-oxidizing populations. Appl Environ Microbiol 63:4704–4712

Sahan E, Muyzer G (2008) Diversity and spatio-temporal distribution of ammonia-oxidizing Archaea and Bacteria in sediments of the Westerschelde estuary. FEMS Microbiol Ecol 64:175–186

Sakami T (2011) Distribution of ammonia-oxidizing archaea and bacteria in the surface sediments of Matsushima Bay in relation to environmental variables. Microb Environ 27:61–66

Sakami T, Andoh T, Morita T, Yamamoto Y (2012) Phylogenetic diversity of ammonia-oxidizing archaea and bacteria in biofilters of recirculating aquaculture systems. Mar Genomics 7:27–31

Santoro AE, Francis CA, de Sieyes NR, Boehm AB (2008) Shifts in the relative abundance of ammonia-oxidizing bacteria and archaea across physicochemical gradients in a subterranean estuary. Environ Microbiol 10:1068–1079

Schloss PD, Westcott SL, Ryabin T, Hall JR, Hartmann M, Hollister EB, Lesniewski RA, Oakley BB, Parks DH, Robinson CJ, Sahl JW, Stres B, Thallinger GG, Van Horn DJ, Weber CF (2009) Introducing MOTHUR: open-source, platform-independent, community-supported software for describing and comparing microbial communities. Appl Environ Microbiol 75:7537–7541

Siswantoa E, Hashima M (2012) A data fusion study on the impacts of the 2011 Japan tsunami on the marine environment of Sendai Bay. Int J Image Data Fusion 3:191–198

Smith JM, Casciotti KL, Chavez FP, Francis CA (2014) Differential contributions of archaeal ammonia oxidizer ecotypes to nitrification in coastal surface waters. ISME J 8:1704–1714

Tait K, Kitidis V, Ward BB, Cummings DG, Jones MR, Somerfield PJ, Widdicombe S (2014) Spatio-temporal variability in ammonia oxidation and ammonia-oxidizing bacteria and archaea in coastal sediments of the western English Channel. Mar Ecol Prog Ser 511:41–58

Tamura K, Dudley J, Nei M, Kumar K (2007) MEGA4: molecular evolutionary genetics analysis (MEGA) software version 4.0. Mol Biol Evol 24:1596–1599

Thandrup BD, Dalsgaard T (2008) Nitrogen cycling in sediment. In: Kirchman DL (ed) Microbial ecology of the oceans, 2nd edn. Wiley, New York, pp 527–568

Tolar BB, King GM, Hollibaugh JT (2013) An analysis of Thaumarchaeota populations from the Northern Gulf of Mexico. Front Microbiol/Aquat Microbiol 72:1–36

Treusch AH, Leininger S, Kletzin A, Schuster SC, Klenk HP, Schleper C (2005) Novel genes for nitrite reductase and Amo-related proteins indicate a role of uncultivated mesophilic crenarchaeota in nitrogen cycling. Environ Microbiol 7:1985–1995

Urakawa H, Kurata S, Fujiwara T, Kuroiwa D, Maki H, Kawabata S, Hiwatari T, Ando H, Kawai T, Watanabe M, Kohata K (2006) Characterization and quantification of ammonia oxidizing bacteria in eutrophic coastal marine sediments using polyphasic molecular approaches and immunofluorescence staining. Environ Microbiol 8:787–803

Wankel SD, Mosier AC, Hansel CM, Paytan A, Francis CA (2011) Spatial variability in nitrification rates and ammonia-oxidizing microbial communities in the agriculturally impacted Elkhorn Slough Estuary, California Scott D. Appl Environ Microbiol 77:269–280

Ward BB, O'Mullan GD (2002) Worldwide distribution of *Nitrosococcus oceani*, a marine ammonia-oxidizing -proteobacterium, detected by PCR and sequencing of 16S rRNA and *amoA* genes. Appl Environ Microbiol 68:4153–4157

Wuchter C, Abbas B, Coolen MJL, Herfort L, van Bleijswijk J, Timmers P, Strous M, Teira E, Herndl GH, Middelburg JJ, Schouten S, Damsté JSS (2006) Archaeal nitrification in the ocean. Proc Natl Acad Sci U S A 103:12317–12322

Zhalnina K, Quadros PD, Camargo FAO, Triplett EW (2012) Drivers of archaeal ammonia-oxidizing communities in soil. Front Microb 3:1–9

Marine Metagenomic Sequence Counts of Reads Assigned to Taxa Consistently Proportionate to Read Counts Obtained for per g of Seawater Sample

12

Toshiaki Kudo, Md. Shaheed Reza, Atsushi Kobiyama, Jonaira Rashid, Yuichiro Yamada, Yuri Ikeda, Daisuke Ikeda, Nanami Mizusawa, Saki Yanagisawa, Kazuho Ikeo, Shigeru Sato, Takehiko Ogata, Shinnosuke Kaga, Shiho Watanabe, Kimiaki Naiki, Yoshimasa Kaga, Satoshi Segawa, Katsuhiko Mineta, Vladimir Bajic, Takashi Gojobori, and Shugo Watabe

Abstract

Development of high-throughput DNA sequencing technologies has enabled scientists to generate vast amounts of genetic information that may provide a comprehensive understanding of key roles played by environmental microorganisms. Generally the microorganisms inhabit a particular niche and correlate well with environmental changes. It is accepted that the read counts obtained through metagenomic analyses correlate semi-quantitatively with the relative abundance of bacterial species. In our marine metagenomic study conducted on

T. Kudo · A. Kobiyama · J. Rashid · Y. Yamada · Y. Ikeda · D. Ikeda · N. Mizusawa
S. Yanagisawa · S. Sato · T. Ogata
Kitasato University School of Marine Biosciences, Sagamihara, Japan

Md. S. Reza
Kitasato University School of Marine Biosciences, Sagamihara, Japan

Department of Fisheries Technology, Bangladesh Agricultural University, Mymensingh, Bangladesh

K. Ikeo
Kitasato University School of Marine Biosciences, Sagamihara, Japan

National Institute of Genetics, Shizuoka, Japan

S. Kaga
Iwate Fisheries Technology Center, Iwate, Japan

Ofunato Fisheries Promotion Center, Iwate Prefectural Government, Iwate, Japan

S. Watanabe · K. Naiki · Y. Kaga · S. Segawa
Iwate Fisheries Technology Center, Iwate, Japan

T. Gojobori et al. (eds.), *Marine Metagenomics*,
https://doi.org/10.1007/978-981-13-8134-8_12

the Ofunato Bay, Iwate Prefecture, Japan, we observed such correlation which exists for bacterioplankton *Candidatus* Pelagibacter ubique, identified as the dominant bacterial species of the bay. Shotgun metagenomic analyses identified three strains of *Ca.* Pelagibacter in the bay, viz., dmdA-HTCC1062, dmdA-HTCC9022, and O19-dmdA, that showed a dynamic change throughout the year particularly in the 10-m depth zone. Interestingly, the total abundances of those strains that fall in the *Ca.* Pelagibacter genus were found to correlate well with the read counts per g seawater samples used for analyses. It is assumed that whole-genome sequence (WGS) reads for members of the metagenome would show similar trend provided that proper precautions are taken to ensure collection of representative sample from the environment.

Keywords
Candidatus Pelagibacter · Dominant strain · Marine bacterioplankton

12.1 Introduction

Bacteria inhabiting the pelagic zone of the ocean are commonly known as pelagic bacteria. These are generally free-living in the ocean and consist of more than 90% of the total bacterial population. Among these heterotrophic bacteria, several genera/species have been reported to dominate. It was previously thought that microorganisms are distributed in the seawater homogenously. However, studies have shown that bacterial abundance varies at mm scale, so as their species richness (Seymour et al. 2000; Long and Azam 2001). All the members inhabiting this zone possess the ability to utilize highly variable but often large fractions of organic matters produced by phytoplankton or by sinking particles, resulting in a dynamic biogeochemical change in bacterial abundance, their composition, diversity, and richness (del Giorgio and Cole 1998; Long and Azam 2001). However, unculturability of the majority of microorganisms (~99% of bacteria and Archaea) under laboratory conditions still remains the challenge. With the advent of the recent developments in next-generation sequencing (NGS), scientists are now able to investigate environmental bacteria using shotgun and 16S rRNA NGS. The decreasing cost of such sequencing technology has also enabled scientists to use it in biomonitoring and routine investigations in the various research fields such as medicine and environment (Reza et al. 2018a).

K. Mineta · V. Bajic · T. Gojobori
Computational Bioscience Research Center, King Abdullah University of Science and Technology, Thuwal, Saudi Arabia

S. Watabe
Kitasato University School of Marine Biosciences, Sagamihara, Japan
e-mail: swatabe@kitasato-u.ac.jp

In an attempt to overcome unculturability and to gather information on microbial diversity and related functions, we selected the Ofunato Bay, a typical enclosed bay located in the northeast part of Japan, and used WGS approach. This approach allows us to examine thousands of organisms in parallel and comprehensively sample all genes, providing insight into community biodiversity and function (Sharpton 2014; Reza et al. 2018b). We previously reported that strains of *Candidatus* Pelagibacter ubique including those of the HTCC1062 type and the Red Sea type changed seasonally and observed a good correlation between chlorophyll *a* concentrations and abundances of the dimethylsulfoniopropionate (DMSP) catabolic genes (Kudo et al. 2018). This time we report that the dominant pelagic bacterioplankton *Ca.* Pelagibacter ubique changed consistently depending on season in the Ofunato Bay.

12.2 Methods

12.2.1 Sample Location

Three sampling stations were selected in the Ofunato Bay along its length named as KSt. 1 (141.73457E, 39.063370 N; avg. depth 10.3 m), KSt. 2 (141.73245E, 39.044612 N; avg. depth 25.3 m), and KSt. 3 (141.72820E, 39.019030 N; avg. depth 38.5 m).

12.2.2 Sample Collection and Processing and Analytical Procedures

Seawater samples were collected monthly from January to December 2015 using a vertical water sampler. Both surface and deep water layers were selected for sampling. Water collected from 1-m depth of the stations was treated as surface sample. On the other hand, deep water samples were collected at depth from 8 (KSt. 1) or 10 m (KSt. 2 and KSt. 3) below the surface. Then the seawater samples were sequentially filtered through 20-μm-pore-sized Nylon Net Filters NY20 with 47-mm diameter (Merck Millipore Ltd., Tullagreen, Ireland) and each of 5-, 0.8-, and 0.2-μm Isopore Membrane Filters with 142-mm diameter (Merck Millipore Ltd.).

Genomic DNA was then extracted from 0.2-μm filters using a PowerWater® DNA Isolation Kit (MO Bio Laboratories Inc., Carlsbad, CA, USA), following the manufacturer's recommendations. The extracted DNA was used for metagenomic shotgun sequencing with an Illumina MiSeq (Illumina Inc., San Diego, CA). The whole-genome sequence (WGS) datasets have been registered in the DDBJ Sequence Read Archive under the accession number DRA005744. Illumina paired reads for each library were first joined by overlapping forward and reverse reads of the same DNA fragment (paired end sequences) using the software FLASH with default parameters (overlap minimum, 10 nt; maximum allowed ratio between the number of mismatched base pairs and the overlap length, 0.25). Quality filtering of these WGS reads was then performed by removing reads<50 bp and quality trimmed to Phred 20 using the Genomics Workbench (CLCbio, Cambridge, MA). BLASTn was performed locally on a server at the laboratory in Kitasato University, Sagamihara, Kanagawa Prefecture, Japan, using the quality- and size-filtered sequences against the NCBI-nt reference database. Taxonomic analysis at the species level was then performed using MEGAN v5.10.3 (Huson et al. 2007) after parsing the BLAST output. Comparative analysis in MEGAN was also performed after normalizing counts per the recommendations of the authors.

12.3 Results and Discussion

12.3.1 Identification of *Ca.* Pelagibacter Strains in the Ofunato Bay

Ca. Pelagibacter ubique is a member of the SAR11 clade of α-proteobacteria and one of the most abundant groups of heterotrophic bacteria in oceans. In the Ofunato Bay, we also observed *Ca.* Pelagibacter ubique to be one of the dominant groups in our WGS data across 7 km long and 2.5 km wide of the bay, covering from KSt. 1 to KSt. 3 in both surface and deep water layers. Our study showed the presence of three strains of *Ca.* Pelagibacter, viz., *Ca.* Pelagibacter ubique dmdA-HTCC1062, dmdA-HTCC7211, and O19-dmdA in the bay, which varied seasonally from January to December in 2015. It has been reported that *Ca.* Pelagibacter reached as high as 60% of all sequences at the genus level during winter (Kobiyama et al. 2018), and the increase was more evident in the deeper waters.

While comparing the WGS read counts for *Ca.* Pelagibacter obtained for per g of seawater from which the sequencing process was conducted using MiSeq, we observed the highest 16,005 reads in March, while lower peaks were observed in June, August, and October (Fig. 12.1).

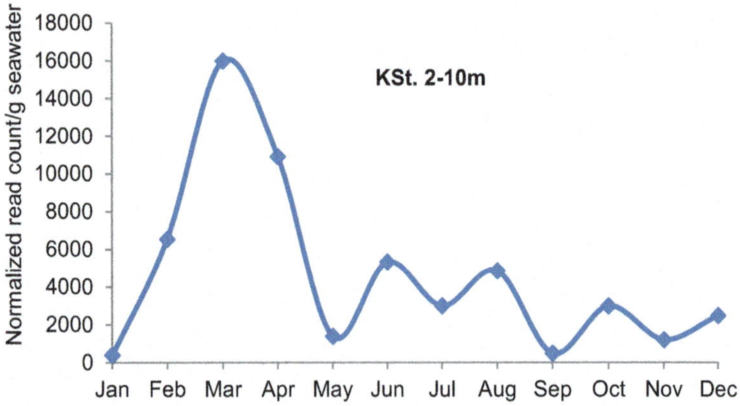

Fig. 12.1 Shotgun metagenomic sequence reads per g seawater assigned to *Ca.* Pelagibacter based on MEGAN analysis

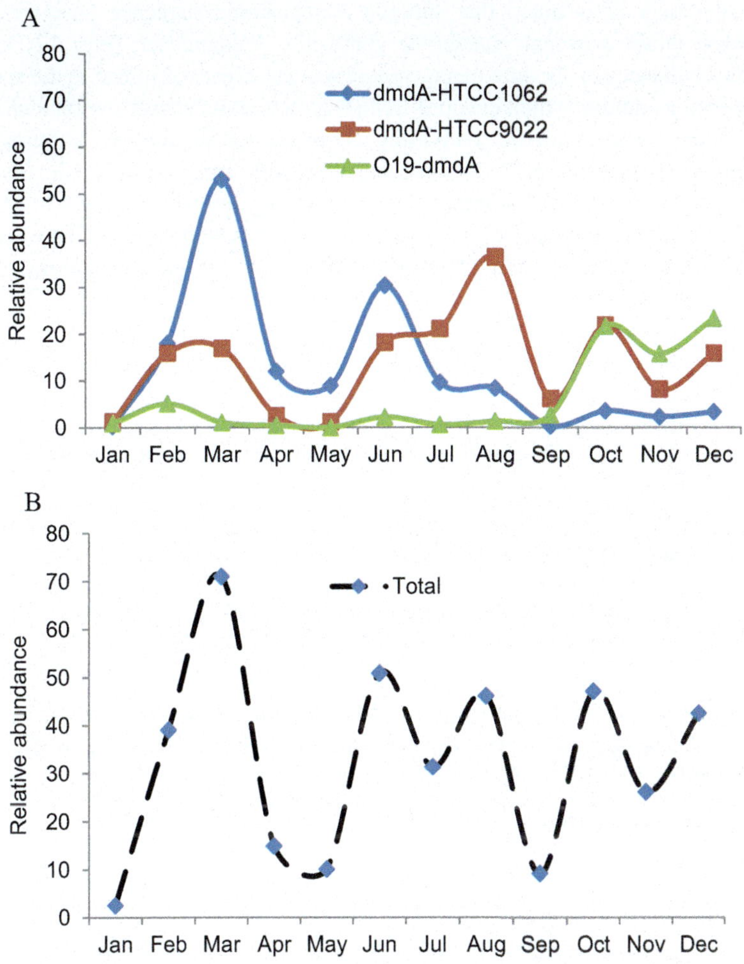

Fig. 12.2 Relative abundance of shotgun metagenomic sequence reads assigned to three strains of *Ca.* Pelagibacter, dmdA-HTCC1062, dmdA-HTCC9022, and O19-dmdA, from January to December 2015 (**a**) and relative abundance of their totality (**b**)

Figure 12.2a showed the relative abundance of shotgun metagenomic sequence reads assigned to three strains. The dmdA-HTCC1062 strain showed the highest peak in spring and small one in June. Interestingly, dmdA-HTCC9022 was most dominant during summer, although its close relatives were comparatively lower in abundance during the same period. The dmdA-HTCC9022 and dmdA-O19 strain both showed small peaks in autumn. When we calculated the seasonal changes in the total abundances of all three strains of *Ca.* Pelagibacter, it was proportionate to the total read counts calculated for per g of seawater sample (Fig. 12.1).

In summary, it is noted that shotgun metagenomic sequence analysis using MEGAN shows seasonal change of genus *Ca.* Pelagibacter (Fig. 12.1). The relative abundance of shotgun metagenomic sequence reads assigned to the specific genes such as dmdA, which encodes DMSP demethylase, shows seasonal changes at the strain level. With the increasing popularity of sequencing technique for environmental monitoring, it has become more and more important to interpret results and correlate them with metadata. Our approaches would benefit to analyze bacterial structure at the genus–strain levels in marine environment. This work was funded in part by King Abdullah University of Science and Technology, in the Kingdom of Saudi Arabia.

References

del Giorgio PA, Cole JJ (1998) Bacterial growth efficiency in natural aquatic systems. Annu Rev Ecol Syst 29:503–541

Huson DH, Auch AF, Qi J, Schuster SC (2007) MEGAN analysis of metagenomic data. Genome Res 17:377–386

Kobiyama A, Ikeo K, Reza MS, Rashid J, Yamada Y, Ikeda Y, Ikeda D, Mizusawa N, Sato S, Ogata T, Kudo T, Kaga S, Watanabe S, Naiki K, Kaga Y, Mineta K, Bajic V, Gojobori T, Watabe S (2018) Metagenome-based diversity analyses suggest a strong locality signal for bacterial communities associated with oyster aquaculture farms in Ofunato Bay. Gene 665:149–154

Kudo T, Kobiyama A, Rashid J, Reza MS, Yamada Y, Ikeda Y, Ikeda D, Mizusawa N, Ikeo K, Sato S, Ogata T, Jimbo M, Kaga S, Watanabe S, Naiki K, Kaga Y, Segawa S, Mineta K, Bajic V, Gojobori T, Watabe S (2018) Seasonal changes in the abundance of bacterial genes related to dimethylsulfoniopropionate catabolism in seawater from Ofunato Bay as revealed by metagenomic analysis. Gene 665:174–184

Long RA, Azam F (2001) Microscale patchiness of bacterioplankton assemblage richness in seawater. Aquat Microb Ecol 26:103–112

Reza MS, Mizusawa N, Kumano A, Oikawa C, Ouchi D, Kobiyama A, Yamada Y, Ikeda Y, Ikeda D, Ikeo K, Sato S, Ogata T, Kudo T, Jimbo M, Yasumoto K, Yoshitake K, Watabe S (2018a) Metagenomic analysis using 16S ribosomal RNA genes of a bacterial community in an urban stream, the Tama River, Tokyo. Fish Sci 84:563–577

Reza MS, Kobiyama A, Yamada Y, Ikeda Y, Ikeda D, Mizusawa N, Ikeo K, Sato S, Ogata T, Jimbo M, Kudo T, Kaga S, Watanabe S, Naiki K, Kaga Y, Mineta K, Bajic V, Gojobori T, Watabe S (2018b) Taxonomic profiles in metagenomic analyses of free-living microbial communities in the Ofunato Bay. Gene 665:192–200

Seymour JR, Mitchell JG, Pearson L, Waters RL (2000) Heterogeneity in bacterioplankton abundance from 4.5 millimetre resolution sampling. Aquat Microb Ecol 22:143–153

Sharpton TJ (2014) An introduction to the analysis of shotgun metagenomic data. Front Plant Sci 5:209

New Aquaculture Technology Based on Host-Symbiotic Co-metabolism

13

Miyuki Mekuchi, Taiga Asakura, and Jun Kikuchi

Abstract

Marine products represent an important source of protein for human consumption. With the expansion of human population, the demand of marine products from aquaculture has been increasing. Aquaculture production currently accounts for over half of all marine products. The development of aquaculture has improved with technological innovation; however, questions related to animals and their symbiotic gut bacteria remain due to the lack of biological knowledge. Therefore, host-symbiotic co-metabolism analyses using high-throughput technologies and next-generation analytics are conducted to improve the current methods of aquaculture. Recently, research has focused on the environmental microbiota in aquaculture facilities and the gut microbiota of aquatic animals. Controlling the microbiological conditions would greatly contribute to further develop aquaculture.

A holistic approach by omics analysis would be the most effective. Omics analyses using high-throughput technologies and next-generation analytics generate massive amounts of data. Conventional hypothesis-driven strategies remain important once the data-driven strategies have produced new insights on pathways not previously identified in the system studied and when a large amount of data was obtained. Combining data-driven and hypothesis-driven strategies to improve scientific knowledge will accelerate innovation and aquaculture technology.

Keywords

Microbiome · Metabolome · Co-metabolism · Host metabolism · Aquaculture products

M. Mekuchi · T. Asakura · J. Kikuchi (✉)
RIKEN Center for Sustainable Resource Science, Yokohama, Kanagawa, Japan
e-mail: jun.kikuchi@riken.jp

T. Gojobori et al. (eds.), *Marine Metagenomics*,
https://doi.org/10.1007/978-981-13-8134-8_13

13.1 Introduction

Aquaculture is becoming an increasingly important source of fish protein for human consumption. Moreover, aquaculture technology has been developed, and culture methods have become more intensive in recent years. However, several issues remain to be resolved, such as aquaculture environment, feeding, and diseases. Aquaculture is seriously hampered by the scarcity of biological knowledge on aquaculture animals and their symbiotic gut bacteria. In recent years, many high-throughput technologies have been developed for investigating various aspects: genome, transcriptome, proteome, metabolome, and microbiome (Fig. 13.1). These technologies, so-called omics, have been gradually applied and utilized in aquaculture. Incorporating big data generated by the high-throughput technologies and next-generation analytics into aquaculture will promote improvement of the current faming technology.

Recently, gut microbiota has received a lot of attention in aquaculture. The gut microbiota affects the host's nutrient acquisition and energy homeostasis by affecting the number of effector molecules. In addition, intestinal environment is thought to be an important factor in remaining healthy. Controlling gut and environmental microbiota is now forced for the successful aquaculture. The gas-

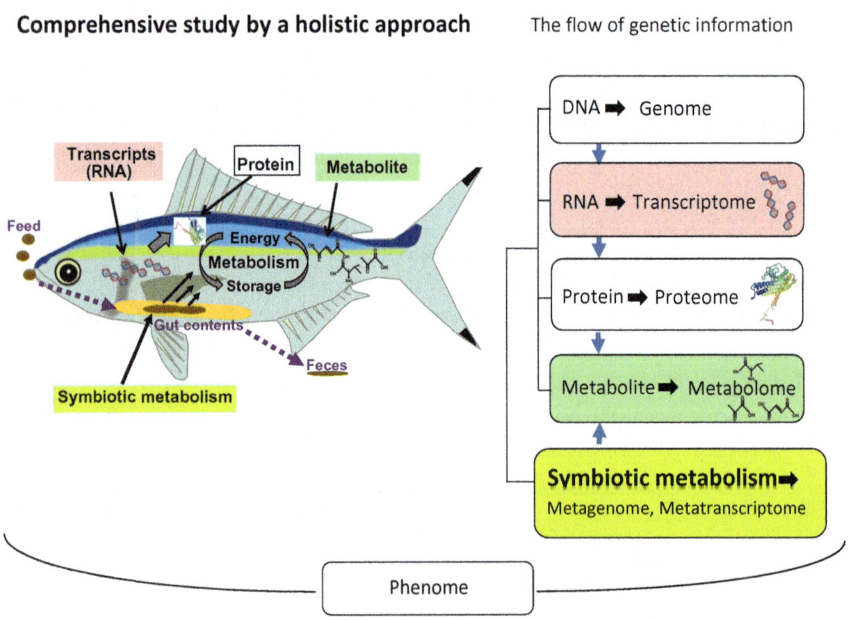

Fig. 13.1 Multi-omics analysis is a comprehensive study by a holistic approach. A multidimensional approach by integrating multi-omics data, including genomic, transcriptomic, proteomic, metabolomic, metabolomic of symbiotic bacterium, and phenomic data. An example of feed intake. Sequential alternation is occurred in many stages. Omics datum are brought together in order to understand the overview of feed intake in fish

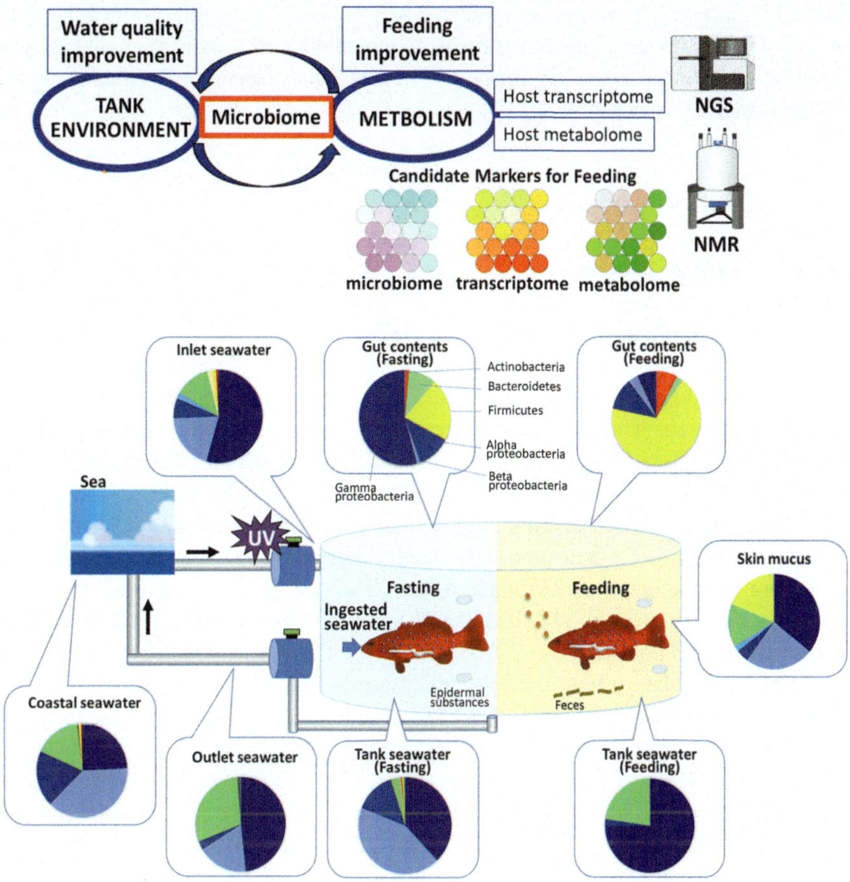

Fig. 13.2 Conceptual model of multi-omics in aquaculture. Multi-omics study and proposed model for monitoring aquaculture environment and symbiotic metabolism upon feeding, as well as timing, via comprehensive analyses of microbiome, host metabolism, and host transcriptome

trointestinal microbiota of fish and shellfish are peculiarly dependent on the external environment, owing to the flow of water through the digestive tract. Most bacterial cells are transient in the gut, with continuous intrusion of microbes from water and food.

Accumulation of biological knowledge on aquaculture animals is essential for the improvement of aquaculture (Fig.13.2). The generation and testing of hypotheses are widely considered the primary method of scientific advancement. However, recently, data-driven advances in scientific knowledge are seen as an overstepping of the limit of hypothesis-driven strategies. Data-driven strategies are efficient when we have no ideas about the role of orphan molecules and when a completely new and/or unpredictable pathway is working. The increasing rate of data generation across all scientific disciplines is providing opportunities for data-driven research.

Data-driven research requires computer-based inductive reasoning to turn the data into hypothesis. In this chapter, we focus on metabolism of aquatic animal hosts and symbiotic gut bacteria, and we introduce "omics" analysis and omics data analyses by data-driven strategies.

13.2 Basic Knowledge

13.2.1 Transcriptome

RNA Expression and the Holistic Approach

Genes encode proteins and proteins control cell functions. Transcription is the process by which the genetic information present in DNA is transferred to mRNA as the first step in protein synthesis. Genes expressed in a particular cell determine how the cell functions. Transcription varies with the external environmental, physiological condition, and developmental stages (Qian et al. 2014). The study of transcriptomics examines the expression levels of RNA in a given cell population. The levels of transcription are often measured individually. However, a transcriptomic analysis enables a holistic approach (Cerda and Manchado 2013). Transcriptomic profiling methods can be used to simultaneously analyze a large number of transcripts quantitatively (Haider and Pal 2013). Currently, the most popular transcriptomic profiling techniques are the cDNA microarray and RNA-Seq based on next-generation sequencing (NGS) (Fig. 13.3).

Microarray

The microarray is a method of analyzing the level of gene expression (see review by (Heller 2002)). Oligonucleotides called probes arrayed on a glass slide are hybridized with fluorescent-labeled cDNA. Genome or transcript sequences are required for designing probes; therefore, opportunities to use microarrays for non-model animals are limited. For model animals or animals for which genomic information is available, the microarray is a powerful tool for transcriptomic analysis. Further, the technology of the microarray has been developed, and the cost of analysis is cheaper than that of RNA-Seq. However, microarray technology remains limited in non-model and wild animals. Probes are designed based on complete coding sequence (CDS) or protein-coding sequences predicted by whole genome sequencing. Microarrays with well-designed probes lead to acceptable results. In fish raised by aquaculture, the microarray was used in various species, such as Atlantic salmon, rainbow trout, channel catfish, flounder, and bluefin tuna. Douglas (2006) summarized fish microarray use. Aquaculture animals are often not pedigreed and are derived from wild stock. Under these circumstances, unexpected single nucleotide polymorphisms (SNPs) exist infrequently in the probe-hybridized area, and the hybridization efficiency is changed.

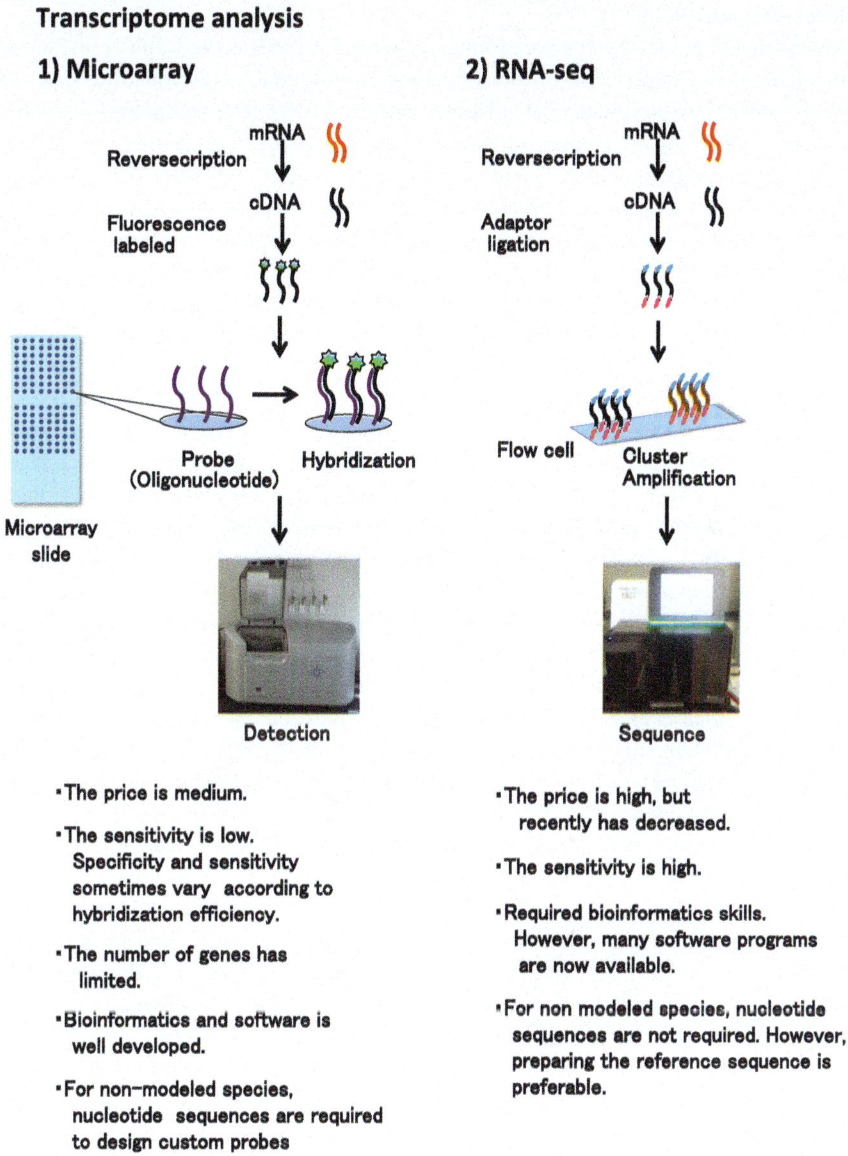

Fig. 13.3 Comparison of two transcriptomic analyses methods. (1) Microarray. Synthesized cDNA is labeled with fluorescence dyes. Labeled probes are spotted on a microarray slide. Target sequences are hybridized, and then glass slide is scanned. Obtained image can be analyzed quantitatively. (2) RNA-seq. Adaptor sequence is ligated to synthesized double-stranded cDNA. Amplified library is sequenced by next-generation sequencer

RNA-Seq and NGS

RNA-sequencing-based transcriptomics is now widely accepted and is the most popular mRNA expression profiling method, which enables the quantification of many transcripts simultaneously (Haider and Pal 2013). The technology for transcriptome profiling provides novel insights into gene expression analysis. Moreover, expression values of RNA-Seq data are more reliable and more accurate than other transcriptomic methods based on calculating and counting each transcript. As compared with other transcriptomic methods, the limitations of the analysis are reduced for RNA-Seq. Moreover, an RNA-Seq-based transcriptome is beneficial for not only for model species but also non-model species. The comparison of different transcriptomic methods have been previously described (Martin et al. 2016).

The construction of a cDNA library is the preparation of DNA-based sequencing forms. RNA is needed to prepare cDNA using reverse transcriptase. Currently, various methods of cDNA library construction have been developed (Podnar et al. 2014; Hrdlickova et al. 2017). Briefly, messenger RNA is extracted using oligo (dT) hybridization and/or rRNA elimination. In eukaryotes, mRNA constitutes approximately 1–5% of the total RNA. The most abundant RNA is rRNA, constituting more than 80% of total RNA. Therefore, removal of rRNA is a key process for analyzing mRNA expression. Subsequently, mRNA is fragmented mostly by alkaline solution or divalent cations. The size of mRNA fragments varies according to the sequencer and the sequencing methods. First-strand cDNA is synthesized by reverse transcriptase and random primers, followed by second-strand cDNA synthesis. Sequencing adapters and barcoding adaptors are ligated. Amplification of the cDNA library is required for detection sensitivity. Currently, RNA-Seq has become popular and its cost has decreased. Moreover, the application of RNA-Seq has been developed and used in various ways (Head et al. 2014).

The high-throughput sequencing technologies and next-generation sequencing (NGS) technology have developed and progressed. Currently, NGS and third-generation sequencing (TGS) have been divided into two categories: short-read and long-read sequencers. The sequencers are selected depending on the usage. For transcriptome analysis, the short-read sequencer (e.g., illumine and proton) is widely used. For de novo assemblage, long-read sequencers are efficient (e.g., PacBio). Recently, low-cost and mobile sequencers have been developed (e.g., Oxford Nanopore Technology). This kind of sequencer will contribute to the acceleration of biological research. The high-throughput sequencers were well described by Goodwin et al. (2016).

One-gigabase throughput is now available per run of NGS. Analysis of the huge amount of RNA-Seq data is often a bottleneck and makes it difficult to understand the transcriptome. Various bioinformatics software packages are now available. The analysis of non-model transcriptomes requires de novo assembling in advance to prepare for reference sequences. For de novo assembling of a transcriptome, de Bruijn graph basis software packages were developed, such as Trinity (Grabherr et al. 2011), Velvet (Zerbino and Birney 2008), SOAPdenovo-Trans, and ABySS (J. T. Simpson et al. 2009). The depth of sequences required is presumed to have

30X coverage for de novo assembly. Sequence read mapping and the calculation of expression levels are then performed by using assembled reference sequences. Calculation of expression levels of each gene involves the estimation of reads per kilobase per million reads (RPKM) (Mortazavi et al. 2008), fragment per kilobase of transcript per million reads (FPKM) (Trapnell et al. 2010), and transcripts per million (TPM) (Wagner et al. 2012). Statistical analysis using software packages such as DEGseq (Wang et al. 2010), edgeR (Robinson et al. 2010), baySeq (Hardcastle and Kelly 2010), and DESeq (Anders and Huber 2010 are then required for inference of differences in expression levels.

Transcriptome tells us and how to apply to aquaculture.

Transcriptome analysis has helped to gain new biological insights for aquaculture studies. In basic science, transcriptomes have contributed to our understanding, as well as to novel transcript discovery, posttranscriptional modification, SNP discovery, and quantification of transcript levels (Qian et al. 2014). The transcriptome is now widely used in various areas, such as immunology, developmental biology, physiology, and toxicology.

Genetically based selective breeding has the potential to increase production efficiency. SNP information is considered beneficial data for conducting genetically based selective breeding. In addition, SNP data are needed to understand the function of genes. The transcriptome is mainly used for quantification of transcript levels.

In aquaculture, fish are reared at a higher density than in the wild. Infection is a critical issue for aquaculture. To address this issue, elucidation of immunological disease resistance mechanisms and stress response mechanisms are required. Comprehensive study using the transcriptomic approach would illuminate the mechanisms and would lead to disease prevention.

Recently full life cycle aquaculture has been encouraged to prevent the decline of wild populations. Broodstock management and seed production are the prerequisites for successful full life cycle aquaculture. The molecular mechanisms of reproduction and development need to be revealed, which might help in developing the technology.

Aquaculture production has impressively increased worldwide. The main aquaculture feed consists of fish meal; however, the supply of fish meal has been limited. Other new ingredients for aquaculture feeds are needed for replacing fish meal. Transcriptome analysis facilitates the identification of new ingredients by monitoring metabolism.

13.2.2 Metabolome

Metabolic profiling or metabolomics is the study of organic metabolites in cells, tissues, and biological fluids (Clayton et al. 2006; Schlipalius et al. 2012; Auro et al. 2014). Metabolomics studies provide comprehensive datasets from target samples (Suhre et al. 2011; Kettunen et al. 2012, 2016). This section describes the range of samples, metabolites, and NMR experiments used in metabolomics studies. Recently, solution NMR has been used for metabolic profiling, in which NMR spectra of mixtures of small biological molecules are subjected to further multivariate analyses for identifying metabolite biomarkers (Brindle et al. 2002; Holmes et al. 2008; Fukuda et al. 2011; Furusawa et al. 2013; Smith et al. 2013) and evaluating nutrients in food (Tomita et al. 2015, 2017; Watanabe et al. 2015; Sekiyama et al. 2017). Innovations in non-targeted approaches of studying biological systems are important for better biomass production and sustainability of such systems.

Homeostasis is the maintenance of an internal physiological environment within a certain range against changes in the external environment (Kikuchi and Yamada 2017). Pathological states lead to collapse of homeostasis and change metabolite compositions and/or profiles. For example, NMR metabolome analysis of humans has provided new insights into interactions between hosts and microorganisms during the establishment of physiological homeostasis in intestines (Li et al. 2008; Morita et al. 2008; Kato et al. 2014; Sugahara et al. 2015). First, it can be applied to a wide variety of sample systems, from crude extracts to intact biological tissues (Beckonert et al. 2007, 2010; Blaise et al. 2007; Kruger et al. 2008; Frost et al. 2014), and from small to macromolecular complexes (Mao et al. 2001; Ward et al. 2011; Choe et al. 2012; Mansfield et al. 2012; Komatsu and Kikuchi 2013; Watanabe et al. 2014; Komatsu et al. 2015), as well as to interaction studies (Feng et al. 2006; Lattao et al. 2012; Simpson et al. 2012; Tuo Wang et al. 2013; Cao et al. 2014). Second, NMR can provide site- and atom-specific information (Eisenreich et al. 2006; Sekiyama and Kikuchi 2007; Ohyama et al. 2009; Peyraud et al. 2009; Komatsu et al. 2014) when coupled with stable isotopic tracing experiments (Kikuchi et al. 2004; Tokuda et al. 2014). Third, uniform stable isotopic labeling of biological samples allows the application of NMR to two-dimensional (2D) and three-dimensional (3D) experiments, similar to protein NMR (Clendinen et al. 2014; Mori et al. 2015; Ito et al. 2016; Komatsu et al. 2016).

Fourth, the exchangeability of spectral data among laboratories worldwide is a marked advantage of NMR-based approaches and enables studies of cross-site analytical validity using careful parameter settings (Dumas et al. 2006; Viant et al. 2009; Ward et al. 2010; Lacy et al. 2014). This advantage arises because chromatography is not involved in the processes for measuring molecular complexity. Similarly, the reproducibility of NMR and its ability to quantify molecular species are major advantages for studies on molecular complexity (Hao et al. 2014; Gallo et al. 2015; Kikuchi et al. 2016). In quantification, absolute measurement values are not always important, but relative measurements within a variety of samples are a cause for concern. However, solution and solid-state NMR produce highly reproducible numerical matrix data (e.g., chemical shift versus intensity) with inter-institution convertibility. Environmental homeostasis evaluation is growing in importance with declining environmental health. Environmental homeostasis can be evaluated at multiple levels using NMR (Bundy et al. 2007; Collette et al. 2010; Samuelsson et al. 2011; Ogata et al. 2012; Whitfield Aslund et al. 2012; Ellis et al. 2014; Ito et al. 2014; Ogawa et al. 2014; Yoshida et al. 2014; Wei et al. 2015; Uchimiya et al. 2017).

In metabolomics studies, the inclusion of multivariate analysis (also known as chemometrics, pattern recognition, and data mining) is imperative. Methods for multivariate analysis fall roughly into two categories: unsupervised (e.g., principle component analysis, PCA) and supervised (e.g., partial least squares, PLS). PCA is the most widely used method in the field of metabolomics. Typically, PCA is used for data overview and trend (or cluster) identification, and it is used for dimension reductions during dataset preprocessing. Other representative, unsupervised, multivariate analyses include hierarchal clustering (HCA) (Date et al. 2012b), K-means clustering (Ogawa et al. 2014), and self-organizing maps (SOM) (Ogura et al. 2016b).

Moreover, correlation-based analysis (identifying a linear relationship between two variables) is widely used in metabolomics studies (Date et al. 2012a). Correlation coefficients are usually visualized as heat map and network diagrams. Market basket analysis is a statistical approach for identifying co-occurrences of variables in datasets; this approach has recently been applied to metabolomics studies (Shiokawa et al. 2016).

PLS is the most commonly used supervised, multivariate analysis in metabolomics studies. PLS and related analyses are the most common classification and regression tools for analyzing metabolomics data. PLS-related analyses are powerful data mining methods because they enable analysis of large, highly complex datasets with collinearity and noise. However, alternative "machine learning" (ML) methods that are suitable for use in the field of metabolomics have been recently introduced (Chatzimichali and Bessant 2016; Date and Kikuchi 2018; Asakura et al. 2018a).

Sample preparation for molecular complexity analysis is a minimum laboratory experimental requirement due to unnecessary feature of column purification (Sekiyama et al. 2010; Kikuchi et al. 2018). Solution NMR detects molecules with low molecular weights, such as polar metabolites in solvents, including deuterated water, methanol-d_4, dimethyl sulfoxide-d_6, and chloroform-d_1 (Sekiyama et al. 2011).

Metabolite identification is one of the most important problems in metabolomics. Standard spectra and chemical shift databases are typically used to assign and annotate metabolic mixtures. Many chemical shift databases are available on the Internet, and using these services is simple and convenient. The Human Metabolome Database (HMDB) (Wishart et al. 2013), Biological Magnetic Resonance Bank (BMRB) (Ulrich et al. 2008), and Madison Metabolomics Consortium Database (MMCD) (Cui et al. 2008) are commonly used chemical shift databases for metabolomics. Our laboratory developed the SpinAssign program, which can annotate metabolites in D_2O and MeOD buffer using peak lists of ^1H-^{13}C HSQC (Chikayama et al. 2010). The p-value has been adopted as a mathematical indicator for annotation evaluation. A metabolomics database for NMR analysis (TOCCATA) has been recently developed (Bingol et al. 2012). In this database, highly reliable annotation is possible by database search using TOCSY peaks for metabolites in a mixture. The Birmingham Metabolite Library (BML) contains ^1H and 2D J-resolved spectral data of metabolites, which can be used to annotate 2D J-resolved spectra of metabolic mixtures (Ludwig et al. 2012). SpinCouple is an annotation

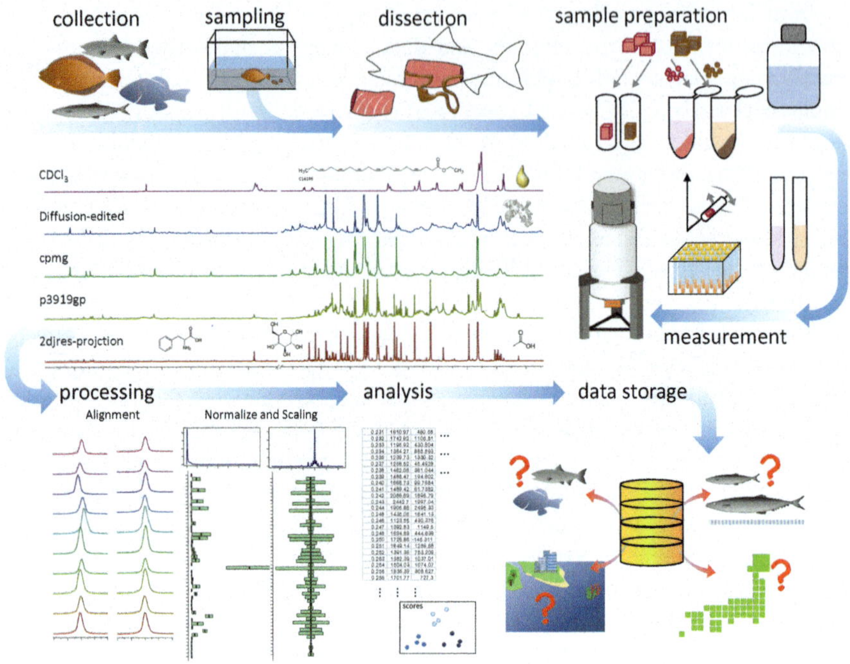

Fig. 13.4 Analytical strategy for NMR-based metabolic profiling. Firstly, fishery products such as wild and cultured fishes and food products are sampled and dissected for the NMR measurements. Polar to nonpolar low molecular weight molecules and polymers in a variety of samples can be extracted and detected by different solution NMR pulse sequences. Optionally, intact tissues can be also analyzed by high-resolution magic angle spinning NMR (From top to middle). The processed metabolic mixture data is sometime shifted due to presence of counter ions; therefore this cases the caution to be made for the peak alignment procedure. After normalization and scaling of all numerical data, the relationships between different metabolites, the similarity between samples, and the presence of biomarkers can be determined using various data mining methods using these multivariable data. Furthermore, because of inter-institution convertibility and high reproducibility of NMR data, data storage and database management can be also significant issues such as origin discrimination, ecosystem implication, etc.

tool for 2D *J*-resolved spectra of metabolic mixtures (Kikuchi et al. 2016), in which the user can locate candidate compounds by entering ^{1}H chemical shifts, spin-spin coupling constants, and intensity values (Fig. 13.4).

13.2.3 Microbiome

With the recent development of the metagenomics, the classification and function analysis of symbiotic bacteria has progressed, and its importance has been reassessed. A microorganism (approximately 100 μm or less) cannot be discriminated by the naked eye based on its structure, and a microbiome or microflora refers to a collection of microorganisms living in a specific environment. The collection includes prokaryotic bacteria, archaea, fungi, and other eukaryotes, among which bacteria are the most diverse and occur in large numbers, whereas prokaryotes supply products naturally produced only by prokaryotes to hosts (Moore and Warren 2012). According to the metagenomics analysis of bovine rumen stomach, 90–95, 2–4, and 1–2% of the entire sequence data is from bacteria, archaea, and eukaryotes, respectively. Prokaryotes have different microbiomes depending on the environment, and a surprisingly large number exist in the environment. In the aquatic environment, this figure is estimated to be 5×10^5 cells/mL at a water depth of 200 m or more, 5×10^4 cells/mL at 200 m or less, and 4.6×10^8 cells/mL in sediment (Whitman et al. 1998). However, it exists in greater densities in the intestines of vertebrates. It is estimated that 3.2×10^{11}, 2.1×10^{10}, 9.5×10^{10}, and 2.7×10^6 prokaryote cells/mL are present in the human colon, bovine lumen, bird (chicken, duck, turkey), and termite hind, respectively (Whitman et al. 1998; Rosselló-Mora and Amann 2001). In fish, the figure is estimated to be 7.0×10^8–3.0×10^{10} cells/mL (Sugita et al. 2005). Microorganisms symbiotic to vertebrates can be transformed into a state in which biopolymers and complex organic substances that cannot be digested by the host to can be used.

Symbiotic microorganisms convert to monomers such as glucose and short-chain fatty acids (SCFAs) by hydrolysis and supplied to the host (Semova et al. 2012). The most abundantly produced SCFAs are acetic acid, propionic acid, and butyric acid, which are easily absorbed by vertebrate intestines; moreover, butyric acid is used as a major energy source of colon epithelial cells in humans (Wong et al. 2006). In addition, studies on the effect of symbiotic microorganisms on host behavior have been reported. Improvement in autistic behavior was observed by controlling the bacterial flora of mice (Hsiao et al. 2013). Further, some intestinal bacteria cause anticancer immunity (Viaud et al. 2013). However, in contrast to several advantages mentioned above, allergic diseases caused by intestinal bacterial flora have been reported (Wang et al. 2015). Thus, it is evident that the symbiotic microbial flora has great influence on the host, which is an important factor determining the growth and control of livestock, and culture, in addition to human health. However, because the symbiotic microbial flora is not described as a host gene, it is taken from the outside and forms a microbial flora by selection. The symbiotic microflora varies from early incorporation owing to lifestyle and other factors. One year after birth, the intestinal bacterial community in a normally delivered baby is similar to that of their mothers, whereas that in babies born by Caesarean section are relatively different (Bäckhed et al. 2015). There are not many studies to analyze functionality using such metagenomics in aquaculture sample compared to human studies. Examples of using fish 16S rRNA gene were described later (Fig. 13.5).

13.3 Host Metabolism

Metabolism

Metabolism is an integrated network of biochemical reactions. There are two categories in metabolism: anabolism and catabolism. Anabolism contains the bio-

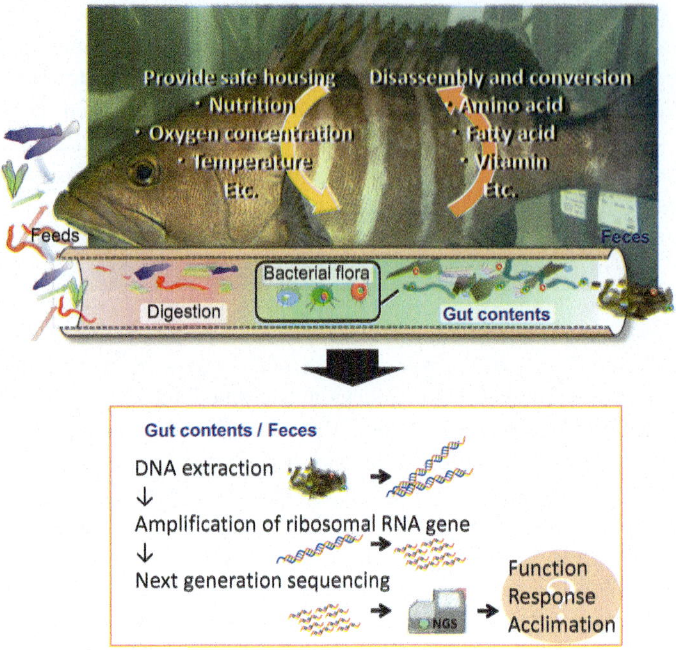

Fig. 13.5 Concept image of microbial analysis. A lot of bacteria are contained in gut contents and feces. The 16S ribosomal RNA genes amplified from the bacterial DNA, and the species and quantity ratio are analyzed by the next generation sequencer (NGS). Evaluate function, response, and acclimation of host and symbiotic microbiome based on information obtained from NGS. This figure is adapted from (Asakura et al. 2014b)

chemical process of transformation and storage of absorbed nutrients. Catabolism is the process of breaking down nutrients and their conversion for energy consumption. The concept of nutritional metabolism has important implications for understanding the efficiency of nutrient utilization and defining nutrition requirements of aquatic animals. Here, we introduce the metabolism of aquatic animals using the transcriptomic analysis and discuss the application of the results obtained.

13.3.1 Oyster Transcriptome

The Pacific oyster, *Crassostrea gigas*, is widely farmed all worldwide and well known for its commercial value. Oysters are filter feeders that graze on plankton, microorganisms, and suspended particles in the surrounding seawater. Feed is transferred to the palpus, followed by passage through the esophagus to reach the stomach. The primary digestive organ of oysters is located in the center of the body and consists of its stomach, hepatopancreas, and crystal sac, which contains digestive enzymes. The hepatopancreas, which is alternatively referred to as the digestive gland or midgut gland, is an organ of the digestive tract and plays the role of liver and pancreas. The metabolic system of oysters changes between spawning time in summer and energy storage time in winter. The pacific oyster accumulates reserves of glycogen and lipids in winter for gametogenesis and spawning in summer. Glycogen and lipids play a role in the energetic and metabolic supply of gametogenesis in many bivalves (Bayne et al. 1982; Ruiz et al. 1992; Mathieu and Lubet 1993; Berthelin et al. 2000). We examined the molecular mechanism of the hepatopancreatic metabolism. Complete genomic information for Pacific oysters has been reported (Zhang et al. 2012). Total genome size was reported to be 559 Mb. A total of 28,027 genes were predicted from the entire genome. Transcriptomic analysis used these predicted genes as a reference for read mapping.

A cDNA library was sequenced by the Illumina sequencing platform. In the hepatopancreas, 15,570 genes were expressed. Regarding glycogen, the expression levels of glycogen in mRNA were higher in winter. The expression gradually increased from the end of summer. The mRNA expression levels of glycogen synthase kinase 3 (GSK3), located upstream of glycogen synthase, were also upregulated in winter. For lipids, the mRNA expression levels of fatty acid synthase were high in the winter. Moreover, gene enrichment analysis showed that the C/EBP expression pathway, which is related to lipid storage, was highly expressed in winter. C/EBP is one of the lipid storage contributors (Rosen et al. 2000). Gene ontology analysis revealed differences in gene profiling between summer and winter (Fig.13.6). Figure 13.6b shows the molecular function categories. The rate of metabolism in summer was higher than in winter. In a detailed hierarchy of metabolism, the proportion of lipid and carbohydrate metabolic process genes were higher in winter. These results accounted for and supported the previous study on molecular mechanisms (Bayne et al. 1982; Berthelin et al. 2000). The proportion of amino acid metabolism was also higher in winter. Protein is stored inside the bivalve as an energy reserve for gametogenetic development (Barber and Blake 1981; Ruiz et al. 1992). On the other hand, the rate of nucleobase-containing compound metabolism process genes was higher in summer than in winter. This is because they were ATP synthesis-related genes. In summer, oysters exhaust and deplete their energy because of spawning and high water temperate stress. Therefore, they are thought to need to generate ATP as an energy source.

13.3.2 Transcriptome of Teleostean Muscle

The swimming power is produced by the segmented myotomal muscles. The skeletal muscles constitute a large part of myotomal muscles. In most fish, the myotomal muscles contain three types of muscle fibers: red, pink, and white. Myotomal muscles primarily contain white muscle and red muscle is located

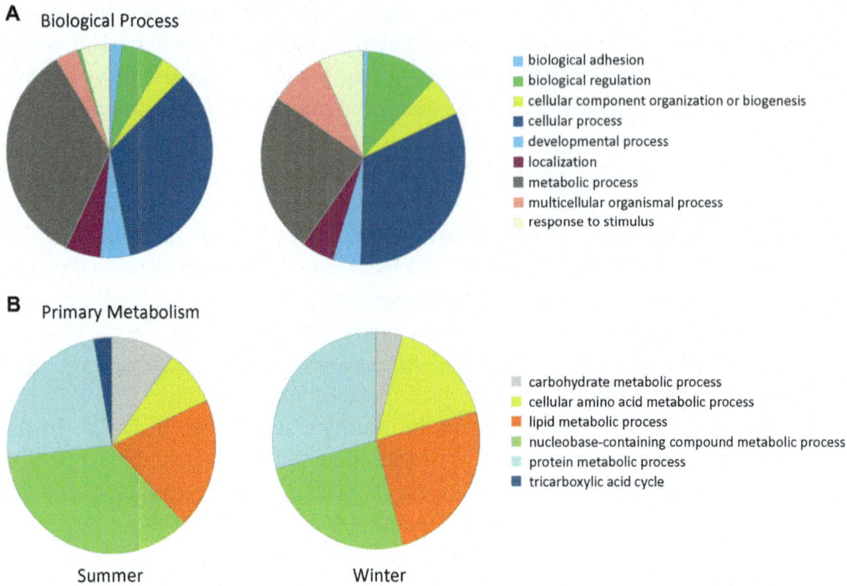

Fig. 13.6 Gene ontology analysis in winter and summer oyster. (**a**) Molecular function category. Metabolic process and cellular process were dominated in both summer and winter. The proportion of gene involved in metabolic process was higher in summer than in winter. (**b**) Primary metabolism category. This category is one of the metabolic process genes. The proportion of carbohydrate and nucleobase-containing compound metabolism was higher in summer. On the other hand, lipid, protein, and cellular amino acid metabolism was lower in summer

laterally. Pink muscle fibers are located between red and white fibers. The muscle color signifies differences in physiological functions and biochemistry. White muscle has anaerobic glycolytic potential and plays a role for fast and short-term swimming. Red muscle works for constant swimming through aerobic metabolism by glucose and fat oxidation (Bone 1978; Guppy et al. 1979; Guppy and Hochachka 1978; Block et al. 2005).

Energy-generating mechanisms depend on enzyme activities. The features of white and red muscles were described by transcriptomic analysis. Two representative fishes, the Pacific bluefin tuna *Thunnus orientalis* and the Pacific cod *Gadus macrocephalus*, were subjected to transcriptome analysis (Shibata et al. 2016). The Pacific bluefin is the fastest swimming pelagic fish with ram ventilation, which requires seawater flow over the gill surfaces (Korsmeyer and Dewar 2001). The Pacific cod was selected as a representative of a fish with a generalist locomotor strategy. A potential trade-off in individual cods between stamina and the ability to use glycolysis-based locomotion was detected (Reidy et al. 2000).

The mRNA expression levels of three energy generation pathways, glycolysis, TCA cycle, and oxidative phosphorylation, were compared between the white and red muscles of two fishes. The proportion of these three pathways is shown

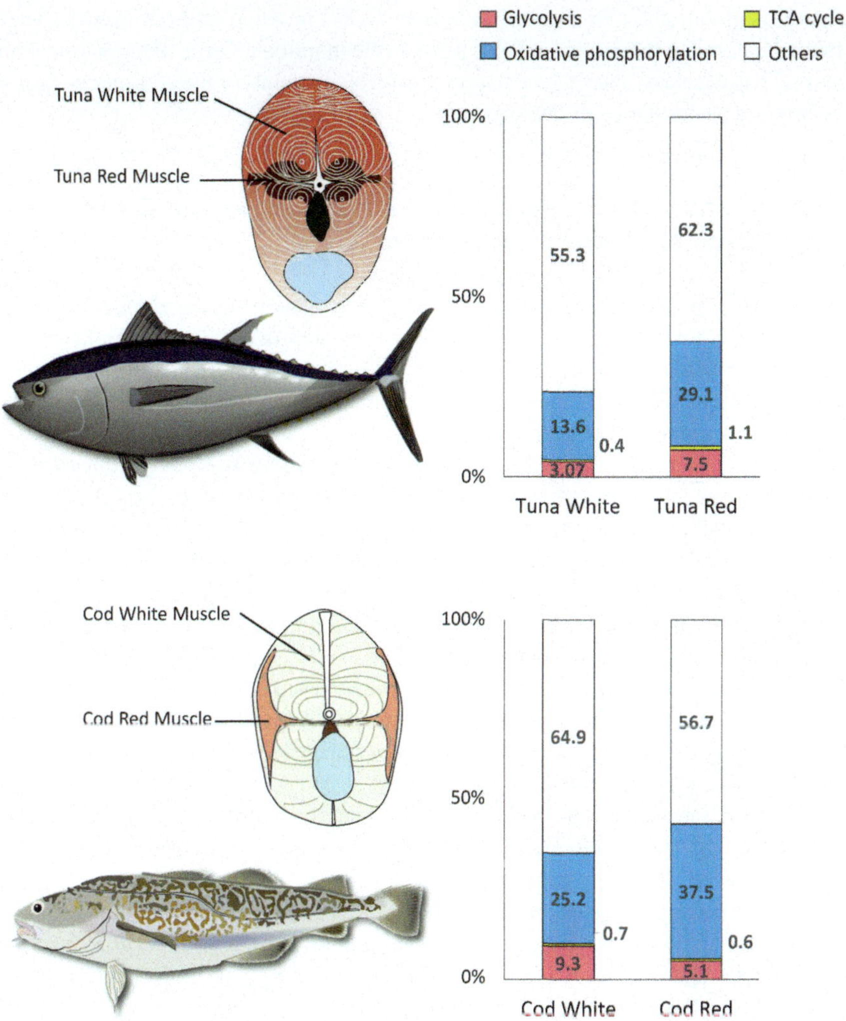

Fig. 13.7 The proportion of white and red muscular gene expression in tuna and cod. Red shows glycolysis, yellow shows TCA cycle, and blue shows oxidative phosphorylation. (Modified figured of Shibata et al. (2016))

in Fig. 13.7. Higher levels of expression of glycolysis genes were detected in white muscles than in red muscles. In white muscles, more glycolysis genes were expressed in tuna than in cod. Moreover, oxidative phosphorylation genes were highly expressed in red muscle. In the white muscles of tuna, it was assumed that highly expressed glycolytic genes played roles in increasing the enzymatic reactions of energy production pathways (Shibata et al. 2016). The white muscles of tuna are red in color, owing to the presence of higher amounts of myoglobin, which transports oxygen to mitochondria (Sannier et al. 1996). The expression levels of tuna myoglobin and cytochrome c oxidase (COX) in white and red muscles were much higher than those of cod. COX plays a role in mitochondrial electron transport and ATP generation. These gene expression levels account for the continuous fastest swimming performance of the tuna.

13.3.3 Transcriptome and Metabolome of the Leopard Coral Grouper *Plectropomus leopardus*

Aquaculture technology of leopard coral groupers has been developed; however, feed efficiency and metabolic control need to be improved. Feed efficiency has a close relationship to the metabolic and biorhythmic characteristics of the cultured fish. The metabolic mechanisms and biorhythmic processes of aquaculture were investigated. (Mekuchi et al. 2017).

Transomic approaches (Ogata et al. 2012), including the integration of transcriptome and metabolome analyses, provide a comprehensive data network of genes and metabolites (Asakura et al. 2014a; Misawa et al. 2016a, b; Ogura et al. 2016a; Samuelsson et al. 2011; Samuelsson and Larsson 2008; Tian et al. 2007; L. Wagner et al. 2014; Yoshida et al. 2014). The association of high-throughput technology analyses have elucidated holistic and multidimensional information on various functions and pathways.

Hatchery-reared leopard coral groupers were fasted for the first 2 days of the experiment and were then fed [Zeitgeber time (ZT)2 and ZT10)] ad libitum for the next two experimental days. Fish were collected every 4 hours for 4 days, with the exception of the evenings of Day 1 and Day 3. Transcriptomic and metabolomics analyses were preformed using a next-generation sequencer (NextSeq 500, Illumina) and NMR (AVANCE II-700, Bruker). In this study, data-driven strategies were cited. Key genes and metabolites were selected by multivariate and correlation pattern analyses conducted using the software package R.

In transcriptomic analysis, the clock genes, *per1a*, *per3*, *cry1aa*, *cry1ab*, *cry1ba*, *cry2*, *clocka*, and *cipc*, were found to exhibit circadian rhythms in the muscle. These clock genes are involved in the perception of time in the pineal gland and central nervous system; however, this study revealed that these genes are also expressed in the skeletal muscle where they exhibited circadian patterns.

In muscle differentiation mechanisms, *myoD* exhibited a 24 h rhythm. The mRNA expression levels of myogenin and *myoD* were highest at ZT6 and ZT2, respectively. These expression patterns for myogenin were identical to that of zebrafish (*Danio rerio*); however, the expression patterns of *myoD* differed from those observed in previous studies (Amaral and Johnston 2012). MyoD and myogenin are known to be the myogenic regulatory factors (MRFs) (Buckingham 1992). Murine CLOCK and BMAL1 have been reported to interact with *MyoD* and maintain the phenotype and functions of the skeletal muscles (Andrews et al. 2010). Among the metabolism-related genes, the expression levels of adiponectin receptor 2 (*adipoR2*) were higher during the night, which was identical to the patterns of myogenin and myoD. Adiponectin and its receptors controlled the amount of body fat (Yamauchi et al. 2003).

During feeding and fasting, hormone-related genes exhibited identical expression patterns. Growth hormone receptor and thyroid hormone receptor genes were highly expressed under fasting conditions. Hormones in the hypothalamic-pituitary-thyroid axis (HPT axis) are responsible for growth and metabolism (Mullur et al. 2014). Metabolomics data were analyzed by projection on a latent structure-discriminant analysis (PLS-DA) and orthogonal PLS-DA (OPLS-DA) using R, as previously described (Asakura et al. 2014a; Ito et al. 2014; Motegi et al. 2015). Discriminant analysis was performed between the groups of circadian time. Loading plot analysis derived from a discriminant analysis revealed that inosine monophosphate (IMP), which was high in the morning, was a prime contributor. Brain IMP of zebrafish exhibited a similar dynamic pattern as did the leopard coral grouper (Li et al. 2015). Malate, fumarate, and lactate are associated with the TCA cycle and glycolysis (Gumbmann and Tappel 1962) (Fig. 13.8). These metabolites were found to be contributors to a daytime pattern. The levels of these three substances were high in the evening and low at night. The dynamics of lactate exhibited an identical pattern as that of zebrafish (Li et al. 2015). In the medaka (*Oryzias latipes*) liver, the amount of malate and fumarate has been reported to decrease during the night (Fujisawa et al. 2016). The TCA cycle and glycolysis are the part of the ATP generation cycles, more of which is assumed to be used during the day than at night. Genes in purine metabolism pathway exhibited the diurnal pattern, which indicated the amount of ATP was higher in daytime than night.

In discrimination analysis between feeding and fasting, leucine and isoleucine appeared to contribute to the fasting side. Leucine and isoleucine are well known to be metabolized in skeletal muscles (van den Thillart 1986). The amounts of leucine and isoleucine were found to increase on the second day of fasting. Leucine and isoleucine are utilized as energy sources under energy-depleting conditions (Gillis and Ballantyne 1996; Holecek et al. 2001), and hence these substances are thought to be important potential sources of energy during fasting.

This comprehensive study provided novel and fundamental metabolic information for the leopard coral grouper. This information pertains not only to the metabolic pathways but also to those of hormonal control.

13.3.4 Metabolome and Ionome

Environmental metabolomics (or metabonomics) provides a method for characterizing the interactions between organisms and their environments. We examined the metabolic and mineral profiles of three kinds of abundant fishes in estuarine ecosystems, yellowfin goby, urohaze-goby, and juvenile Japanese seabass sampled from Tsurumi River estuary, Japan (Yoshida et al. 2014) (Fig.13.9). Multivariate analyses, including NMR-based metabolomics and ICP-OES-based ionomics approaches, revealed that the profiles were clustered according to differences among body tissues rather than differences in body size, sex, and species. The metabolic and mineral

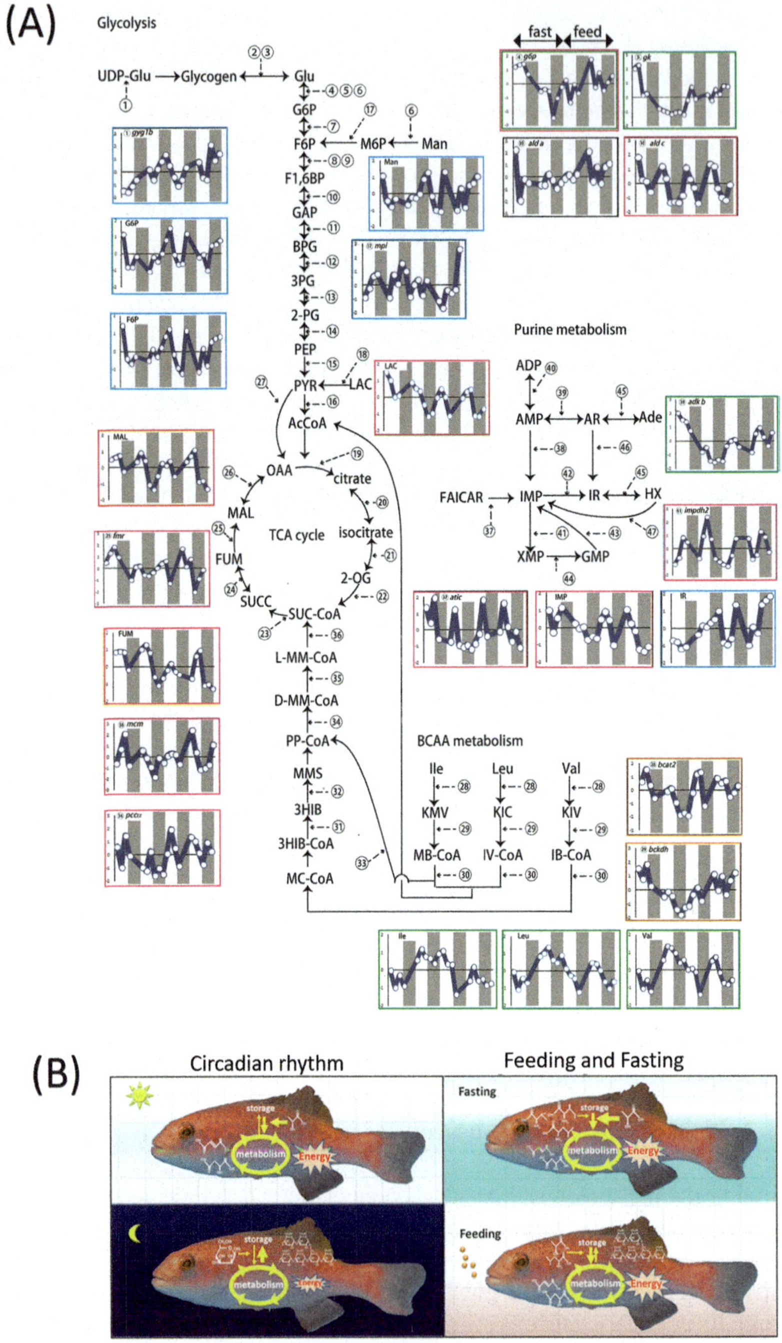

Fig. 13.8 (**a**) Glycolysis, TCA cycles, and purine metabolism pathway. Small italic letter indicated genes. Yellow-colored genes and metabolites exhibited diurnal circadian fashion. Green

profiles of the muscle and fin tissues, respectively, suggested that these tissues are the most appropriate for evaluating environmental perturbations.

Recent studies have demonstrated that NMR spectrum of biospecimens, such as tissue and blood, is useful for offering a rich-metabolite profiling (Nishiyama et al. 2015). Previously, we used a novel analytical strategy, named signal enhancement by spectral integration (SENSI), to resolve the low S/N problem in ^{13}C NMR by the integration of plural spectra without additional measurements for analysis and annotation (Misawa et al. 2016a). Because SENSI involves the integration of plural spectra measured according to varied principles from different samples, we proposed that it could be applied to similar types of NMR spectral data independent of their dimensions. Therefore, we introduced an SENSI method designed for 2D NMR spectra, named SENSI 2D, and described its performance with respect to assisting assignment and extracting population characteristics in 2D *J*-resolved NMR spectroscopy.

Over 1000 (n = 1022) 2D *J*-resolved spectra from individual yellowfin goby, *Acanthogobius flavimanus*, were acquired as described previously (Misawa et al. 2016b). We have reported an NMR database based on 2D *J*-resolved measurement (Kikuchi et al. 2016). The SENSI 2D method introduced here with enhanced S/N ratios, additional peak detection, and CV value calculation function will provide further clues and references for the assignment of 2D *J*-resolved spectra. Furthermore, the SENSI 2D spectra with a high S/N ratio enabled successful observation of body size-dependent population characteristics, including metabolite levels, by showing differences in the average intensity of peaks derived from plural spectra before integration, even for low-intensity peaks. Moreover, with recent technological advancements, increasing attention has been paid to NMR-based metabolomics for mass-limited samples or submillimeter-size organisms, or when using highly miniaturized NMR instruments (Chikayama et al. 2016; Nishiyama et al. 2015; Wong et al. 2014; Zalesskiy et al. 2014). Additionally, we believe that SENSI can be adapted to these new technologies because it greatly improves the low S/N ratio problem encountered in low magnetic fields that occur in miniaturized NMR systems.

Fig. 13.8 (continued) colored showed nocturnal circadian fashion. Blue and pink show the dynamics patterns varied according to the fasting and feeding. Glycogenin synthase 1b (gyg1b), glucose-6-phosphate dehydrogenase (g6p), glucokinase (gk), aldolase (ald), phosphate isomerase (mpi), malate dehydrogenase (mdh), branched chain aminotransferase2 (bcat2), branched chain keto acid dehydrogenase (bckdh), propionyl-CoA carboxylase (pcc), methylmalonyl-CoA mutase (mcm), IMP cyclohydrase (atic), adenosine kinase b (adk b), IMP dehydrogenase 1b (impdh 1b) glucose (Glu), glucose-6-phosphate (G6P), fructose 6-phosphate (F6P), mannose-6-phosphate (M6P), mannose (Man), fructose-1,6-biphosphate (F1,6BP), glyceraldehyde-3-phosphate (GAP), 1,3-bisphosphoglycerate (BPG), 3-phosphoglyceric acid (3PG), 2-phosphoglycerate (2PG), phosphoenolpyruvic acid (PEP), pyruvic Acid (PYR), acetyl-CoA (AcCOA), oxaloacetic acid (OAA), 2-oxoglutarate (2-OG), succinyl-CoA (SUC-CoA), succinic acid (SUCC). (**b**) Summary of multi-omics analysis of the leopard coral grouper. Metabolism changes in the circadian rhythm and nutritional condition

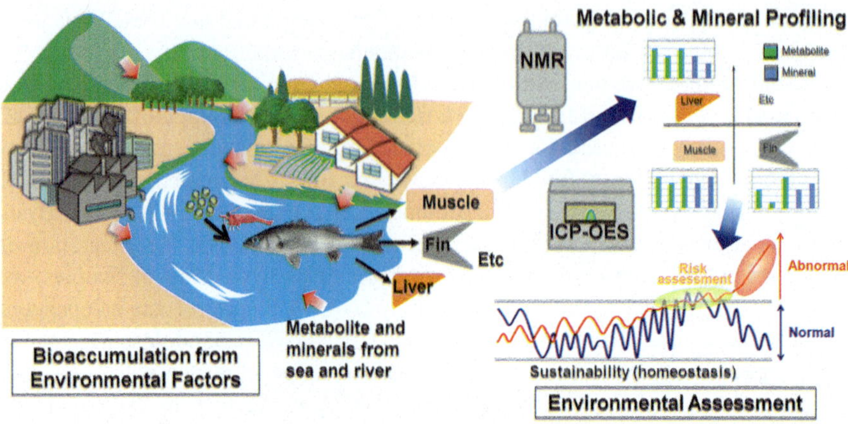

Fig. 13.9 Concept diagram illustrating the evaluation of homeostatic states of natural samples with metabolic profiling by NMR and ICP-OES. The metabolic and mineral perturbations in wild fishes can be occurred from river waters as well as sea tide through bioaccumulation of natural food web system. Any statistically enough number of environmental samples, such as fish tissues, water, algae, and sediments, can be possibly analyzed (left). High-throughput, unpurified NMR, as well as ICP-OES data showed homeostatic changes in the balance of metabolite and mineral balances (top right). The multivariate analysis showed the classification of metabolic balance difference, and loading plots indicated metabolic marker signals owing to this homeostatic changes (bottom second). Comparative analysis of these perturbations can assess range of normal and risk of environmental threats (bottom right)

13.4 Symbiotic Metabolism

The intestinal microbiota has been linked to a wide range of biological processes that benefit the host, including nutritional conditions and the immune system (Yeoman et al. 2011; Maynard et al. 2012; Xing et al. 2013). In addition, energy production is presumably affected by the composition and interaction of the gut microbiota (Ley et al. 2005; Tremaroli and Backhed 2012). Microbiota composition drastically changes in response to environmental and biological conditions and stimuli (Carmona-Antonanzas et al. 2014; Kohl et al. 2014; Dethlefsen et al. 2006; Sugahara et al. 2015; Misawa et al. 2015; Kato et al. 2014). Recent studies have reported that diet conditions strongly affect the gut microbiota in many animals, including fish (Kohl et al. 2014; Brown et al. 2012; Xia et al. 2014; Waite and Taylor 2014; Asakura et al. 2014b). Knowledge regarding the gut microbiota of teleostean fish has been gradually accumulating; however, information regarding the dynamics and interaction with host fish remains limited and appears to differ according to condition and species.

13.4.1 Microbiome and Metabolism of Coral Leopard Grouper

The composition of the gut microbiome of the leopard coral grouper was investigated during feeding and fasting conditions (Mekuchi et al. 2018). Subsequently, the microbiota functional capacity was estimated and compared with that of the host transcriptome and metabolome data.

The DNA extracted from gut contents and tank seawater was amplified using the universal bacterial primers of the V3–V4 regions of the 16S rRNA gene. The amplicons were sequenced by the next-generation sequencer (Miseq, Illumina). Sequencing reads were analyzed using QIIME software (Caporaso et al. 2010), and an operational taxonomic unit was calculated.

In gut content, at the phylum level, four phyla, *Proteobacteria*, *Firmicutes*, *Actinobacteria*, and *Bacteroidetes*, dominated (>97%). *Proteobacteria* was the dominant phylum (Fig. 13.10a). *Firmicutes* gradually increased based on feeding, whereas *Bacteroidetes* gradually decreased and became difficult to detect by the end of the feeding period.

In tank seawater, *Proteobacteria* and *Bacteroidetes* accounted for >99.7% of bacteria. *Proteobacteria* was the dominant phylum in both gut and seawater. At the class level, *Alphaproteobacteria* and *Gammaproteobacteria* were the dominants in *Proteobacteria* (Fig. 13.10b). *Gammaproteobacteria* was the major class in gut and seawater.

Multivariate analysis of discriminant analysis (PLS-DA) and an artificial neural network self-organizing map (SOM) analysis revealed that the gut contents of the feeding group were characterized by *Firmicutes* and *Fusobacteria*, whereas that of the fasting group were characterized by *Proteobacteria.*

During fasting, *Gammaproteobacteria* was the dominant bacteria derived from ingested ambient seawater. Fish drink seawater to compensate for dehydration (Grosell and Genz 2006; Miyuki Mekuchi et al. 2010). Ocean microbiome data published by Tara Ocean Project have demonstrated that *Alphaproteobacteria* constitute the predominant class (Sunagawa et al. 2015); in contrast, in most cases, *Gammaproteobacteria* have been reported as the dominant bacteria in rearing tanks. Fish skin and gut are rich in *Gammaproteobacteria* (Larsen et al. 2013; Xia et al. 2014; Etyemez and Balcazar 2015; Schmidt et al. 2015; Tarnecki et al. 2016; Ramirez and Romero 2017). The density of fish-derived organic matter was higher in rearing tanks than in the natural seawater; therefore, *Gammaproteobacteria* were more abundant in rearing tanks.

Proteobacteria and *Bacteroidetes* were dominant in the fasting group, whereas *Firmicutes* and *Fusobacteria* were dominant in the feeding group. Subsequent correlation analysis revealed that *Firmicutes* and *Fusobacteria* exhibited a positive correlation, whereas *Firmicutes* and *Bacteroidetes* showed a negative correlation. In mammals, *Firmicutes* is known to increase under conditions of high-fat food intake, and it is also known that *Firmicutes* and *Bacteroidetes* are negatively correlated (Turnbaugh et al. 2006, 2009; Turnbaugh and Gordon 2009). Fish feed was thought to contain an excess amount of fat; therefore, *Firmicutes* easily increased.

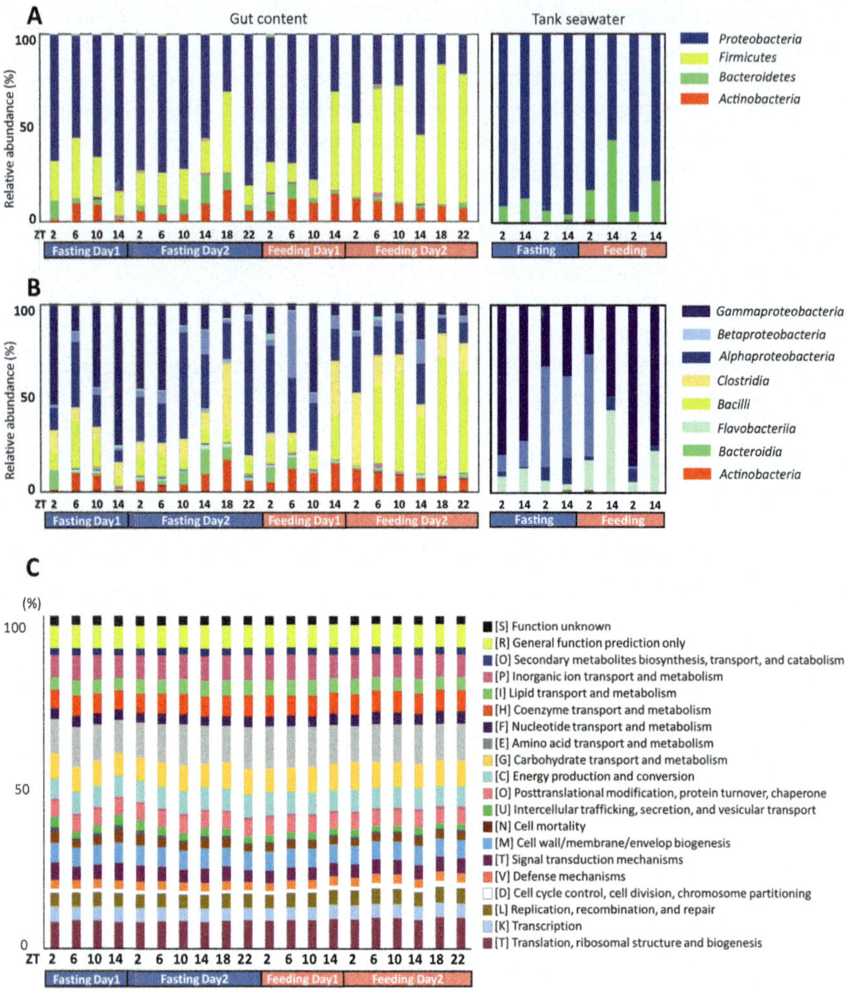

Fig. 13.10 (**a**) Microbial taxonomic composition in phylum level. Left graph shows gut content and right graph shows seawater. Each circle indicates samples collected at the same time. (**b**) Microbial taxonomic composition in class level. (**c**) Functional structure of intestinal microbiota exhibited by COG categories

The functional variability of the fish gut microbiome was investigated (Fig. 13.10c). The functional structure was stable during fasting and feeding. The abundance distribution of the basic functions was stable in the human gut and in the ocean (Turnbaugh et al. 2009; Motegi et al. 2015). However, the detailed clusters of orthologous group (COG) classification dynamics pattern showed that energy production, lipid metabolism, and inorganic metabolism increased under fasting conditions. The microbiota was predicted to utilize host gut nutrients and/or energy during feeding condition; however, it needed to generate energy independently under fasting conditions. Microbiota analysis aided in the comprehensive understanding of the metabolism of the leopard coral grouper.

13.4.2 Sevenband Grouper

When studying co-metabolic modulation by hosts and their microbial symbionts, metabolic profiling is a key approach for characterization and evaluation of metabolism and physiology. Using NMR-based metabolomic (or metabonomics) approaches provide technical insights for characterizing the interactions of hosts and symbionts (Nicholson et al. 1999; Brindle et al. 2002; Holmes et al. 2008), and this method is capable of generating comparable data between laboratories, thus supporting its continued use (Viant et al. 2009). This approach has many advantages for studying host-microbial interactions and assessing metabolic function and homeostasis at the molecular fingerprinting level (Nicholson et al. 1999; Dumas et al. 2006). Therefore, an NMR-based metabolomic approach has been successfully applied to evaluate animal homeostasis, including that in human beings (Clayton et al. 2006; Holmes et al. 2008; Fukuda et al. 2011; Claesson et al. 2012; Schlipalius et al. 2012; Furusawa et al. 2013). In addition, this NMR-based metabolomic approach is valuable in aquatic ecosystems for studying the environmental effects of pharmaceuticals and other chemicals on fishes (Samuelsson et al. 2006, 2011 Samuelsson and Larsson 2008). Such studies have contributed knowledge of basic physiology and development of fish, disease, water pollution, and other aspects (Southam et al. 2008, 2011; Williams et al. 2009; Picone et al. 2011; Bilandzic et al. 2011; Dove et al. 2012; Wagner et al. 2014). Therefore, little information is available on metabolic variations associated with fluctuations in microbial composition and structures in diverse fish species affected by the crosstalk between hosts and their microbial symbionts.

Microbial DNA extraction was performed according to previous studies with slight modifications (Date et al. 2012a). Each DNA sample was amplified by polymerase chain reaction using universal bacterial primers 954f and 1369r targeted to the V6–V8 regions of the 16S rRNA gene according to a previous report (Date et al. 2010). The sequencing analysis and the data processing were outsourced to Operon Biotechnologies Co. Ltd. (Tokyo, Japan). The categorizations of bacterial taxa were performed using a ribosomal database project [RDP; http://rdp.cme.msu.edu/seqmatch/seqmatch_intro.jsp; classifier (Wang et al. 2007)].

The microbiota and metabolic profiles in feces showed trends of clustering based on feeding conditions, whereas those in gut contents showed no clear clustering under the feeding conditions. This result indicated that the feces as an aggregate of final metabolic products were influenced by feedings compared with gut contents.

Time series data differences in feeding behavior in the rearing environment of the same species were evaluated by profiling methods described above. Results of PCA show that feces profile varied greatly, but temporarily, due to a change in feed, and tended to return to the original profile after a period of time (Fig. 13.11a). Based

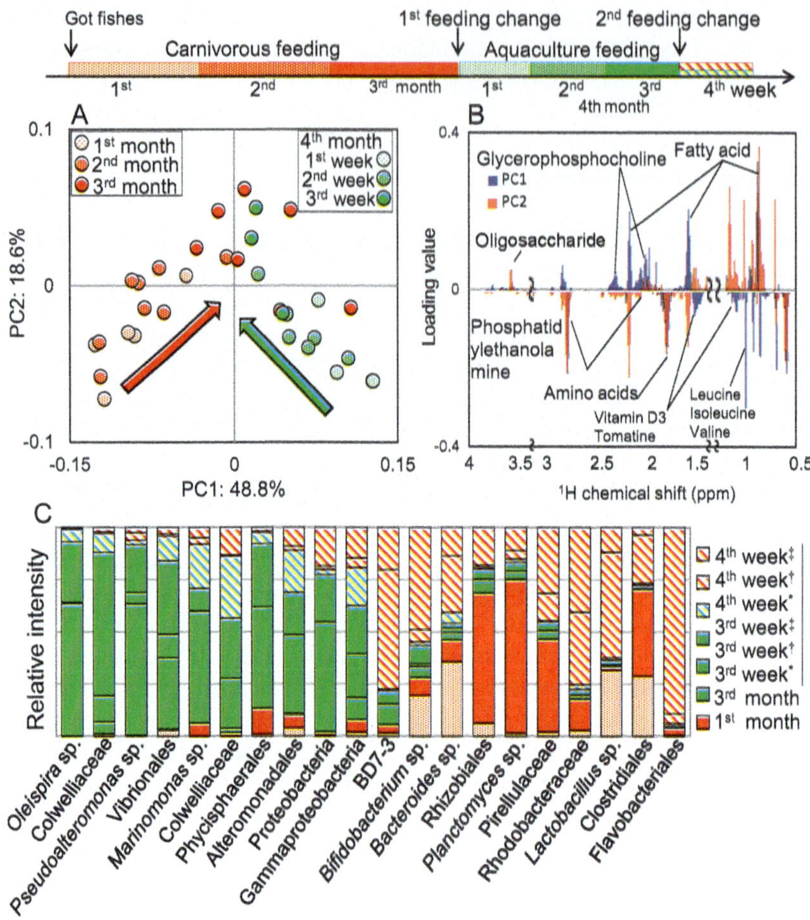

Fig. 13.11 Metabolic and microbial profiles of feeding response in *Epinephelus septemfasciatus* feces. Symbols representing individual communities are colored by diet (red, carnivorous feeding; green, aquaculture feeding). Three *Epinephelus septemfasciatus* were bred by carnivorous feeding for 3 months. They were then divided into three tanks (*, †, ‡) and bred for 3 weeks by aquaculture feeding. One *E. septemfasciatus* was then bred with aquaculture feeding, and the others were bred with carnivorous feeding. (A) PCA score plot on 1H NMR profile from feces (n = 38, k = 808, R2X = 0.488, R2Y = 0. 187, Q2 = 0.442). (B) PCA loading plots on 1H NMR profiles. (C) Bacterial profile that changed characteristically during the 4 months of rearing. These data were first normalized by total reads for each sample (i.e., the ratio in each fecal sample) and then secondarily normalized on the basis of the sum of the ratios of the same bacteria. This figure is adapted from (Asakura et al. 2014b)

on the loading plot analysis, some metabolites, such as phospholipids and fatty acids, contributed to the variations owing to the effect of aquaculture feeding (Fig. 13.11b). In addition, microbiota profiles varied widely with the effect of change in aquaculture feed, but the community was likely to be restored when carnivorous feeding was again available to the fishes (Fig. 13.11c). This result suggests that feeding changes in *Epinephelus septemfasciatus* affected the variations in metabolic and microbiota profiles of feces. In these results, a difference in response to time series of feeding associated with fecal metabolic and microbiota profiles can be observed. Based on the information that the intestinal environment is reflected in the feces of fish, we propose that this could be an informative noninvasive technique for cultured fish.

Recently, Kikuchi's group have been also published host and symbiotic metabolic analysis to the wild yellowfin goby (Wei et al. 2018) and 24 natural fish diversities living in Kanto and Tohoku region in Japan (Asakura et al. 2018b).

13.4.3 Comparative Analysis of Chemical and Microbial Profiles in Estuarine Sediments Sampled from Kanto and Tohoku Regions in Japan

Eutrophic estuarine environments are rich in biodiversity, and humanity greatly benefits from the ecosystem services they provide (Palmer and Filoso 2009). In addition, the response of benthic communities to organic enrichment affects the aquaculture industry (Borja et al. 2009). Microbial flora are present in a variety of environments, including soils, oceans, and symbiotic ecosystems, and these microorganisms are responsible for driving the biogeochemical cycling of elements on earth (Hugenholtz et al. 1998). For example, the biodegradation of aromatic compounds, which are the most prevalent and persistent environmental pollutants, by microorganisms is a major mechanism by which organic pollutants are removed from contaminated sites (Seo et al. 2009). One of the challenges that microbial ecologists faced is the identification of microorganisms that perform specific metabolic processes in the natural environment (Dumont and Murrell 2005; Neufeld et al. 2007). The characteristics of organic matter present in estuarine sediments are determined by its original configuration and the geochemical environment of the sediment (Santín et al. 2008) and by microbial inputs; therefore, methods for evaluating how the composition and structure of sediment organic and inorganic matter is related to the community structure and function of microorganisms are of great interest. To this end, NMR-based organic component analyses were combined with ICP-OES-based ionomics and pyrosequencing-based microbial community analyses of estuarine and coastal environments in the Tohoku and Kanto regions of Japan.

An amplicon analysis of 16S rRNA genes was performed using a next-generation pyrosequencing technique based on the FLX system (this analysis was outsourced to Operon Biotechnology Co. Ltd., Tokyo, Japan).

A correlation and network analysis was performed for identifying the elemental, microbial community, nitrogen ions, and organic component profiles, and these profiles were compared between the Kanto and Tohoku regions (Fig. 13.12).

In the network analysis, which was performed using the open-source software Gephi, microbial and chemical relationships and networks were visualized accord-

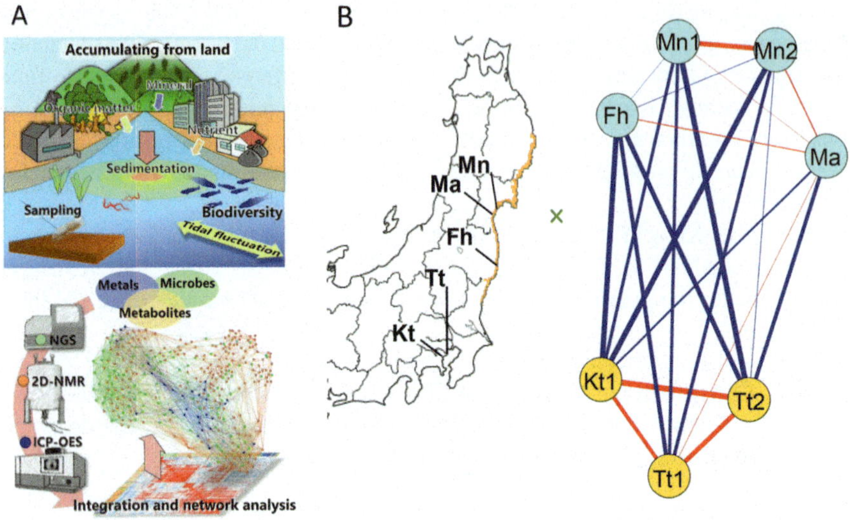

Fig. 13.12 (**a**) Conceptual image of integration and network analysis. Various substances accumulate at the estuarine and coastal bottom. The sampled sediments are analyzed by a plurality of apparatuses, and the integrated results are visualized on the network. (**b**) Correlation network analysis computed for all measured numerical values from seven geographically varied sediments. When sediment information is similar, nodes connect with red edges and become thicker with higher similarity. Blue edges indicate not similar. It is shown that the sediment in the Kanto region (orange) of Japan has a high similarity and greatly differs from sediments in the Tohoku region (cyan) Adapted with permission from (Asakura et al. 2014a). Copyright (2014) American Chemical Society

ing to a homogeneous correlation (Fig. 13.12a). Figure 13.12a shows individual data based on regional differences, and Fig. 13.12b shows correlations at each site based on data variation. Figure 13.12b shows that estuaries located in the same region had a positive correlation and that Kanto and Tohoku were separated by a negative correlation. Site Fh showed a relatively weak positive correlation with site Ma but negative correlation with the Kanto region. The network analysis captured the trends and characteristics of the individual sampling sites, although the individual sampling sites showed relatively weak correlations among them. The analytical strategy, including the microbial-gated correlation and network analysis, described in this study, is a powerful approach for analyzing, visualizing, and evaluating the complex metabolic dynamics and networks in sediment microbial ecosystems. Particularly, the microbial-gated correlation analysis is useful for comparing the relationships among (two or more) groups and categories. In this study, we adopted a "microbial"-gated correlation analysis; however, elemental- and sediment organic matter-gated correlation analyses are possible by modifying the "gating" (i.e., deploying) of elements to ICP data or organic matters in sediments from NMR data, respectively, and it can be used to characterize, visualize, and compare the relationships among the components of sediment ecosystems. This analytical strategy will enable us to evaluate and characterize not only estuarine sediment ecosystems but also other ecosystems, such as deep and shallow sea sediments and agronomic and forest soils. The analytical strategy promises to be useful for deciphering complicated metabolic dynamics, networks, and interactions in a variety of microbial ecosystems.

13.5 Conclusion and Future Perspectives

Aquaculture has increased its contribution to the per capita seafood consumption (FAO 2016), because of the increasing global population and limited fishery resources. Aquaculture has rapidly expanded worldwide and incorporated new species. Because global aquaculture fish production continues to expand, an improved understanding of how environmental factors interact with fish health and production is needed. Metagenomics research has a great potential to contribute to aquaculture development.

Metabolism and Intestinal Microbiota Control

Feed is one of the largest production costs associated with aquaculture, and the price of fish meal is predicted to increase as aquaculture expands worldwide. Therefore, an improvement in feed efficiency is required, and feeding concerns will be major factors limiting aquaculture development. Metabolism of host fish and their microbiota constitutes the key for developing novel feeding. The development of alternative ingredients of fish meal is urgently required. Recently significant advances have been made toward alternatives to fish-based diets, such as grain and worms (Wong et al. 2013). However, the adaptability to grain-based diet is different according to species because of their constitutional abnormalities. Co-metabolism by gut microbiota has been gradually become a focus. Gut microbiota compensate for host metabolism, which might in turn affect fish health and production. By controlling the intestinal microbiota, novel aquaculture strategies could be developed.

Effects of Microbiota on Disease Defense and Immune System

Fish are sometimes exposed to lethal diseases caused by bacterial, fungal, viral, and parasitic agents. Maintaining high-stock density, overfeeding, and the use of destructive fishing techniques lead to disease outbreaks in aquaculture. Aquaculture damage caused by diseases involves huge losses. Currently, utilization of medicine, such as antibiotics, is the most effective way to prevent fish diseases (Banerjee and Ray 2017). However, aquatic animals subjected to medicine sometimes do not allow for their shipment as a food product. Moreover, overuse of medicine could cause selective pressure for the evolution of drug-resistant bacteria. Probiotics are live microorganisms that confer several beneficial effects to the host. For example, they enhance immunity, help in digestion, provide protection against pathogens, improve water quality, and promote growth and reproduction. Recently, probiotics have become popular in the aquaculture industry (Hai 2015; Giatsis et al. 2016; Goncalves and Gallardo-Escarate 2017). Probiotics are now focusing on being an alternative method to antibiotics, such that they can be used instead of antibiotics. Recently, a wide range of bacteria has been reported as potential probiotic candidates in the fish farming sectors.

Aquaculture System and Microbiota

The demand for fish as a protein source has increased worldwide. The research on intensive aquaculture has been imperative. Recirculating aquaculture systems (RAS) and shallow raceway systems (SRS) (Øiestad 1999) technologies have been developed. These intensive aquaculture systems become more important in continuous and sustainable aquaculture practices. The organic matter derived from unconsumed food and fish are filtered. Microbe communities are developed in filtering systems in RAS and play important roles in clarification. Research regarding bacterial communities have been conducted on effectivity of biological filters (Sugita et al. 2005; Schreier et al. 2010). Understanding and controlling the bacterial communities help to develop aquaculture production systems and reduce the disease outbreaks.

Scientific analysis of aquatic products and processing by a holistic approach would be effective in many fields, such as environmental monitoring, quality control, safety, security, and resource protection. The data-driven approach will help find important clues from the massive data generated by omics analysis (Fig. 13.13).

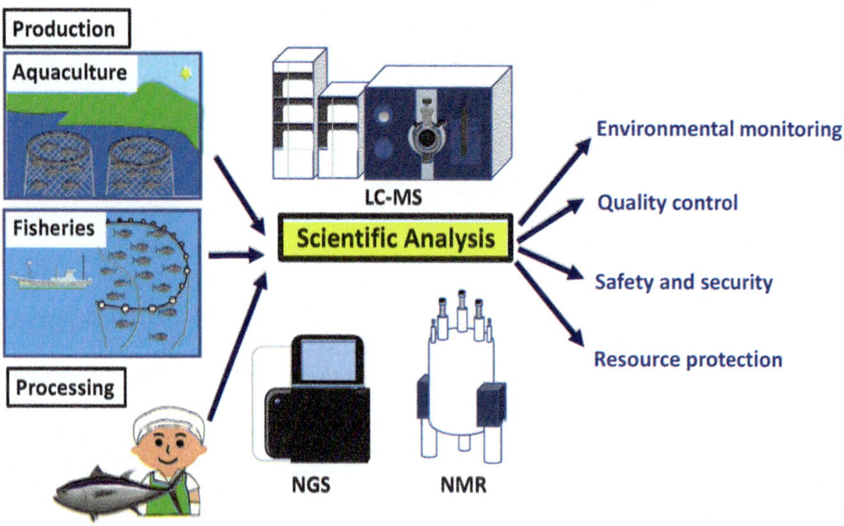

Fig. 13.13 Perspective of future aquaculture utilizing scientific analysis. Big data in omics analysis can be utilized in aquaculture, fisheries, and processing of marine products. Association analysis and machine learning on big data would help developing in environmental monitoring, quality control, resource production, food safety, and security

References

Amaral IP, Johnston IA (2012) Circadian expression of clock and putative clock-controlled genes in skeletal muscle of the zebrafish. Am J Physiol Regul Integr Comp Physiol 302(1):R193–R206. https://doi.org/10.1152/ajpregu.00367.2011

Anders S, Huber W (2010) Differential expression analysis for sequence count data. Genome Biol 11(10):R106. https://doi.org/10.1186/gb-2010-11-10-r106

Andrews JL, Zhang X, McCarthy JJ, McDearmon EL, Hornberger TA, Russell B et al (2010) CLOCK and BMAL1 regulate MyoD and are necessary for maintenance of skeletal muscle phenotype and function. Proc Natl Acad Sci U S A 107(44):19090–19095. https://doi.org/10.1073/pnas.1014523107

Asakura T, Date Y, Kikuchi J (2014a) Comparative analysis of chemical and microbial profiles in estuarine sediments sampled from Kanto and Tohoku regions in Japan. Anal Chem 86(11):5425–5432. https://doi.org/10.1021/ac5005037

Asakura T, Sakata K, Yoshida S, Date Y, Kikuchi J (2014b) Noninvasive analysis of metabolic changes following nutrient input into diverse fish species, as investigated by metabolic and microbial profiling approaches. Peer J 2:e550. https://doi.org/10.7717/peerj.550

Asakura T, Sakata K, Date Y Kikuchi J (2018a) Application of ensemble deep neural network to metabolomics studies Anal. Chim Acta. (in press)

Asakura T, Sakata K, Date Y, Kikuchi J (2018b) Regional feature extraction of various fishes based on chemical and microbial variable selection using machine learning. Anal Meth 10:2160–2168. https://doi.org/10.1039/c8ay00377g

Auro K, Joensuu A, Fischer K, Kettunen J, Salo P, Mattsson H et al (2014) A metabolic view on menopause and ageing. Nat Commun 5:4708. https://doi.org/10.1038/ncomms5708

Bäckhed F, Roswall J, Peng Y, Feng Q, Jia H, Kovatcheva-Datchary P et al (2015) Dynamics and stabilization of the human gut microbiome during the first year of life. Cell Host Microbe 17(5):690 703

Banerjee G, Ray AK (2017) The advancement of probiotics research and its application in fish farming industries. Res Vet Sci 115:66–77. https://doi.org/10.1016/j.rvsc.2017.01.016

Barber BJ, Blake NJ (1981) Energy storage and utilization in relation to gametogenesis in *Argopecten irradians concentricus* (Say). J Exp Mar Biol Ecol 52(2–3):121–134

Bayne B, Bubel A, Gabbott P, Livingstone D, Lowe D, Moore M (1982) Glycogen utilisation and gametogenesis in *Mytilus edulis* L. Mar Biol Lett 3(2):98–105

Beckonert O, Keun HC, Ebbels TM, Bundy J, Holmes E, Lindon JC et al (2007) Metabolic profiling, metabolomic and metabonomic procedures for NMR spectroscopy of urine, plasma, serum and tissue extracts. Nat Protoc 2(11):2692–2703. https://doi.org/10.1038/nprot.2007.376

Beckonert O, Coen M, Keun HC, Wang Y, Ebbels TM, Holmes E et al (2010) High-resolution magic-angle-spinning NMR spectroscopy for metabolic profiling of intact tissues. Nat Protoc 5(6):1019–1032. https://doi.org/10.1038/nprot.2010.45

Berthelin C, Kellner K, Mathieu M (2000) Storage metabolism in the Pacific oyster (*Crassostrea gigas*) in relation to summer mortalities and reproductive cycle (west coast of France). Comp Biochem Physiol B Biochem Mol Biol 125(3):359–369

Bilandzic N, Dokic M, Sedak M (2011) Metal content determination in four fish species from the Adriatic Sea. Food Chem 124(3):1005–1010. https://doi.org/10.1016/j.foodchem.2010.07.060

Bingol K, Zhang F, Bruschweiler-Li L, Brüschweiler R (2012) TOCCATA: a customized carbon total correlation spectroscopy NMR metabolomics database. Anal Chem 84(21):9395–9401

Blaise BJ, Giacomotto J, Elena B, Dumas ME, Toulhoat P, Segalat L et al (2007) Metabotyping of *Caenorhabditis elegans* reveals latent phenotypes. Proc Natl Acad Sci U S A 104(50):19808–19812. https://doi.org/10.1073/pnas.0707393104

Block BA, Teo SL, Walli A, Boustany A, Stokesbury MJ, Farwell CJ et al (2005) Electronic tagging and population structure of Atlantic bluefin tuna. Nature 434(7037):1121–1127. https://doi.org/10.1038/nature03463

Bone Q (1978) Locomotor muscle. Fish Physiol 7:361–424

Borja Á, Rodríguez JG, Black K, Bodoy A, Emblow C, Fernandes TF et al (2009) Assessing the suitability of a range of benthic indices in the evaluation of environmental impact of fin and shellfish aquaculture located in sites across Europe. Aquaculture 293(3):231–240

Brindle JT, Antti H, Holmes E, Tranter G, Nicholson JK, Bethell HW et al (2002) Rapid and noninvasive diagnosis of the presence and severity of coronary heart disease using 1H-NMR-based metabonomics. Nat Med 8(12):1439–1444. https://doi.org/10.1038/nm802

Brown K, DeCoffe D, Molcan E, Gibson DL (2012) Diet-induced dysbiosis of the intestinal microbiota and the effects on immunity and disease. Nutrients 4(8):1095–1119. https://doi.org/10.3390/nu4081095

Buckingham M (1992) Making muscle in mammals. Trends Genet 8(4):144–148. https://doi.org/10.1016/0168-9525(92)90373-c

Bundy JG, Keun HC, Sidhu JK, Spurgeon DJ, Svendsen C, Kille P et al (2007) Metabolic profile biomarkers of metal contamination in a sentinel terrestrial species are applicable across multiple sites. Environ Sci Technol 41(12):4458–4464

Cao X, Lattao C, Pignatello JJ, Mao J, Schmidt-Rohr K (2014) Sorption selectivity in natural organic matter probed with fully deuterium-exchanged and carbonyl-13C-labeled benzophenone and 1H-13C NMR spectroscopy. Environ Sci Technol 48(15):8645–8652. https://doi.org/10.1021/es501129f

Caporaso JG, Kuczynski J, Stombaugh J, Bittinger K, Bushman FD, Costello EK et al (2010) QIIME allows analysis of high-throughput community sequencing data. Nat Methods 7(5):335–336. https://doi.org/10.1038/nmeth.f.303

Carmona-Antonanzas G, Tocher DR, Martinez-Rubio L, Leaver MJ (2014) Conservation of lipid metabolic gene transcriptional regulatory networks in fish and mammals. Gene 534(1):1–9. https://doi.org/10.1016/j.gene.2013.10.040

Cerda J, Manchado M (2013) Advances in genomics for flatfish aquaculture. Genes Nutr 8(1):5–17. https://doi.org/10.1007/s12263-012-0312-8

Chatzimichali EA, Bessant C (2016) Novel application of heuristic optimisation enables the creation and thorough evaluation of robust support vector machine ensembles for machine learning applications. Metabolomics 12(1):16. https://doi.org/10.1007/s11306-015-0894-4

Chikayama E, Sekiyama Y, Okamoto M, Nakanishi Y, Tsuboi Y, Akiyama K et al (2010) Statistical indices for simultaneous large-scale metabolite detections for a single NMR spectrum. Anal Chem 82(5):1653–1658. https://doi.org/10.1021/ac9022023

Chikayama E, Yamashina R, Komatsu K, Tsuboi Y, Sakata K, Kikuchi J et al (2016) FoodPro: a web-based tool for evaluating covariance and correlation NMR spectra associated with food processes. Metabolites 6(4):36. https://doi.org/10.3390/metabo6040036

Choe A, Chuman T, von Reuss SH, Dossey AT, Yim JJ, Ajredini R et al (2012) Sex-specific mating pheromones in the nematode Panagrellus redivivus. Proc Natl Acad Sci U S A 109(51):20949–20954. https://doi.org/10.1073/pnas.1218302109

Claesson MJ, Jeffery IB, Conde S, Power SE, O'Connor EM, Cusack S et al (2012) Gut microbiota composition correlates with diet and health in the elderly (Research Support, Non-U.S. Gov't). Nature 488(7410):178–184. https://doi.org/10.1038/nature11319

Clayton TA, Lindon JC, Cloarec O, Antti H, Charuel C, Hanton G et al (2006) Pharmaco-metabonomic phenotyping and personalized drug treatment. Nature 440(7087):1073–1077. https://doi.org/10.1038/nature04648

Clendinen CS, Lee-McMullen B, Williams CM, Stupp GS, Vandenborne K, Hahn DA et al (2014) 13C NMR Metabolomics: applications at natural abundance. Anal Chem 86(18):9242–9250. https://doi.org/10.1021/ac502346h

Collette TW, Teng Q, Jensen KM, Kahl MD, Makynen EA, Durhan EJ et al (2010) Impacts of an anti-androgen and an androgen/anti-androgen mixture on the metabolite profile of male fathead minnow urine. Environ Sci Technol 44(17):6881–6886. https://doi.org/10.1021/es1011884

Cui Q, Lewis IA, Hegeman AD, Anderson ME, Li J, Schulte CF et al (2008) Metabolite identification via the madison metabolomics consortium database. Nat Biotech 26(2):162–164. https://doi.org/10.1038/nbt0208-162

Date Y, Kikuchi J (2018) Application of a deep neural network to metabolomics studies and its performance in determining important variables Anal. Chem 90:1805–1810. https://doi.org/10.1021/acs.analchem.7b03795

Date Y, Nakanishi Y, Fukuda S, Kato T, Tsuneda S, Ohno H et al (2010) New monitoring approach for metabolic dynamics in microbial ecosystems using stable-isotope-labeling technologies. J Biosci Bioeng 110(1):87–93. https://doi.org/10.1016/j.jbiosc.2010.01.004

Date Y, Iikura T, Yamazawa A, Moriya S, Kikuchi J (2012a) Metabolic sequences of anaerobic fermentation on glucose-based feeding substrates based on correlation analyses of microbial and metabolite profiling. J Proteome Res 11(12):5602–5610. https://doi.org/10.1021/pr3008682

Date Y, Sakata K, Kikuchi J (2012b) Chemical profiling of complex biochemical mixtures from various seaweeds. Polym J 44(8):888–894. https://doi.org/10.1038/Pj.2012.105

Dethlefsen L, Eckburg PB, Bik EM, Relman DA (2006) Assembly of the human intestinal microbiota. Trends Ecol Evol 21(9):517–523

Douglas SE (2006) Microarray studies of gene expression in fish. Omics J Integ Biol 10(4):474–489

Dove AD, Leisen J, Zhou M, Byrne JJ, Lim-Hing K, Webb HD et al (2012) Biomarkers of whale shark health: a metabolomic approach. PLoS One 7(11):e49379. https://doi.org/10.1371/journal.pone.0049379

Dumas ME, Maibaum EC, Teague C, Ueshima H, Zhou B, Lindon JC et al (2006) Assessment of analytical reproducibility of 1H NMR spectroscopy based metabonomics for large-scale epidemiological research: the INTERMAP study. Anal Chem 78(7):2199–2208. https://doi.org/10.1021/ac0517085

Dumont MG, Murrell JC (2005) Stable isotope probing—linking microbial identity to function. Nat Rev Microbiol 3(6):499–504

Eisenreich W, Slaghuis J, Laupitz R, Bussemer J, Stritzker J, Schwarz C et al (2006) 13C isotopologue perturbation studies of *Listeria monocytogenes* carbon metabolism and its modulation by the virulence regulator PrfA. Proc Natl Acad Sci U S A 103(7):2040–2045. https://doi.org/10.1073/pnas.0507580103

Ellis RP, Spicer JI, Byrne JJ, Sommer U, Viant MR, White DA et al (2014) (1)H NMR metabolomics reveals contrasting response by male and female mussels exposed to reduced seawater pH, increased temperature, and a pathogen. Environ Sci Technol 48(12):7044–7052. https://doi.org/10.1021/es501601w

Etyemez M, Balcazar JL (2015) Bacterial community structure in the intestinal ecosystem of rainbow trout (*Oncorhynchus mykiss*) as revealed by pyrosequencing-based analysis of 16S rRNA genes. Res Vet Sci 100:8–11. https://doi.org/10.1016/j.rvsc.2015.03.026

FAO (2016) FAO Fisheries and Aquaculture Report eng no 1133

Feng X, Simpson AJ, Simpson MJ (2006) Investigating the role of mineral-bound humic acid in phenanthrene sorption. Environ Sci Technol 40(10):3260–3266

Frost G, Sleeth ML, Sahuri-Arisoylu M, Lizarbe B, Cerdan S, Brody L et al (2014) The short-chain fatty acid acetate reduces appetite via a central homeostatic mechanism. Nat Commun 5:3611. https://doi.org/10.1038/ncomms4611

Fujisawa K, Takami T, Kimoto Y, Matsumoto T, Yamamoto N, Terai S et al (2016) Circadian variations in the liver metabolites of medaka (*Oryzias latipes*). Sci Rep 6:20916. https://doi.org/10.1038/srep20916

Fukuda S, Toh H, Hase K, Oshima K, Nakanishi Y, Yoshimura K et al (2011) Bifidobacteria can protect from enteropathogenic infection through production of acetate. Nature 469(7331):543–547. https://doi.org/10.1038/nature09646. nature09646 [pii]

Furusawa Y, Obata Y, Fukuda S, Endo TA, Nakato G, Takahashi D et al (2013) Commensal microbe-derived butyrate induces the differentiation of colonic regulatory T cells. Nature 504(7480):446–450. https://doi.org/10.1038/nature12721. nature12721 [pii]

Gallo V, Intini N, Mastrorilli P, Latronico M, Scapicchio P, Triggiani M et al (2015) Performance assessment in fingerprinting and multi component quantitative NMR analyses. Anal Chem 87(13):6709–6717. https://doi.org/10.1021/acs.analchem.5b00919

Giatsis C, Sipkema D, Ramiro-Garcia J, Bacanu GM, Abernathy J, Verreth J et al (2016) Probiotic legacy effects on gut microbial assembly in tilapia larvae. Sci Rep 6:33965. https://doi.org/10.1038/srep33965

Gillis T, Ballantyne J (1996) The effects of starvation on plasma free amino acid and glucose concentrations in lake sturgeon. J Fish Biol 49(6):1306–1316

Goncalves AT, Gallardo-Escarate C (2017) Microbiome dynamic modulation through functional diets based on pre- and probiotics (mannan-oligosaccharides and *Saccharomyces cerevisiae*) in juvenile rainbow trout (*Oncorhynchus mykiss*). J Appl Microbiol 122(5):1333–1347. https://doi.org/10.1111/jam.13437

Goodwin S, McPherson JD, McCombie WR (2016) Coming of age: ten years of next-generation sequencing technologies. Nat Rev Genet 17(6):333–351. https://doi.org/10.1038/nrg.2016.49

Grabherr MG, Haas BJ, Yassour M, Levin JZ, Thompson DA, Amit I et al (2011) Full-length transcriptome assembly from RNA-Seq data without a reference genome. Nat Biotechnol 29(7):644–652. https://doi.org/10.1038/nbt.1883

Grosell M, Genz J (2006) Ouabain-sensitive bicarbonate secretion and acid absorption by the marine teleost fish intestine play a role in osmoregulation. Am J Phys Regul Integr Comp Phys 291(4):R1145–R1156

Gumbmann M, Tappel AL (1962) The tricarboxylic acid cycle in fish. Arch Biochem Biophys 98(2):262–270

Guppy M, Hochachka PW (1978) Controlling the highest lactate dehydrogenase activity known in nature. Am J Phys 234(3):R136–R140

Guppy M, Hulbert WC, Hochachka PW (1979) Metabolic sources of heat and power in tuna muscles. II Enzyme and metabolite profiles. J Exp Biol 82(1):303–320

Hai NV (2015) The use of probiotics in aquaculture. J Appl Microbiol 119(4):917–935. https://doi.org/10.1111/jam.12886

Haider S, Pal R (2013) Integrated analysis of transcriptomic and proteomic data. Curr Genomics 14(2):91–110. https://doi.org/10.2174/1389202911314020003

Hao J, Liebeke M, Astle W, De Iorio M, Bundy JG, Ebbels TM (2014) Bayesian deconvolution and quantification of metabolites in complex 1D NMR spectra using BATMAN. Nat Protoc 9(6):1416–1427. https://doi.org/10.1038/nprot.2014.090

Hardcastle TJ, Kelly KA (2010) baySeq: empirical Bayesian methods for identifying differential expression in sequence count data. BMC Bioinformatics 11(1):422. https://doi.org/10.1186/1471-2105-11-422

Head SR, Komori HK, LaMere SA, Whisenant T, Van Nieuwerburgh F, Salomon DR et al (2014) Library construction for next-generation sequencing: overviews and challenges. BioTechniques 56(2):61. https://doi.org/10.2144/000114133

Heller MJ (2002) DNA microarray technology: devices, systems, and applications. Annu Rev Biomed Eng 4(1):129–153. https://doi.org/10.1146/annurev.bioeng.4.020702.153438

Holecek M, Sprongl L, Tilser I (2001) Metabolism of branched-chain amino acids in starved rats: the role of hepatic tissue. Physiol Res 50(1):25–33

Holmes E, Loo RL, Stamler J, Bictash M, Yap IK, Chan Q et al (2008) Human metabolic phenotype diversity and its association with diet and blood pressure. Nature 453(7193):396–400. https://doi.org/10.1038/nature06882

Hrdlickova R, Toloue M, Tian B (2017) RNA-seq methods for transcriptome analysis. Wiley Interdiscip Rev RNA 8(1):e1364. https://doi.org/10.1002/wrna.1364

Hsiao EY, McBride SW, Hsien S, Sharon G, Hyde ER, McCue T et al (2013) Microbiota modulate behavioral and physiological abnormalities associated with neurodevelopmental disorders. Cell 155(7):1451–1463

Hugenholtz P, Goebel BM, Pace NR (1998) Impact of culture-independent studies on the emerging phylogenetic view of bacterial diversity. J Bacteriol 180(18):4765–4774

Ito K, Sakata K, Date Y, Kikuchi J (2014) Integrated analysis of seaweed components during seasonal fluctuation by data mining across heterogeneous chemical measurements with network visualization. Anal Chem 86(2):1098–1105. https://doi.org/10.1021/ac402869b

Ito K, Tsutsumi Y, Date Y, Kikuchi J (2016) Fragment assembly approach based on graph/network theory with quantum chemistry verifications for assigning multidimensional NMR signals in metabolite mixtures. ACS Chem Biol 11(4):1030–1038. https://doi.org/10.1021/acschembio.5b00894

Kato T, Fukuda S, Fujiwara A, Suda W, Hattori M, Kikuchi J et al (2014) Multiple omics uncovers host-gut microbial mutualism during prebiotic fructooligosaccharide supplementation. DNA Res 21(5):469–480. https://doi.org/10.1093/dnares/dsu013

Kettunen J, Tukiainen T, Sarin AP, Ortega-Alonso A, Tikkanen E, Lyytikainen LP et al (2012) Genome-wide association study identifies multiple loci influencing human serum metabolite levels. Nat Genet 44(3):269–276. https://doi.org/10.1038/ng.1073

Kettunen J, Demirkan A, Wurtz P, Draisma HH, Haller T, Rawal R et al (2016) Genome-wide study for circulating metabolites identifies 62 loci and reveals novel systemic effects of LPA. Nat Commun 7:11122. https://doi.org/10.1038/ncomms11122

Kikuchi J, Yamada S (2017) NMR window of molecular complexity showing homeostasis in superorganisms. Analyst 142(22):4161–4172. https://doi.org/10.1039/c7an01019b

Kikuchi J, Shinozaki K, Hirayama T (2004) Stable isotope labeling of *Arabidopsis thaliana* for an NMR-based metabolomics approach. Plant Cell Physiol 45(8):1099–1104. https://doi.org/10.1093/pcp/pch117

Kikuchi J, Tsuboi Y, Komatsu K, Gomi M, Chikayama E, Date Y (2016) SpinCouple: development of a web tool for analyzing metabolite mixtures via two-dimensional J-resolved NMR database. Anal Chem 88(1):659–665. https://doi.org/10.1021/acs.analchem.5b02311

Kikuchi J, Ito K, Date Y (2018) Environmental metabolomics with data science as investigation of ecosystem homeostasis. Prog NMR Spectroscopy 104:56–88. https://doi.org/10.1016/j.pnmrs.2017.11.003

Kohl KD, Amaya J, Passement CA, Dearing MD, McCue MD (2014) Unique and shared responses of the gut microbiota to prolonged fasting: a comparative study across five classes of vertebrate hosts. FEMS Microbiol Ecol 90(3):883–894. https://doi.org/10.1111/1574-6941.12442

Komatsu T, Kikuchi J (2013) Comprehensive signal assignment of 13C-labeled lignocellulose using multidimensional solution NMR and 13C chemical shift comparison with solid-state NMR. Anal Chem 85(18):8857–8865. https://doi.org/10.1021/ac402197h

Komatsu T, Ohishi R, Shino A, Akashi K, Kikuchi J (2014) Multi-spectroscopic analysis of seed quality and 13C-stable-iotopologue monitoring in initial growth metabolism of *Jatropha curcas* L. Metabolites 4(4):1018–1033. https://doi.org/10.3390/metabo4041018

Komatsu T, Kobayashi T, Hatanaka M, Kikuchi J (2015) Profiling planktonic biomass using element-specific, multicomponent nuclear magnetic resonance spectroscopy. Environ Sci Technol 49(11):7056–7062. https://doi.org/10.1021/acs.est.5b00837

Komatsu T, Ohishi R, Shino A, Kikuchi J (2016) Structure and metabolic-flow analysis of molecular complexity in a 13C-labeled tree by 2D and 3D NMR. Angew Chem Int Ed 55(20):6000–6003. https://doi.org/10.1002/anie.201600334

Korsmeyer KE, Dewar H (2001) Tuna metabolism and energetics. Fish Physiol 19:35–78

Kruger NJ, Troncoso-Ponce MA, Ratcliffe RG (2008) 1H NMR metabolite fingerprinting and metabolomic analysis of perchloric acid extracts from plant tissues. Nat Protoc 3(6):1001–1012. https://doi.org/10.1038/nprot.2008.64

Lacy P, McKay RT, Finkel M, Karnovsky A, Woehler S, Lewis MJ et al (2014) Signal intensities derived from different NMR probes and parameters contribute to variations in quantification of metabolites. PLoS One 9(1):e85732. https://doi.org/10.1371/journal.pone.0085732

Larsen A, Tao Z, Bullard SA, Arias CR (2013) Diversity of the skin microbiota of fishes: evidence for host species specificity. FEMS Microbiol Ecol 85(3):483–494. https://doi.org/10.1111/1574-6941.12136

Lattao C, Cao X, Li Y, Mao J, Schmidt-Rohr K, Chappell MA et al (2012) Sorption selectivity in natural organic matter studied with nitroxyl paramagnetic relaxation probes. Environ Sci Technol 46(23):12814–12822. https://doi.org/10.1021/es302157j

Ley RE, Backhed F, Turnbaugh P, Lozupone CA, Knight RD, Gordon JI (2005) Obesity alters gut microbial ecology. Proc Natl Acad Sci U S A 102(31):11070–11075. https://doi.org/10.1073/pnas.0504978102

Li M, Wang BH, Zhang MH, Rantalainen M, Wang SY, Zhou HK et al (2008) Symbiotic gut microbes modulate human metabolic phenotypes. Proc Natl Acad Sci U S A 105(6):2117–2122. https://doi.org/10.1073/pnas.0712038105

Li Y, Li G, Gorling B, Luy B, Du J, Yan J (2015) Integrative analysis of circadian transcriptome and metabolic network reveals the role of de novo purine synthesis in circadian control of cell cycle. PLoS Comput Biol 11(2):e1004086. https://doi.org/10.1371/journal.pcbi.1004086

Ludwig C, Easton JM, Lodi A, Tiziani S, Manzoor SE, Southam AD et al (2012) Birmingham Metabolite Library: a publicly accessible database of 1-D 1H and 2-D 1H J-resolved NMR spectra of authentic metabolite standards (BML-NMR). Metabolomics 8(1):8–18. https://doi.org/10.1007/s11306-011-0347-7

Mansfield SD, Kim H, Lu FC, Ralph J (2012) Whole plant cell wall characterization using solution-state 2D NMR. Nat Protoc 7(9):1579–1589. https://doi.org/10.1038/nprot.2012.064

Mao JD, Xing B, Schmidt-Rohr K (2001) New structural information on a humic acid from two-dimensional 1H-13C correlation solid-state nuclear magnetic resonance. Environ Sci Technol 35(10):1928–1934

Martin SAM, Dehler CE, Król E (2016) Transcriptomic responses in the fish intestine. Dev Comp Immunol 64(Supplement C):103–117. https://doi.org/10.1016/j.dci.2016.03.014

Mathieu M, Lubet P (1993) Storage tissue metabolism and reproduction in marine bivalves—a brief review. Invertebr Reprod Dev 23(2–3):123–129

Maynard CL, Elson CO, Hatton RD, Weaver CT (2012) Reciprocal interactions of the intestinal microbiota and immune system. Nature 489(7415):231

Mekuchi M, Hatta T, Kaneko T (2010) Mg-calcite, a carbonate mineral, constitutes Ca precipitates produced as a byproduct of osmoregulation in the intestine of seawater-acclimated Japanese eel *Anguilla japonica*. Fish Sci 76(2):199–205

Mekuchi M, Sakata K, Yamaguchi T, Koiso M, Kikuchi J (2017) Trans-omics approaches used to characterise fish nutritional biorhythms in leopard coral grouper (*Plectropomus leopardus*). Sci Rep 7(1):9372. https://doi.org/10.1038/s41598-017-09531-4

Mekuchi M, Asakura T, Sakata K, Yamaguchi T, Teruya K, Kikuchi J (2018) Intestinal microbiota composition is altered according to nutritional biorhythms in the leopard coral grouper (*Plectropomus leopardus*). Plos one X:XX

Misawa T, Date Y, Kikuchi J (2015) Human metabolic, mineral, and microbiota fluctuations across daily nutritional intake visualized by a data-driven approach. J Proteome Res 14(3):1526–1534. https://doi.org/10.1021/pr501194k

Misawa T, Komatsu T, Date Y, Kikuchi J (2016a) SENSI: signal enhancement by spectral integration for the analysis of metabolic mixtures (10.1039/C5CC09442A). Chem Commun 52(14):2964–2967. https://doi.org/10.1039/c5cc09442a

Misawa T, Wei F, Kikuchi J (2016b) Application of two-dimensional nuclear magnetic resonance for signal enhancement by spectral integration using a large dataset of metabolic mixtures. Anal Chem 88:6130–6134. https://doi.org/10.1021/acs.analchem.6b01495

Moore SJ, Warren MJ (2012) The anaerobic biosynthesis of vitamin B12. Biochem Soc Trans 40(3):581–586. https://doi.org/10.1042/bst20120066

Mori T, Tsuboi Y, Ishida N, Nishikubo N, Demura T, Kikuchi J (2015) Multidimensional high-resolution magic angle spinning and solution-state NMR characterization of 13C-labeled plant metabolites and lignocellulose (Article). Sci Rep 5:11848. https://doi.org/10.1038/srep11848. http:/\penalty\z@/\penalty\z@www.nature.com/\penalty\z@articles/\penalty\z@srep11848#supplementary-\penalty\z@information

Morita H, Toh H, Fukuda S, Horikawa H, Oshima K, Suzuki T et al (2008) Comparative genome analysis of *Lactobacillus reuteri* and *Lactobacillus fermentum* reveal a genomic island for reuterin and cobalamin production. DNA Res 15(3):151–161. https://doi.org/10.1093/dnares/dsn009. dsn009 [pii]

Mortazavi A, Williams BA, McCue K, Schaeffer L, Wold B (2008) Mapping and quantifying mammalian transcriptomes by RNA-Seq. Nat Methods 5(7):621–628. https://doi.org/10.1038/nmeth.1226

Motegi H, Tsuboi Y, Saga A, Kagami T, Inoue M, Toki H et al (2015) Identification of reliable components in Multivariate Curve Resolution-Alternating Least Squares (MCR-ALS): a data-driven approach across metabolic processes. Sci Rep 5:15710. https://doi.org/10.1038/srep15710

Mullur R, Liu YY, Brent GA (2014) Thyroid hormone regulation of metabolism. Physiol Rev 94(2):355–382. https://doi.org/10.1152/physrev.00030.2013

Neufeld JD, Wagner M, Murrell JC (2007) Who eats what, where and when? Isotope-labelling experiments are coming of age. ISME J 1(2):103–110

Nicholson JK, Lindon JC, Holmes E (1999) 'Metabonomics': understanding the metabolic responses of living systems to pathophysiological stimuli via multivariate statistical analysis of biological NMR spectroscopic data (Review). Xenobiotica 29(11):1181–1189. https://doi.org/10.1080/004982599238047

Nishiyama Y, Endo Y, Nemoto T, Bouzier-Sore A-K, Wong A (2015) High-resolution NMR-based metabolic detection of microgram biopsies using a 1 mm HRμMAS probe. Analyst 140(24):8097–8100

Ogata Y, Chikayama E, Morioka Y, Everroad RC, Shino A, Matsushima A et al (2012) ECOMICS: a web-based toolkit for investigating the biomolecular web in ecosystems using a trans-omics approach. PLoS One 7(2):e30263. https://doi.org/10.1371/journal.pone.0030263

Ogawa DMO, Moriya S, Tsuboi Y, Date Y, Prieto-da-Silva AR, Radis-Baptista G et al (2014) Biogeochemical typing of paddy field by a data-driven approach revealing sub-systems within a complex environment – a pipeline to filtrate, organize and frame massive dataset from multi-omics analyses. PLoS One 9(10):e110723. https://doi.org/10.1371/journal.pone.0110723

Ogura T, Date Y, Masukujane M, Coetzee T, Akashi K, Kikuchi J (2016a) Improvement of physical, chemical, and biological properties of aridisol from Botswana by the incorporation of torrefied biomass. Sci Rep 6:28011. https://doi.org/10.1038/srep28011

Ogura T, Hoshino R, Date Y, Kikuchi J (2016b) Visualization of microfloral metabolism for marine waste recycling. Metabolites 6(1):7. https://doi.org/10.3390/metabo6010007

Ohyama K, Suzuki M, Kikuchi J, Saito K, Muranaka T (2009) Dual biosynthetic pathways to phytosterol via cycloartenol and lanosterol in Arabidopsis. Proc Natl Acad Sci U S A 106(3):725–730. https://doi.org/10.1073/pnas.0807675106

Øiestad V (1999) Shallow raceways as a compact, resource-maximizing farming procedure for marine fish species. Aquac Res 30(11–12):831–840. https://doi.org/10.1046/j.1365-2109.1999.00408.x

Palmer MA, Filoso S (2009) Restoration of ecosystem services for environmental markets. Science 325(5940):575–576

Peyraud R, Kiefer P, Christen P, Massou S, Portais JC, Vorholt JA (2009) Demonstration of the ethylmalonyl-CoA pathway by using 13C metabolomics. Proc Natl Acad Sci U S A 106(12):4846–4851. https://doi.org/10.1073/pnas.0810932106

Picone G, Engelsen SB, Savorani F, Testi S, Badiani A, Capozzi F (2011) Metabolomics as a powerful tool for molecular quality assessment of the fish *Sparus aurata*. Nutrients 3(2):212–227. https://doi.org/10.3390/nu3020212

Podnar J, Deiderick H, Huerta G, Hunicke-Smith S (2014) Next-generation sequencing RNA-seq library construction. Curr Protoc Mol Biol 106:4.21.1–4.2119. https://doi.org/10.1002/0471142727.mb0421s106

Qian X, Ba Y, Zhuang Q, Zhong G (2014) RNA-Seq technology and its application in fish transcriptomics. OMICS 18(2):98–110. https://doi.org/10.1089/omi.2013.0110

Ramirez C, Romero J (2017) Fine flounder (*Paralichthys adspersus*) microbiome showed important differences between wild and reared specimens. Front Microbiol 8:271. https://doi.org/10.3389/fmicb.2017.00271

Reidy SP, Kerr SR, Nelson JA (2000) Aerobic and anaerobic swimming performance of individual Atlantic cod. J Exp Biol 203(2):347–357

Robinson MD, McCarthy DJ, Smyth GK (2010) edgeR: a Bioconductor package for differential expression analysis of digital gene expression data. Bioinformatics 26(1):139–140. https://doi.org/10.1093/bioinformatics/btp616

Rosen ED, Walkey CJ, Puigserver P, Spiegelman BM (2000) Transcriptional regulation of adipogenesis. Genes Dev 14(11):1293–1307

Rosselló-Mora R, Amann R (2001) The species concept for prokaryotes. FEMS Microbiol Rev 25(1):39–67

Ruiz C, Abad M, Sedano F, Garcia-Martin L, Lopez JS (1992) Influence of seasonal environmental changes on the gamete production and biochemical composition of *Crassostrea gigas* (Thunberg) in suspended culture in El Grove, Galicia, Spain. J Exp Mar Biol Ecol 155(2):249–262

Samuelsson LM, Larsson DG (2008) Contributions from metabolomics to fish research. Mol BioSyst 4(10):974–979. https://doi.org/10.1039/b804196b

Samuelsson LM, Forlin L, Karlsson G, Adolfsson-Eric M, Larsson DGJ (2006) Using NMR metabolomics to identify responses of an environmental estrogen in blood plasma of fish. Aquatic Toxicol 78(4):341–349. https://doi.org/10.1016/j.aquatox.2006.04.008

Samuelsson LM, Bjorlenius B, Forlin L, Larsson DG (2011) Reproducible (1)H NMR-based metabolomic responses in fish exposed to different sewage effluents in two separate studies. Environ Sci Technol 45(4):1703–1710. https://doi.org/10.1021/es104111x

Sannier F, Lecoeur C, Zhao Q, Garreau I, Piot JM (1996) Separation of hemoglobin and myoglobin from yellowfin tuna red muscle by ultrafiltration: effect of pH and ionic strength. Biotechnol Bioeng 52(4):501–506. https://doi.org/10.1002/(sici)1097-0290(19961120)52:4<501::aid-bit5>3.0.co;2-t

Santín C, González-Pérez M, Otero X, Vidal-Torrado P, Macías F, Álvarez M (2008) Characterization of humic substances in salt marsh soils under sea rush (*Juncus maritimus*). Estuar Coast Shelf Sci 79(3):541–548

Schlipalius DI, Valmas N, Tuck AG, Jagadeesan R, Ma L, Kaur R et al (2012) A core metabolic enzyme mediates resistance to phosphine gas (Research Support, N.I.H., Extramural Research Support, Non-U.S. Gov't). Science 338(6108):807–810. https://doi.org/10.1126/science.1224951

Schmidt VT, Smith KF, Melvin DW, Amaral-Zettler LA (2015) Community assembly of a euryhaline fish microbiome during salinity acclimation. Mol Ecol 24(10):2537–2350. https://doi.org/10.1111/mec.13177

Schreier HJ, Mirzoyan N, Saito K (2010) Microbial diversity of biological filters in recirculating aquaculture systems. Curr Opin Biotechnol 21(3):318–325. https://doi.org/10.1016/j.copbio.2010.03.011

Sekiyama Y, Kikuchi J (2007) Towards dynamic metabolic network measurements by multi-dimensional NMR-based fluxomics. Phytochemistry 68(16–18):2320–2329. https://doi.org/10.1016/j.phytochem.2007.04.011

Sekiyama Y, Chikayama E, Kikuchi J (2010) Profiling polar and semipolar plant metabolites throughout extraction processes using a combined solution-state and high-resolution magic angle spinning NMR approach. Anal Chem 82(5):1643–1652. https://doi.org/10.1021/ac9019076

Sekiyama Y, Chikayama E, Kikuchi J (2011) Evaluation of a semipolar solvent system as a step toward heteronuclear multidimensional NMR-based metabolomics for 13C-labeled bacteria, plants, and animals. Anal Chem 83(3):719–726. https://doi.org/10.1021/ac102097u

Sekiyama Y, Okazaki K, Kikuchi J, Ikeda S (2017) NMR-based metabolic profiling of field-grown leaves from sugar beet plants harbouring different levels of resistance to cercospora leaf spot disease. Metabolites 7(1):4. https://doi.org/10.3390/metabo7010004

Semova I, Carten JD, Stombaugh J, Mackey LC, Knight R, Farber SA et al (2012) Microbiota regulate intestinal absorption and metabolism of fatty acids in the zebrafish. Cell Host Microbe 12(3):277–288. https://doi.org/10.1016/j.chom.2012.08.003

Seo J-S, Keum Y-S, Li QX (2009) Bacterial degradation of aromatic compounds. Int J Environ Res Public Health 6(1):278–309

Shibata M, Mekuchi M, Mori K, Muta S, Chowdhury VS, Nakamura Y et al (2016) Transcriptomic features associated with energy production in the muscles of Pacific bluefin tuna and Pacific cod. Biosci Biotechnol Biochem 80(6):1114–1124. https://doi.org/10.1080/09168451.2016.1151341

Shiokawa Y, Misawa T, Date Y, Kikuchi J (2016) Application of market basket analysis for the visualization of transaction data based on human lifestyle and spectroscopic measurements. Anal Chem 88(5):2714–2719. https://doi.org/10.1021/acs.analchem.5b04182

Simpson JT, Wong K, Jackman SD, Schein JE, Jones SJ, Birol I (2009) ABySS: a parallel assembler for short read sequence data. Genome Res 19(6):1117–1123. https://doi.org/10.1101/gr.089532.108

Simpson AJ, Simpson MJ, Soong R (2012) Nuclear magnetic resonance spectroscopy and its key role in environmental research. Environ Sci Technol 46(21):11488–11496. https://doi.org/10.1021/es302154w

Smith MI, Yatsunenko T, Manary MJ, Trehan I, Mkakosya R, Cheng J et al (2013) Gut microbiomes of Malawian twin pairs discordant for kwashiorkor. Science 339(6119):548–554. https://doi.org/10.1126/science.1229000

Southam AD, Easton JM, Stentiford GD, Ludwig C, Arvanitis TN, Viant MR (2008) Metabolic changes in flatfish hepatic tumours revealed by NMR-based metabolomics and metabolic correlation networks. J Proteome Res 7(12):5277–5285. https://doi.org/10.1021/pr800353t

Southam AD, Lange A, Hines A, Hill EM, Katsu Y, Iguchi T et al (2011) Metabolomics reveals target and off-target toxicities of a model organophosphate pesticide to roach (*Rutilus rutilus*): implications for biomonitoring. Environ Sci Technol 45(8):3759–3767. https://doi.org/10.1021/Es103814d

Sugahara H, Odamaki T, Fukuda S, Kato T, Xiao JZ, Abe F et al (2015) Probiotic Bifidobacterium longum alters gut luminal metabolism through modification of the gut microbial community. Sci Rep 5:13548. https://doi.org/10.1038/srep13548

Sugita H, Nakamura H, Shimada T (2005) Microbial communities associated with filter materials in recirculating aquaculture systems of freshwater fish. Aquaculture 243(1):403–409. https://doi.org/10.1016/j.aquaculture.2004.09.028

Suhre K, Wallaschofski H, Raffler J, Friedrich N, Haring R, Michael K et al (2011) A genome-wide association study of metabolic traits in human urine. Nat Genet 43(6):565–569. https://doi.org/10.1038/ng.837

Sunagawa S, Coelho LP, Chaffron S, Kultima JR, Labadie K, Salazar G et al (2015) Ocean plankton. Structure and function of the global ocean microbiome. Science 348(6237):1261359. https://doi.org/10.1126/science.1261359

Tarnecki AM, Patterson WF 3rd, Arias CR (2016) Microbiota of wild-caught Red Snapper *Lutjanus campechanus*. BMC Microbiol 16(1):245. https://doi.org/10.1186/s12866-016-0864-7

Tian C, Chikayama E, Tsuboi Y, Kuromori T, Shinozaki K, Kikuchi J et al (2007) Top-down phenomics of *Arabidopsis thaliana*: metabolic profiling by one- and two-dimensional nuclear magnetic resonance spectroscopy and transcriptome analysis of albino mutants. J Biol Chem 282(25):18532–18541. https://doi.org/10.1074/jbc.M700549200

Tokuda G, Tsuboi Y, Kihara K, Saitou S, Moriya S, Lo N et al (2014) Metabolomic profiling of 13C-labelled cellulose digestion in a lower termite: insights into gut symbiont function. Proc Biol Sci 281(1789):20140990. https://doi.org/10.1098/rspb.2014.0990

Tomita S, Nemoto T, Matsuo Y, Shoji T, Tanaka F, Nakagawa H et al (2015) A NMR-based, non-targeted multistep metabolic profiling revealed l-rhamnitol as a metabolite that characterised apples from different geographic origins. Food Chem 174:163–172. https://doi.org/10.1016/j.foodchem.2014.11.028

Tomita S, Ikeda S, Tsuda S, Someya N, Asano K, Kikuchi J et al (2017) A survey of metabolic changes in potato leaves by NMR-based metabolic profiling in relation to resistance to late blight disease under field conditions. Magn Reson Chem 55(2):120–127. https://doi.org/10.1002/mrc.4506

Trapnell C, Williams BA, Pertea G, Mortazavi A, Kwan G, van Baren MJ et al (2010) Transcript assembly and quantification by RNA-Seq reveals unannotated transcripts and isoform switching during cell differentiation. Nat Biotechnol 28(5):511–515. https://doi.org/10.1038/nbt.1621

Tremaroli V, Backhed F (2012) Functional interactions between the gut microbiota and host metabolism. Nature 489(7415):242–249. https://doi.org/10.1038/nature11552

Turnbaugh PJ, Gordon JI (2009) The core gut microbiome, energy balance and obesity. J Physiol 587(17):4153–4158. https://doi.org/10.1113/jphysiol.2009.174136

Turnbaugh PJ, Ley RE, Mahowald MA, Magrini V, Mardis ER, Gordon JI (2006) An obesity-associated gut microbiome with increased capacity for energy harvest. Nature 444(7122):1027–1031. https://doi.org/10.1038/nature05414

Turnbaugh PJ, Ridaura VK, Faith JJ, Rey FE, Knight R, Gordon JI (2009) The effect of diet on the human gut microbiome: a metagenomic analysis in humanized gnotobiotic mice. Sci Transl Med 1(6):6ra14. https://doi.org/10.1126/scitranslmed.3000322

Uchimiya M, Tsuboi Y, Ito K, Date Y, Kikuchi J (2017) Bacterial substrate transformation tracked by stable-isotope-guided NMR metabolomics: Application in a natural aquatic microbial community. Metabolites 7(4):52. https://doi.org/10.3390/metabo7040052

Ulrich EL, Akutsu H, Doreleijers JF, Harano Y, Ioannidis YE, Lin J et al (2008) BioMagResBank. Nucleic Acid Res 36(suppl 1):D402–D408

van den Thillart G (1986) Energy metabolism of swimming trout (Salmo gairdneri) (journal article). J Comp Physiol B 156(4):511–520. https://doi.org/10.1007/bf00691037

Viant MR, Bearden DW, Bundy JG, Burton IW, Collette TW, Ekman DR et al (2009) International NMR-based environmental metabolomics intercomparison exercise. Environ Sci Technol 43(1):219–225

Viaud S, Saccheri F, Mignot G, Yamazaki T, Daillère R, Hannani D et al (2013) The intestinal microbiota modulates the anticancer immune effects of cyclophosphamide. Science 342(6161):971–976

Wagner GP, Kin K, Lynch VJ (2012) Measurement of mRNA abundance using RNA-seq data: RPKM measure is inconsistent among samples. Theory Biosci 131(4):281–285. https://doi.org/10.1007/s12064-012-0162-3

Wagner L, Trattner S, Pickova J, Gomez-Requeni P, Moazzami AA (2014) H-1 NMR-based metabolomics studies on the effect of sesamin in Atlantic salmon (*Salmo salar*). Food Chemistry 147:98–105. https://doi.org/10.1016/j.foodchem.2013.09.128

Waite DW, Taylor MW (2014) Characterizing the avian gut microbiota: membership, driving influences, and potential function. Front Microbiol 5:223. https://doi.org/10.3389/fmicb.2014.00223

Wang Q, Garrity GM, Tiedje JM, Cole JR (2007) Naive Bayesian classifier for rapid assignment of rRNA sequences into the new bacterial taxonomy (Research Support, U.S. Gov't, Non-P.H.S.). Appl Environ Microbiol 73(16):5261–5267. https://doi.org/10.1128/AEM.00062-07

Wang L, Feng Z, Wang X, Wang X, Zhang X (2010) DEGseq: an R package for identifying differentially expressed genes from RNA-seq data. Bioinformatics 26(1):136–138. https://doi.org/10.1093/bioinformatics/btp612

Wang T, Park YB, Caporini MA, Rosay M, Zhong L, Cosgrove DJ et al (2013) Sensitivity-enhanced solid-state NMR detection of expansin's target in plant cell walls. Proc Natl Acad Sci U S A 110(41):16444–16449. https://doi.org/10.1073/pnas.1316290110

Wang S, Charbonnier L-M, Rivas MN, Georgiev P, Li N, Gerber G et al (2015) MyD88 adaptor-dependent microbial sensing by regulatory T cells promotes mucosal tolerance and enforces commensalism. Immunity 43(2):289–303

Ward JL, Baker JM, Miller SJ, Deborde C, Maucourt M, Biais B et al (2010) An inter-laboratory comparison demonstrates that [H]-NMR metabolite fingerprinting is a robust technique for collaborative plant metabolomic data collection. Metabolomics 6(2):263–273. https://doi.org/10.1007/s11306-010-0200-4

Ward JL, Baker JM, Llewellyn AM, Hawkins ND, Beale MH (2011) Metabolomic analysis of Arabidopsis reveals hemiterpenoid glycosides as products of a nitrate ion-regulated, carbon flux overflow. Proc Natl Acad Sci U S A 108(26):10762–10767. https://doi.org/10.1073/pnas.1018875108

Watanabe T, Shino A, Akashi K, Kikuchi J (2014) Chemical profiling of Jatropha tissues under different torrefaction conditions: application to biomass waste recovery. PLoS One 9(9):e106893. https://doi.org/10.1371/journal.pone.0106893

Watanabe M, Ohta Y, Licang S, Motoyama N, Kikuchi J (2015) Profiling contents of water-soluble metabolites and mineral nutrients to evaluate the effects of pesticides and organic and chemical fertilizers on tomato fruit quality. Food Chem 169:387–395. https://doi.org/10.1016/j.foodchem.2014.07.155

Wei F, Ito K, Sakata K, Date Y, Kikuchi J (2015) Pretreatment and integrated analysis of spectral data reveal seaweed similarities based on chemical diversity. Anal Chem 87(5):2819–2826. https://doi.org/10.1021/ac504211n

Wei F, Sakata K, Asakura T, Date Y, Kikuchi J (2018) Systemic homeostasis in metabolome, ionome and microbiome of wild yellowfin goby in estuarine ecosystem. Sci Rep 8:3478. https://doi.org/10.1038/s41598-018-20120-x

Whitfield Aslund ML, McShane H, Simpson MJ, Simpson AJ, Whalen JK, Hendershot WH et al (2012) Earthworm sublethal responses to titanium dioxide nanomaterial in soil detected by (1)H NMR metabolomics. Environ Sci Technol 46(2):1111–1118. https://doi.org/10.1021/es202327k

Whitman WB, Coleman DC, Wiebe WJ (1998) Prokaryotes: the unseen majority. Proc Natl Acad Sci 95(12):6578–6583

Williams TD, Wu HF, Santos EM, Ball J, Katsiadaki I, Brown MM et al (2009) Hepatic transcriptomic and metabolomic responses in the stickleback (*Gasterosteus aculeatus*) exposed to environmentally relevant concentrations of dibenzanthracene. Environ Sci Technol 43(16):6341–6348. https://doi.org/10.1021/Es9008689

Wishart DS, Jewison T, Guo AC, Wilson M, Knox C, Liu Y et al (2013) HMDB 3.0 – the human metabolome database in 2013. Nucleic Acids Res 41(D1):D801–D807. https://doi.org/10.1093/nar/gks1065

Wong JM, de Souza R, Kendall CW, Emam A, Jenkins DJ (2006) Colonic health: fermentation and short chain fatty acids. J Clin Gastroenterol 40(3):235–243

Wong S, Waldrop T, Summerfelt S, Davidson J, Barrows F, Kenney PB et al (2013) Aquacultured rainbow trout (*Oncorhynchus mykiss*) possess a large core intestinal microbiota that is resistant to variation in diet and rearing density. Appl Environ Microbiol 79(16):4974–4984. https://doi.org/10.1128/aem.00924-13

Wong A, Li X, Molin L, Solari F, Elena-Herrmann B, Sakellariou D (2014) muHigh resolution-magic-angle spinning NMR spectroscopy for metabolic phenotyping of *Caenorhabditis elegans*. Anal Chem 86(12):6064–6070. https://doi.org/10.1021/ac501208z

Xia JH, Lin G, Fu GH, Wan ZY, Lee M, Wang L et al (2014) The intestinal microbiome of fish under starvation. BMC Genomics 15(1):266

Xing M, Hou Z, Yuan J, Liu Y, Qu Y, Liu B (2013) Taxonomic and functional metagenomic profiling of gastrointestinal tract microbiome of the farmed adult turbot (*Scophthalmus maximus*). FEMS Microbiol Ecol 86(3):432–443. https://doi.org/10.1111/1574-6941.12174

Yamauchi T, Kamon J, Ito Y, Tsuchida A, Yokomizo T, Kita S et al (2003) Cloning of adiponectin receptors that mediate antidiabetic metabolic effects. Nature 423(6941):762–769. https://doi.org/10.1038/nature01705

Yeoman CJ, Chia N, Yildirim S, Miller MEB, Kent A, Stumpf R et al (2011) Towards an evolutionary model of animal-associated microbiomes. Entropy 13(3):570–594

Yoshida S, Date Y, Akama M, Kikuchi J (2014) Comparative metabolomic and ionomic approach for abundant fishes in estuarine environments of Japan. Sci Rep 4:7005. https://doi.org/10.1038/srep07005

Zalesskiy SS, Danieli E, Blümich B, Ananikov VP (2014) Miniaturization of NMR systems: desktop spectrometers, microcoil spectroscopy, and "NMR on a chip" for chemistry, biochemistry, and industry. Chem Rev 114(11):5641–5694

Zerbino DR, Birney E (2008) Velvet: algorithms for de novo short read assembly using de Bruijn graphs. Genome Res 18(5):821–829. https://doi.org/10.1101/gr.074492.107

Zhang G, Fang X, Guo X, Li L, Luo R, Xu F et al (2012) The oyster genome reveals stress adaptation and complexity of shell formation. Nature 490(7418):49–54. https://doi.org/10.1038/nature11413

Part IV

Applications in Ocean and Fisheries Sciences: Analysis of the Red Tide

Influences of Diurnal Sampling Bias on Fixed-Point Monitoring of Plankton Biodiversity Determined Using a Massively Parallel Sequencing-Based Technique

14

Satoshi Nagai, Noriko Nishi, Shingo Urushizaki, Goh Onitsuka, Motoshige Yasuike, Yoji Nakamura, Atushi Fujiwara, Seisuke Tajimi, Takanori Kobayashi, Takashi Gojobori, and Mitsuru Ototake

Abstract

In this study, we investigated the influence of diurnal sampling bias on the community structure of plankton by comparing the biodiversity among seawater samples (n = 9) obtained every 3 h for 24 h by using massively parallel sequencing (MPS)-based plankton monitoring at a fixed point conducted at Himedo seaport in Yatsushiro Sea, Japan. During seawater sampling, the semidiurnal tidal current having an amplitude of 0.3 m s^{-1} was dominant, and the westward residual current driven by the northeasterly wind was continuously observed during the 24-h monitoring. Therefore, the relative abundance of plankton

S. Nagai (✉) · M. Yasuike · Y. Nakamura · A. Fujiwara
National Research Institute of Fisheries Science, Japan Fisheries Research and Education Agency, Yokohama, Kanagawa, Japan
e-mail: snagai@affrc.go.jp

N. Nishi · S. Urushizaki
AXIOHELIX Co. Ltd., Tokyo, Japan

G. Onitsuka
National Research Institute of Fisheries and Environment of Inland Sea, Japan Fisheries Research and Education Agency, Hiroshima, Japan

S. Tajimi
Kumamoto Prefectural Fisheries Research Center, Japan Fisheries Research and Education Agency, Kumamoto, Japan

T. Kobayashi
Kitasato University School of Marine Biosciences, Sagamihara, Japan

T. Gojobori
Center for Information Biology, National Institute of Genetics, Shizuoka, Japan

M. Ototake
National Research Institute of Aquaculture, Japan Fisheries Research and Education Agency, Minami-ise, Mie, Japan

T. Gojobori et al. (eds.), *Marine Metagenomics*,
https://doi.org/10.1007/978-981-13-8134-8_14

species apparently fluctuated among the samples, but no significant difference was noted according to G-test ($p > 0.05$). Significant differences were observed between the samples obtained from a different locality and at different dates, suggesting that the influence of diurnal sampling bias on plankton diversity is acceptable and data taken at a certain time in a day can be used as the representative one.

Keywords
Diurnal sampling · Metagenome · MPS-based method · Operational taxonomic units · Plankton biodiversity · Yatsushiro Sea

Abbreviations

MPS	Massively parallel sequencing
NMDS	Non-metric multidimensional scaling
OTUs	Operational taxonomy units

14.1 Introduction

Phytoplankton succession is a complex process resulting from local-scale and transitory blooms associated with spatial and temporal environmental heterogeneities (Margalef 1967; Smayda 1973). Succession is a change in species composition within a given water mass resulting from changing physical (light and temperature), chemical (nutrients, water quality, and allelopathy), and biological (competition and grazers) factors within that water mass (Smayda 1980). Growth interactions between phytoplankton species involve allelopathy, which is defined as any direct or indirect inhibitory or stimulatory effect that one plant has on another by the production of chemical secretions (Maestrini and Bonin 1981). Plankton communities are very sensitive to variation in environmental conditions (Smayda 1998; Stomp et al. 2008). In changing environments, phytoplankton compete with each other to acquire their own ecological niche and to adapt to local environmental conditions. Although dense monospecific blooms are common, many species occasionally co-occur as fine-scale blooms in shallow coastal waters. However, the species composition and abundance fluctuate greatly based on their environmental preferences (Smayda 1980), and dense blooms may suddenly disappear within 1–3 days because of turbulence in water columns caused by winds or tidal currents, resulting in hourly fluctuation in microalgal communities (Margalef 1978; Kemp and Mitsch 1979; Brunet and Lizon 2003; Sharples 2008; Blauw et al. 2012). The abundance of dinoflagellates, one of the major microalgal components in shallow coastal waters, changes significantly along the water columns within a day because of their swimming behavior, the so-called diel vertical migration (Eppley et al. 1968; Heaney and Eppley 1981). These environmental and biological factors might greatly influence the reliability of a fixed-point observation during a weekly or monthly monitoring of plankton.

The Yatsushiro Sea is a semi-enclosed, narrow coastal water body located off the western coast of Kyusyu Island, Japan. Its length is ca. 50 km from the northeast to the southwest coast. The bay connects through several narrow straits with the Ariake Sea and the East China Sea. The northeastern area is shallow (< 10 m), and it gradually deepens southwestward to about 70 m near the Ushibuka Strait, through which seawater is mainly exchanged between the Yatsushiro Sea and the East China Sea. The midwest-southwest section of the Yatsushiro Sea is a large fish-farming area in Japan. In the summer of 2009 and 2010, a noxious red tide causative species *Chattonella antiqua* formed a dense red tide in the Yatsushiro Sea and caused severe damage to aquacultured fish (Onitsuka et al. 2011). The red tide killed about 1.0 and 1.8 million cultured yellowtail, striped jack, and great amberjack in 2009 and 2010, respectively, and the amount of damage accounted for losses of \2,900,000,000 and \5,300,000,000 (US$290,000,000 and 53,000,000, respectively; exchange rate of 100 \ = $1; Fisheries Agency, 2010, 2011). Intensive field observations and different laboratory experiments have been conducted to determine the growth mechanism of *Chattonella* under oceanographic conditions in the Yatsushiro Sea (e.g., Sakurada et al. 2008; Shikata et al. 2011; Onitsuka et al. 2011; Aoki et al. 2012). To elucidate the biological interaction between *Chattonella* species and other plankton species or bacteria during the development of the red tide, i.e., appearance, exponential growth, decline, and disappearance, we conducted a massively parallel sequencing (MPS)-based monitoring of plankton and bacteria at four sampling sites in the Yatsushiro Sea once a week in the summer between 2011 and 2014. We investigated the influence of diurnal sampling bias on the community structure of plankton if the monitoring data is applicable to further analyses by comparing the biodiversity among seawater samples obtained every 3 h for 24 h ($n = 9$) by using MPS-based plankton monitoring at a fixed point in Himedo seaport in the Yatsushiro Sea, Japan.

14.2 Analytical Methods

14.2.1 Sampling and DNA Extraction

Seawater samples were obtained using a 5-m-long hose (diameter, 25 mm) every 3 h from 1630 on Oct 3, 2012, to 1630 on Oct 4, 2012 (nine samples in total) from Himedo seaport (32°43.8′N, 130°41.2′E) in the Yatsushiro Sea (Kumamoto Prefecture, Japan; Fig. 14.1). The variation in plankton biodiversity at Himedo seaport was evaluated by obtaining two samples from Kusuura (station KM1, 32°23.8′N, 130°13.5′E) in the Yatsushiro Sea (Fig. 14.1) on Aug 17 and 24, 2011. All the plankton in the seawater samples was trapped by filtering 500-mL seawater through a 8-μm pore-size polycarbonate filters (Nuclepore membrane; GE Healthcare, Tokyo, Japan), followed by filtering through 1-μm pore-size filters (GE Healthcare, Japan). The filters were stored in a deep freezer (−80 °C) until use. For DNA extraction, a 5% Chelex® suspension (Chelex 100 Molecular Biology Grade Resin; Bio-Rad Laboratories Inc., Richmond, CA, USA) was prepared by dispersing the resin into ultrapure water. For effective DNA extraction from

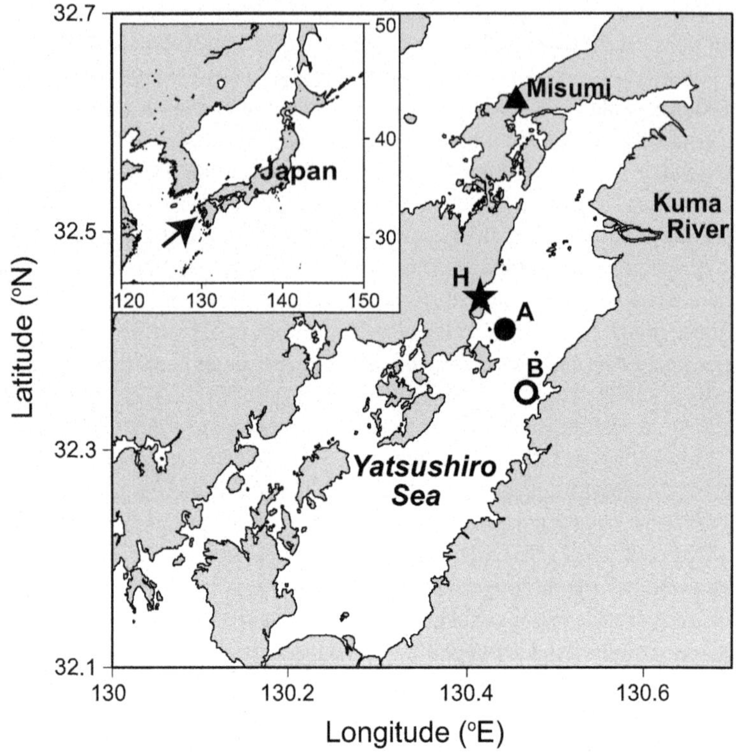

Fig. 14.1 Sampling locations at the Himedo seaport and Kusuura (station KM1) in Yatsushiro Sea, Japan

plankton components trapped on the filters, the filters were cut in half, placed in 1.5-mL tubes (A.150; Assist, Tokyo, Japan), and extracted in 150 μL of 5% Chelex buffer. The plankton cells were crushed using a pellet pestle motor (Kontes Glass, Vineland, NJ, USA) for 60 s, and 350 μL of buffer was added to make the final volume of 500 μL. DNA was extracted by heating the 1.5-mL tubes at 95 °C for 20 min (Nagai et al. 2012). DNA extracted from the 8-μm and 1-μm filters was mixed in equal amounts (50 μL + 50 μL) and used as template DNA.

14.2.2 Polymerase Chain Reaction Amplification and 454 Pyrosequencing

A set of primer pairs for 18S rRNA gene (SSU-F1,289, F: TGGAGY-GATTTGTCTGGTTDATTCCG; SSU-R1,772, R: TCACCTACGGAAACCTTGT-TACG; modified from Nishitani et al. 2012) were used as universal primers to amplify the V7–9 hypervariable regions; unique 6-bp tags were added to the 5′-end of the primers for recognition of each sample in the metagenome analysis. Polymerase chain reaction (PCR) was performed using a thermal cycler (PC-808;

ASTEC, Fukuoka, Japan) in a reaction mixture (50 μL) containing 2.0-μL template DNA (< 1 ng); 0.2 mM of each dNTP; 1× PCR buffer; 1.5 mM Mg^{2+}; 1.0 U KOD-Plus-ver.2 (TOYOBO, Osaka, Japan), which has intensive 3′ → 5′ exonuclease activity; and 1.0 μM of each primer. The PCR cycling conditions were as follows: initial denaturation at 94 °C for 2 min, followed by 25 cycles at 94 °C for 15 s, 56 °C for 30 s, and 68 °C for 1 min. PCR amplification was verified by agarose gel electrophoresis. The PCR products were purified using a High-Pure PCR Product Purification Kit (Roche, Basel, Switzerland) and eluted in 35 μL of elution buffer following the manufacturer's protocol. Amplified PCR products were quantified using a NanoDrop system (NanoDrop Technologies, Wilmington, USA) and pooled together in equal quantities. Amplicon sequencing was performed using a Roche 454 GS-XLR70 Titanium sequencing platform. A 454-pyrosequencing library was constructed using a GS Titanium Rapid Library Preparation Kit (Roche Diagnostics, Branford, CT, USA), and pyrosequencing was performed using a Roche 454 GS FLX^{+} sequencer (Roche Diagnostics, USA) according to the manufacturer's protocol.

14.2.3 MPS Data Treatment Processes and Operational Taxonomic Unit Picking

Nucleotide sequences were demultiplexed depending on the 5′-multiplex identifier (MID) tag and primer sequences. Since the sequencing adapters were ligated after PCR, each sample sequence contained both strand sequences. Both strand sequences shared the same MID tag sequence, but the 5′-end primers were different. Therefore, different strand sequences were also sorted. The sequences containing (1) >1 bp mismatch in the MID tag, (2) >2 bp mismatches in the primer sequences, (3) unknown base in both the MID tag and primer sequences, and (4) monopolymers longer than 9 bp were trimmed from the sequences at both ends. The 3′ tails with an average quality score of less than 25 at the end of the last 50-bp window were also trimmed from each sequence. Sequences longer than 430 bp were truncated to 430 bp by trimming the 3′ tails. The trimmed sequences shorter than 100 bp were filtered out. Subsequently, each sequence was aligned with template sequences to create multiple sequence alignment in SILVA release 111 (http://www.arb-silva.de/) (Quast et al. 2013), and only the sequences that were aligned in appropriate positions were extracted. The demultiplexing, trimming, filtering, and multiple alignment processes were performed using trim.seqs, align.seqs, screen.sea, and filter.seqs commands in Mothur according to the standard operating procedure (SOP) (http://www.mothur.org.\/wiki/454_SOP) (Schloss et al. 2011). The reverse-strand sequences were converted to their reverse-complement sequences. Erroneous and chimeric sequences were detected and removed using pre.cluster (diffs = 4) and chimera.uchime (minh = 0.1; http://drive5.com/usearch/manual/uchime_algo.html) (Edgar et al. 2011) commands in Mothur, respectively. In addition, singletons were removed. The forward-strand sequences and reverse-complement sequences of the reverse strand for each variable region were processed separately. The remaining sequences were assembled using the clmergeclass command in Claident (Tanabe

2012a) and Assams-assembler v 0.1.2013.07.19 (Tanabe 2012b), with sequence identity thresholds of 0.99 (0.01 distance radius in sequence space) and a minimum overlap length of 100 bp. The contig sequences were counted as operational taxonomic units (OTUs) and used for the following taxonomic identification analysis. Sequence assembly of homologous amplicon sequences by Claident and Assams is equivalent to complete linkage clustering, except that the resulting sequences are not representative sequences but consensus sequences. Demultiplexed, filtered, but untrimmed sequence data are deposited in the DDBJ Sequence Read Archive under accession no. DRA002425.

14.2.4 Taxonomic Identification of the OTUs

The selected OTUs were taxonomically identified as follows. A subset of nucleotide databases consisting of sequences that satisfied the below-mentioned conditions were prepared for BLAST search. One keyword was selected from among "ribosomal," "rrna," and "rdna," but "protein" was not included in the title. For taxonomy search, the keywords "metagenome," "uncultured," or "environmental" were not included. The sequences of retrieved GenBank IDs from the nucleotide database, downloaded from the NCBI FTP server, were extracted on July 2, 2012, and used to construct a template sequence database. Subsequently, we performed the taxonomic identification of each OTU by using BLAST search (Cheung et al. 2010). The BLAST search was conducted with NCBI BLAST+ 2.2.26+ (Camacho et al. 2009) by using default parameters, the nucleotide subset described above as database, and all OTU representative sequences as query; subsequently, the taxonomic information was obtained from the BLAST hit with top bit score for each query sequence.

14.2.5 Statistical Analyses

The relative abundance of OTUs at the supergroup levels among samples collected at different times was compared by counting the number of sequences at each supergroup level for the nine samples and by creating pie charts. In this study, supergroups were defined as *Alveolata*, *Amoebozoa*, *Apusozoa*, *Archaeplastida*, *Excavata*, *Hacrobia*, *Metazoa*, *Opisthokonta*, *Rhizaria*, *Stramenopila*, and *Viridiplantae* according to Adl et al. (2005), Burki et al. (2007), Frommolt et al. (2008), Archibald (2009), Hampl et al. (2009), and Okamoto et al. (2009). The relative abundance of the top 20 OTUs was depicted in pie charts and compared among the samples. The significance of variation was evaluated using the G-test of independence with Williams' correction. Further, the relative abundance of all species appearing in the samples was compared from a heatmap created by using heatmap.2 implemented in the statistical software R (R Development Core Team 2014) with the "vegan" package. Two samples obtained from different sampling points in the Yatsushiro Sea (32°43.8′N, 130°41.2′E) were added as the outgroup to evaluate the degree of variation. The biodiversity of plankton communities in the samples was analyzed by applying similarity indices (Jaccard and Chao indices) in R with "vegan" package.

14.2.6 Meteorological and Oceanographic Data

Hourly data on sea level elevation at Misumi port were collected and provided by the Japan Meteorological Agency (http://www.data.jma.go.jp/gmd/kaiyou/db/tide/genbo/index.php). Meteorological and oceanographic data were continuously monitored from the mooring station A (32°24.7′N, 130°26.6′E) and station B (32°21.5′N, 130°28.9′) (Fig. 14.1). Wind speed and direction at station A were measured using a wind sensor (Model 05106; R.M. Young Company, Traverse City, MI, USA). Vertical profiles of horizontal currents with 1-m vertical resolution at station A were observed using an acoustic Doppler current profiler (Aquadopp Profiler Z-cell, Nortek AS, Rud, Norway). These data were recorded and stored hourly, and water temperature and salinity in the surface layer at station B were measured every 20 min. The median values calculated for 1 h were processed and used in the present study.

14.3 How to Interpretate the Data

14.3.1 MPS-Based Plankton Diversity Survey

Summary of MPS-based plankton community survey of nine samples obtained from the Himedo seaport and two samples from Kusuura in the Yatsushiro Sea is presented in Table 14.1. The number of raw MPS was 14,583–41,739 (29,825 ± 8814, mean ± standard deviation), and the number of MPS after the removal of sequence errors and chimeras ranged from 9878 to 29,569 (20,621 ± 5839). The maximum number of MPS was 3 times larger than the minimum number of MPS obtained from the 11 samples. There was no significant positive correlation between the number of MPS and OTUs ($r = 0.48$, n = 11). However, when this relationship was investigated using only the samples from Himedo seaport, the correlation was significant ($r = 0.79$, n = 9), indicating that when the number of MPS in these samples is smaller than that in other samples, the detected number of OTUs for the nine samples also tends to be smaller (data not shown). Random resampling of the MPS was conducted from the samples having the minimum number of MPS (9878) to minimize this bias for further analysis, and then the OTUs were determined. The number of raw OTUs and OTUs after resampling was 507–658 (558 ± 104) and 448–544 (467 ± 81), respectively (Table 14.1). Many recent studies have explored eukaryotic diversity by using the pyrosequencing technology. The observed richness (= OTUs) per eukaryotic environmental sample estimated using the MPS-based method was 383–2133 (Amaral-Zettler et al. 2009), 1229–1823 (Cheung et al. 2010), and 942–1756 (Pawlowski et al. 2011). These values are significantly higher than those detected in the present study (Table 14.1).

Table 14.1 Basic information on massive parallel sequences (MPS)-based plankton community survey in Himedo seaport and other location (KM1) in Yatsushiro Sea, Japan

Numbers of MPS and OTUs	Himedo seaport									Yatsushiro Sea	
	Sampling time									Sampling date	
	16:30	19:30	22:30	1:30	4:30	7:30	10:30	13:30	16:30	Aug. 17	Aug. 24
Forward MPS	7192	13,770	17,931	10,560	14,353	11,014	20,618	17,781	19,964	17,140	9177
Reverse MPS	7391	15,026	18,096	12,457	16,741	12,365	21,121	20,941	20,341	15,796	8295
MPS (b)*	14,583	28,796	36,027	23,017	31,094	23,379	41,739	38,722	40,305	32,936	17,472
MPS (a)**	9878	20,342	23,420	15,522	20,706	16,686	29,569	24,666	26,756	25,737	13,553
OTUs (be)***	507	615	658	578	628	580	626	600	610	458	280
OTUs (af)****	507	534	544	514	504	506	480	472	448	373	254

* (b), before data treatment; ** (a), after data treatment; removal of errors and chimeric sequences; *** (be), before correction; **** (af), after correction. Higher numbers of operational taxonomic units (OTUs) were detected in the samples containing higher numbers of MPS; therefore, resampling of MPS was conducted in accordance with the minimum number of MPS (9878 at 1630), and OTUs were determined

The pyrosequencing technology enables the detection of thousands of OTUs of eukaryotes from various ecosystems (Nolte et al. 2010; Cheung et al. 2010). However, high rate of sequencing errors leads to overestimation of microbial diversity and mislabeling of sequences as new species (Huse et al. 2010); reduction in sequencing errors vary depending on the platform used for MPS data treatment, i.e., elimination of errors and chimeric sequences. In addition, the 18S rRNA gene has commonly been used as a meta-barcoding marker, but the resulting sequences originate from different hypervariable regions (Amaral-Zettler et al. 2009; Cheung et al. 2010; Pawlowski et al. 2011; Monchy et al. 2012). Therefore, the plankton richness or diversity in different ecosystems estimated using different platforms cannot be compared directly. An international standardization of these platforms for MPS-based survey for plankton diversity is necessary to allow the comparison among different ecosystems.

The relative abundance of identified OTUs at the supergroup levels was compared in the Himedo seaport samples. The relative abundances in eight of the nine samples, excluding the sample obtained at 1330 hours on Oct 4, 2012, were similar (Fig. 14.2), and there was no significant difference in the relative abundance among samples, as revealed by G-test ($p > 0.05$). However, the composition of the supergroup was remarkably different from that of the two samples obtained from Kusuura, which were collected on different dates (Fig. 14.2). The relative abundance of the top 20 OTUs was also compared among the samples. The abundances of the top 20 OTUs in the samples from Himedo seaport were 48.8–67.7% (58.0 ± 5.8%), and they showed little fluctuation (Fig. 14.3). The relative abundance was similar among the samples obtained at 1630 on Oct 3 and those obtained at 0130, 0430, 0730, 1030, and 1330 on Oct 4 and among the samples obtained at 1930 and 2230 on Oct 3 and those obtained at 1630 on Oct 4; there was no significant difference in the relative abundance among the samples, as revealed by G-test ($p > 0.05$). However, the abundances of the top 20 OTUs in the samples were remarkably different from those of the two samples obtained from Kusuura. Incidentally, the highest-ranked OTU in the Himedo seaport samples was *Pseudo-nitzschia* species (*Bacillariophyta*) with 17.3–39.2%, followed by *Oithona* sp. 1 and *Oithona* sp. 2

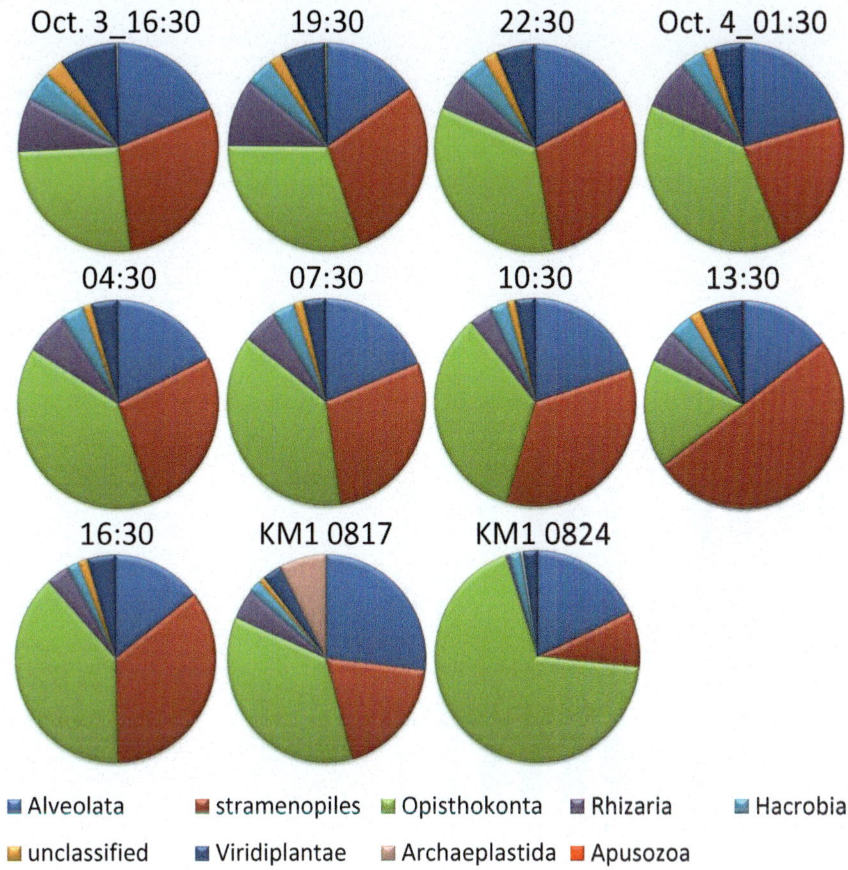

Fig. 14.2 Relative abundances of operational taxonomic units (OTUs) at the supergroup level detected using the massively parallel sequencing (MPS)-based plankton survey in seawater samples of Himedo seaport and Kusuura in Yatsushiro Sea, Japan

(*Arthropoda*). Dinoflagellate and diatom species were the most dominant microalgal groups within the top 20 OTUs, and the data showed common composition of the predominant species in coastal waters, such as the genus *Chaetoceros* and *Pseudo-nitzschia* in diatoms and the genus *Karenia* in dinoflagellates (Manabe et al. 1994; Nishikawa et al. 2010). In contrast, the genus *Oithona* (*Copepoda*, *Cyclopoida*) has been described as the most ubiquitous and abundant copepod in the oceans worldwide (Gallienne and Robins 2001). *Oithona* spp. were also detected with a relatively high abundance of 7.1–16.4% in Kagoshima Bay, Japan (Minowa et al. 2011). This suggests that this genus is one of the most dominant copepod groups in the Yatsushiro Sea.

The heatmap and cluster dendrogram clearly showed that the top 20 OTUs were predominant, but the abundance of the remaining OTUs was rather low (Fig. 14.4). Interestingly, several OTUs were prominent in only one sample, indicating the patchy distribution of the species on a fine spatial scale. This is also a

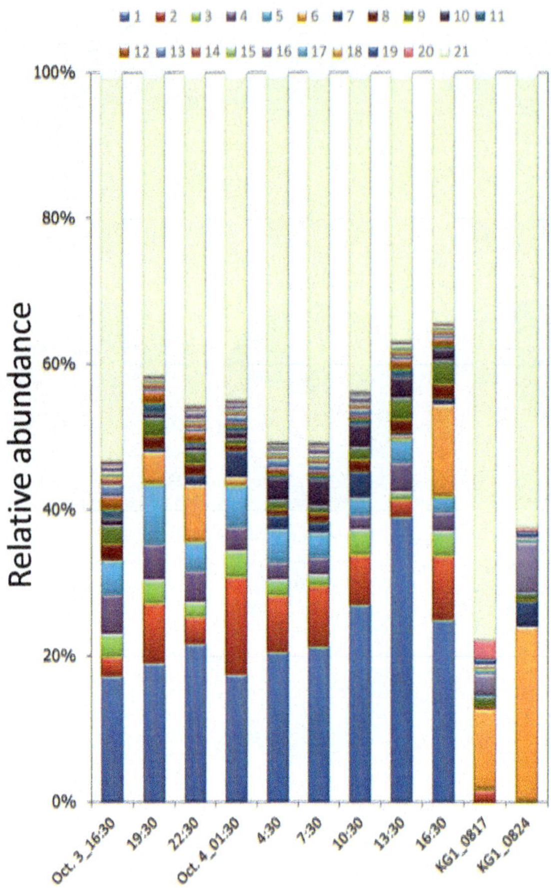

No.	species
1	*Pseudo-nitzschia australis* *Pseudo-nitzschia lineola*
2	*Oithona* sp. 1 New Caledonia-RJH-2001
3	*Oithona* sp. 2 New Caledonia-RJH-2004
4	*Micromonas* sp. RCC299 *Micromonas pusilla*
5	*Larcopyle butschlii*
6	*Paracalanus parvus*
7	*Gymnodinium aureolum* *Gymnodinium* sp. MUCC284
8	*Cyclotella atomus* *Cyclotella striata*
9	*Chaetoceros* sp.
10	*Neocalanus cristatus* *Neocalanus flemingeri*
11	*Karlodinium micrum*
12	*Micromonas pusilla* *Mantoniella squamata*
13	*Crassostrea gigas*
14	*Cercomonas* sp. ATCC 50367
15	*Leucocryptos marina* *Prorocentrum micans* *Prorocentrum rhathymum* *Prorocentrum mexicanum* *Peridinium cf. umbonatum* *Amphidinium semilunatum*
16	*Bradya* sp. Greenland-RJH-2004
17	*Azadinium spinosum* *Karlodinium micrum* *Karenia bidigitata* *Gymnodinium aureolum*
18	*Eucampia zoodiacus* *Eucampia antarctica*
19	*Thalassiosira lundiana*
20	*Karlodinium micrum*
21	other OTU

Fig. 14.3 Relative abundances of the top 20 operational taxonomic units (OTUs) detected using the massively parallel sequencing (MPS)-based plankton survey in seawater samples from Himedo seaport and Kusuura in Yatsushiro Sea, Japan. The list of detected species is shown on the right side. More than one species is listed in several ranks because the top bit scores were detected in Blast search

common feature of coastal waters that experience hourly fluctuations in microalgal communities because of the turbulence of water columns caused by winds or tidal currents (Margalef 1978; Kemp and Mitsch 1979).

The variability of plankton biodiversity among samples was evaluated by applying similarity indices (Jaccard and Chao indices) in the non-metric multidimensional scaling (NMDS) analysis. The NMDS plots and dendrograms inferred from the two indices were considerably similar to each other. Therefore, only the data calculated using the Jaccard method are shown in Figs. 14.5 and 14.6. All nine samples from Himedo seaport were grouped together but apart from the two samples from Kusuura (station KM1; Fig. 14.5). In the dendrogram inferred from the similarity indices, the nine samples from Himedo seaport formed a separate clade distant from the one with the two samples from Kusuura and both received high multiscale

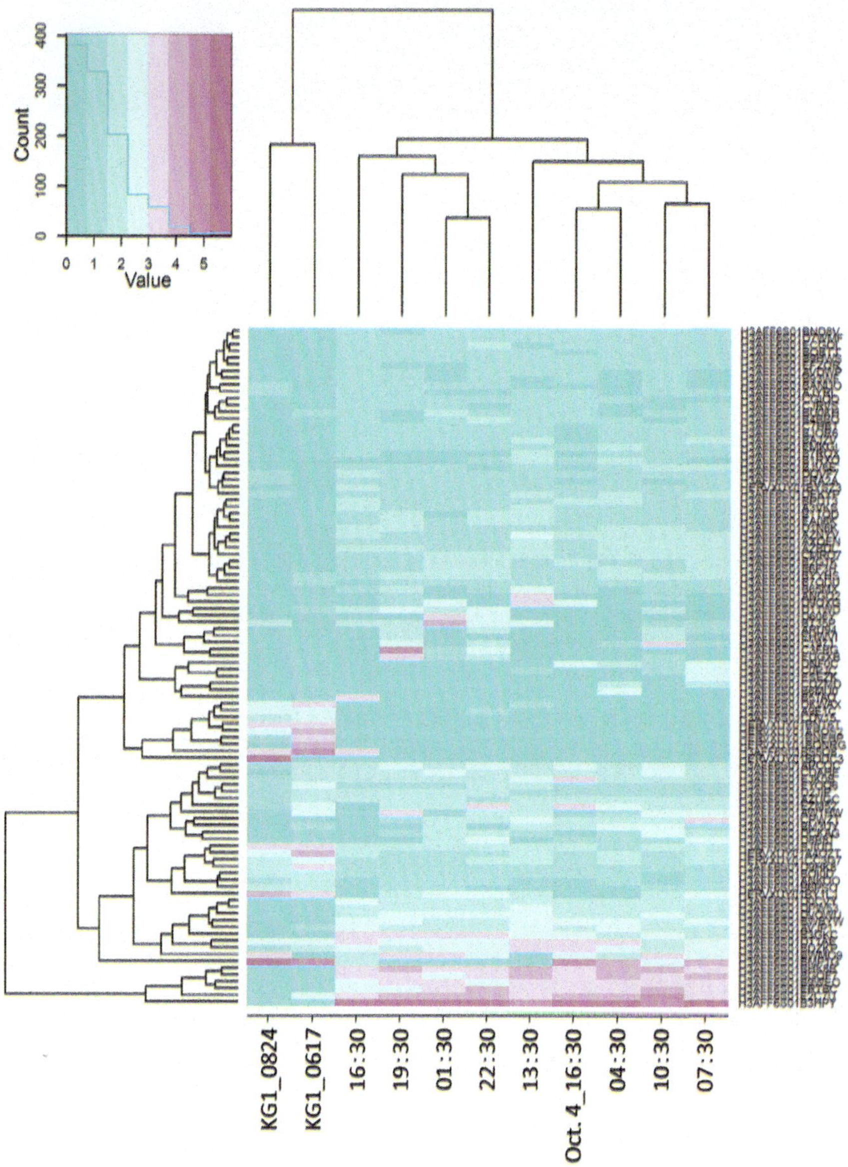

Fig. 14.4 Heatmap and cluster dendrogram of the plankton biodiversity survey conducted using massively parallel sequencing (MPS)-based plankton survey in seawater samples of Himedo seaport and Kusuura in Yatsushiro Sea, Japan

bootstrap values (Fig. 14.6). Four samples from Himedo seaport obtained at 0130 to 1030 and three samples obtained at 1930, 2230, and 1630 on Oct 4 formed sister clades; this could be the result of the species composition as shown in Fig. 14.3.

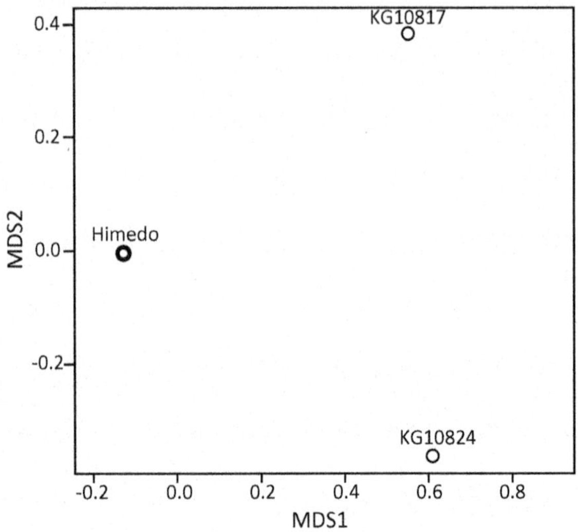

Fig. 14.5 A plot of non-metric multidimensional scaling (NMDS) analysis for plankton biodiversity survey conducted using massively parallel sequencing (MPS)-based plankton survey in seawater samples of Himedo seaport and Kusuura in Yatsushiro Sea, Japan. The similarity index was calculated using the Jaccard method

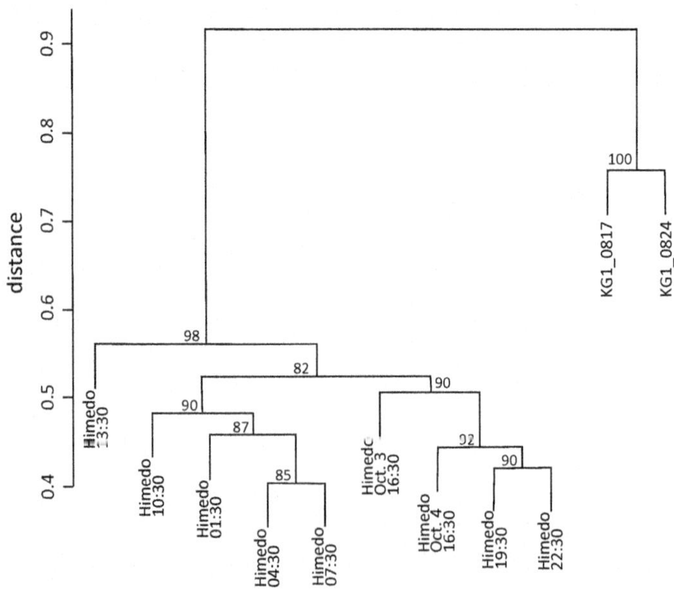

Fig. 14.6 A dendrogram of the results of massively parallel sequencing (MPS)-based plankton biodiversity survey conducted using seawater samples from Himedo seaport and Kusuura in Yatsushiro Sea, Japan. The similarity index was calculated using the Jaccard method

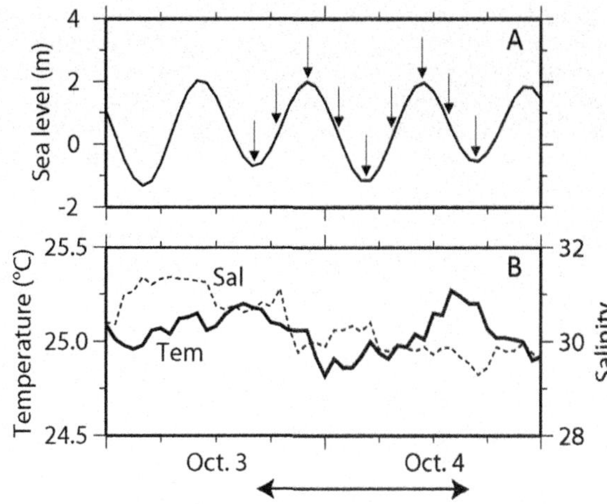

Fig. 14.7 Temporal changes in sea level at Misumi; water temperature and salinity in the surface layer at station B from Oct 3 to Oct 4, 2012. (**a**) Sea levels at Misumi; (**b**) water temperature and salinity in the surface layer. Arrows show the sampling times (above). A two-headed arrow indicates the sampling period (below)

14.3.2 Environmental Conditions During Sampling

Semidiurnal tide showing two peaks per day was predominant, and the range of sea level varied between 2 and 3 m during the sampling period in the northern part of the Yatsushiro Sea (Fig. 14.7A). The sampling was conducted from spring tide to neap tide (fool moon was observed on Sep 30 and quarter moon on Oct 10). Water temperature varied daily as much as 0.5 °C, reaching a peak once a day, in the afternoon. The diurnal temperature change was not consistent with the semidiurnal tidal cycles, probably due to the sea surface heat flux (Fig. 14.7B). Salinity decreased gradually during the sampling period irrespective of the tidal cycles; this might be affected by residual currents driven by the predominant northeasterly winds (5–10 m·s^{-1}; Fig. 14.8A). The north-south currents were largely dominated by a semidiurnal constituent at depths of 1.5 m and 10 m (Fig. 14.8B). The highest current speed exceeded 0.3 m·s^{-1} at a depth of 1.5 m. Residual current at 1.5-m depth leaned toward west likely due to the relatively strong northeasterly wind (Fig. 14.8B). The tidal excursion was estimated to be about 3 or 4 km in the north-south direction during the 24-h monitoring, but the transport of water mass caused a westward residual current of 10 km in the surface layer (Fig. 14.9).

14.3.3 Influence of Environmental Conditions on Diurnal Sampling Bias in Plankton Monitoring

Hourly changes in plankton community structures of coastal waters have frequently been observed because of the turbulence of water columns caused by tidal currents and cycles (Brunet and Lizon 2003; Sharples 2008; Blauw et al. 2012), wind mixing (Chen et al. 2010; Zhang et al. 2014), and vertical migrations of planktons

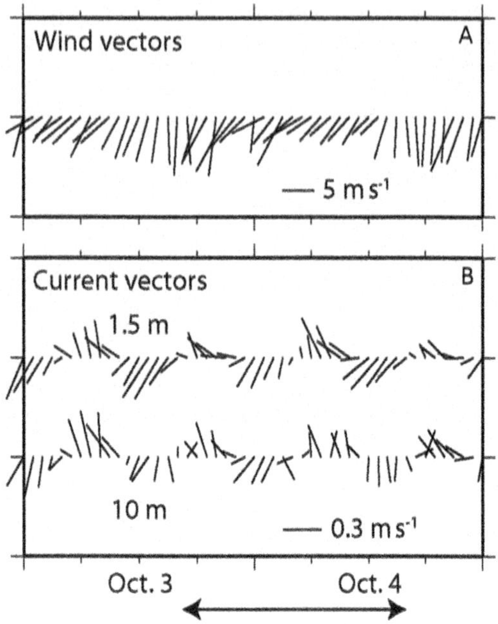

Fig. 14.8 Stick diagrams of hourly wind and hourly currents at depths of 1.5 m and 10 m at station A from Oct 3 to Oct 4, 2012. A, wind; B, current. A two-headed arrow indicates the sampling period

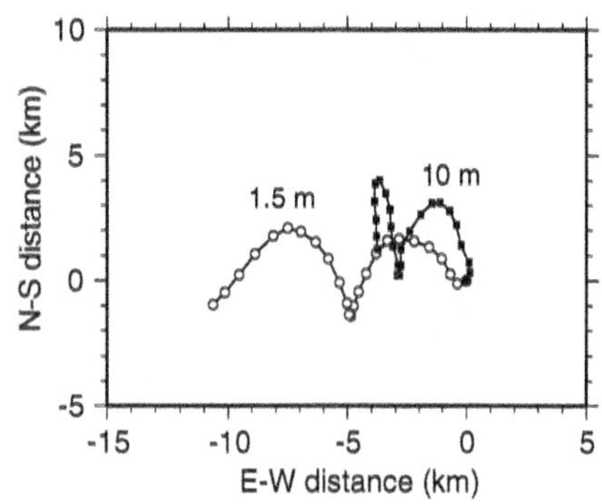

Fig. 14.9 A progressive vector diagram of hourly currents at depths of 1.5 m and 10 m measured at station A. Open circles (1.5 m) and closed squares (10 m) indicate 1-h interval

(Eppley et al. 1968; Heaney and Eppley 1981; Anderson and Stolzenbach 1985; Watanabe et al. 1991; Yamamoto et al. 2010).

During the vertical migration, microalgal species such as *Dinophyceae* and *Raphidophyceae* swim toward the surface before dawn and toward the deeper layers at dusk (Shikata et al. 2014). This behavior enables flagellates to optimize photosynthesis regardless of weather conditions or water clarity (Ault 2000) and to acquire nutrients over a wide depth range (Watanabe et al. 1991). Field observations have revealed that these species can migrate vertically 10 m within a day (Takeuchi 1988; Fauchot et al. 2005). In this study, a 5-m-long hose, which was the depth of water column at sampling location, was used to obtain seawater samples from Himedo seaport; therefore, the influence of the vertical migration of flagellate species on the sampling bias might have been minimized.

Environmental forcing by the tidal cycle is an important driver of phytoplankton variability in coastal waters (Brunet and Lizon 2003; Sharples 2008; Blauw et al. 2012). Systems with a semidiurnal tide, such as the North Sea, show horizontal displacement of water masses with a periodicity of 12 h and 25 min, resulting in the fluctuation of phytoplankton assemblages (Blauw et al. 2012). Aoki et al. (2012) showed that, under favorable conditions of river discharge and wind, massive blooms of harmful microalgal species formed in the northeastern area are rapidly transported southwestward within a few days in the Yatsushiro Sea. Similarly, Yoshida and Numata (1982) indicated the synergistic effect of tidal current and wind, i.e., the velocity and direction, on the accumulation and movement of harmful microalgal blooms within a short period (1–3 days) in Kagoshima Bay, Japan. Nagai also observed that *Karenia mikimotoi* bloomed patchily at relatively low concentrations with no coloration on a small seaport, stretching from east to west in the eastern Seto-Inland Sea, Japan. The colorless blooms, however, rapidly formed a dense red tide within 3 h by transportation from west to east due to the residual current driven by the strong westerly wind, resulting in their concentration in considerably high density of up to 150,000 cells per mL (Nagai, unpublished data). These observations confirmed that the diurnal changes of community structures of plankton in coastal waters frequently occur by advection and turbulence of water columns due to tidal currents, wind mixing, and/or density currents. Meanwhile, these observations suggest that the admixture events may also cause proliferation of surface water monopolized by one or two microalgal species and homogenization of species diversity due to the proliferation.

The circulation behavior of currents can be simulated by analyzing the residual currents developed by the admixture of tidal currents, winds, and/or density currents. In this study, during seawater sampling, the semidiurnal tidal current was dominant at 0.3 $m \cdot s^{-1}$, but the westward residual current driven by the northeasterly wind (5–10 $m \cdot s^{-1}$) was continuously observed during the 24-h-monitoring period (Fig. 14.8). The tidal excursion was about 3 or 4 km in the north-south direction, but the water mass was transported 10 km westward in the surface layer by residual current (Fig. 14.9). Therefore, the relative abundance of plankton species apparently fluctuated among the samples (Figs. 14.3, 14.4), but there was no significant difference according to G-test ($p > 0.05$). The scale of water mass showing similar species diversity of plankton (i.e., the bloom scale) was probably larger horizontally than that of the circulation behavior of the current during the survey period. Further, the diversity of plankton was remarkably different from that of samples obtained from a different locality and at different dates. Taken together, these findings suggest that the influence of diurnal sampling biases on plankton diversity research by MPS-based monitoring is not significant, and it is within the acceptable levels.

Acknowledgments This study was supported by a Grant-in-Aid (Marine Metagenomics for Monitoring the Coastal Microbiota) from the Ministry of Agriculture, Forestry and Fisheries of Japan.

References

Adl SM, Simpson AG, Farmer MA, Andersen RA, Anderson OR, Barta JR et al (2005) The new higher level classification of eukaryotes with emphasis on the taxonomy of protists. J Euk Microbiol 52:399–451

Amaral-Zettler L, McCliment E, Ducklow H, Huse S (2009) A method for studying Protistan diversity using massively parallel sequencing of V9 Hypervariable regions of small-subunit ribosomal RNA genes. PLoS One 4:e6372

Anderson DM, Stolzenbach KD (1985) Selective retention of two dinoflagellates in a wall-mixed estuarine embayment: the important of diel vertical migration and surface avoidance. Mar Ecol Prog Ser 25:39–50

Aoki K, Goh O, Shimizu M, Kuoda H, Matsuyama Y et al (2012) Factors controlling the spatio-temporal distribution of the 2009 *Chattonella antiqua* bloom in the Yatsushiro Sea, Japan. Est Coast Shelf Sci 114:148–155

Archibald JM (2009) The puzzle of plastid evolution. Curr Biol 19:R81–R88

Ault TR (2000) Vertical migration by the marine dinoflagellate *Prorocentrum triestinum* maximises photosynthetic yield. Oecol 125:466–475

Blauw AN, Benincà E, Laane RWPM, Greenwood N, Huisman J (2012) Dancing with the tides: fluctuations of coastal phytoplankton orchestrated by different oscillatory modes of the tidal cycle. PLoS One 7:e49319

Brunet C, Lizon F (2003) Tidal and diel periodicities of size-fractionated phytoplankton pigment signatures at an offshore station in the southeastern English Channel. Est Coast Shelf Sci 56:833–843

Burki F, Shalchian-Tabrizi K, Minge M, Skjæveland Å, Nikolaev SI et al (2007) Phylogenomics reshuffles the eukaryotic supergroups. PLoS One 2:e790

Camacho C, Coulouris G, Avagyan V, Ma N, Papadopoulos J et al (2009) BLAST+: architecture and applications. BMC Bioinformatics 10:421

Chen Z, Hu C, Muller-Karger FE, Luther ME (2010) Short-term variability of suspended sediment and phytoplankton in Tampa Bay, Florida: observations from a coastal oceanographic tower and ocean color satellites. Est Coast Shelf Sci 89:62–72

Cheung M, Au C, Chu K, Kwan H, Wong C (2010) Composition and genetic diversity of picoeukaryotes in subtropical coastal waters as revealed by 454 pyrosequencing. ISME J 4:1053–1059

Edgar RC, Haas BJ, Clemente JC, Quince C, Knight R (2011) UCHIME improves sensitivity and speed of chimera detection. Bioinformatics 27:2194–2200

Eppley RW, Holm-Hansen O, Strickland JDH (1968) Some observation on the vertical migration of dinoflagellates. J Phycol 4:330–340

Fauchot J, Levasseur M, Roy S (2005) Daytime and nighttime vertical migrations of *Alexandrium tamarense* in the St. Lawrence estuary (Canada). Mar Ecol Prog Ser 296:241–250

Fisheries Agency (2010) Information on the occurrence of red tide in Kyushu area. Kyushu Fisheries Coordination Office, Annual Report 2009 p. 95. (in Japanese)

Fisheries Agency (2011). Information on the occurrence of red tide in Kyushu area. Kyushu Fisheries Coordination Office, Annual Report 2010 pp. 102–103. (in Japanese)

Frommolt R, Werner S, Paulsen H, Goss R, Wilhelm C et al (2008) Ancient recruitment by chromists of green algal genes encoding enzymes for carotenoid biosynthesis. Mol Biol Evol 25:2653–2667

Gallienne CP, Robins DB (2001) Is *Oithona* the most important copepod in the world's oceans? J Plankton Res 23:1421–1432

Hampl V, Hug L, Leigh JW, Dacks JB, Fanz-Lang B et al (2009) Phylogenomic analyses support the monophyly of Excavata and resolve relationships among eukaryotic "supergroups". Proc Natl Acad Sci U S A 106:3859–3864

Heaney SI, Eppley RW (1981) Light, temperature and nitrogen as interacting factors affecting diel vertical migrations of dinoflagellates in culture. J Plankton Res 3:331–344

Huse SM, Welch DM, Morrison HG, Sogin ML (2010) Ironing out the wrinkles in the rare biosphere through improved OTU clustering. Environ Microbiol 12:1889–1898

Kemp WM, Mitsch WJ (1979) Turbulence and phytoplankton diversity: a general model of the "paradox of plankton". Ecol Model 7:201–222

Maestrini S, Bonin D (1981) Allelopathic relationships between phytoplankton species. Can Bull Fish Aquat Sci 210:323–338

Manabe T, Tanda M, Hori Y, Nagai S, Nakamura Y (1994) Changes in eutrophication and phytoplankton in Harima-Nada—results of environmental monitoring for 20 years. Bull Coast Oceanogr 31:169–181. (in Japanese with English abstract)

Margalef R (1967) Some concepts relative to the organization of plankton. Oceanogr Mar Biol Annu Rev 5:257–289

Margalef R (1978) Life-forms of phytoplankton as survival alternatives in an unstable environment. Oceanologia 1:493–509

Minowa M, Kobari T, Akamatsu H, Ichikawa T, Fukuda R et al (2011) Seasonal changes in abundance, biomass and depth distribution of mesozooplankton community in Kagoshima Bay. Bull Japan Soc Fish Oceanogr 75:71–81

Monchy S, Grattepanche JD, Breton E, Meloni D, Sanciu G et al (2012) Microplanktonic community structure in a coastal system relative to a Phaeocystis bloom inferred from morphological and tag pyrosequencing methods. PLoS One 7:e39924

Nagai S, Yamamoto K, Hata N, Itakura S (2012) Study of DNA extraction methods for use in loop-mediated isothermal amplification detection of single resting cysts in the toxic dinoflagellates *Alexandrium tamarense* and *A. catenella*. Mar Genomics 7:51–56

Nishikawa T, Hori Y, Nagai S, Miyahara K, Nakamura Y et al (2010) Nutrient and phytoplankton dynamics in Harima-Nada, eastern Seto Inland Sea, Japan during a 35-year period from 1973 to 2007. Estuar Coast 33:417–427

Nishitani G, Nagai S, Hayakawa S, Kosaka Y, Sakurada K et al (2012) Multiple plastids collected by the dinoflagellate *Dinophysis mitra* through kleptoplastidy. App Environ Microbiol 78:813–821

Nolte V, Pandey RV, Jost S (2010) Contrasting seasonal niche separation between rare and abundant taxa conceals the extent of protist diversity. Mol Ecol 19:2908–2915

Okamoto N, Chantangsi C, Horak A, Leander BS, Keeling PJ (2009) Molecular phylogeny and description of the novel katablepharid Roombia truncata gen. Et sp. nov., and establishment of the Hacrobia taxon nov. PLoS One 4:e7080

Onitsuka G, Aoki K, Shimizu M, Matsuyama Y, Kimoto K et al (2011) Short-term dynamics of a *Chattonella antiqua* bloom in the Yatsushiro Sea, Japan, in summer 2010: characteristics of its appearance in the southern area. Bull Japan Soc Fish Oceanogr 75:143–153. (in Japanese with English abstract)

Pawlowski J, Christen R, Lecroq B, Bachar D, Shahbazkia HR et al (2011) Eukaryotic richness in the abyss: insights from pyrotag sequencing. PLoS One 6:e18169

Quast C, Pruesse E, Yilmaz P, Gerken J, Schweer T, Yarza P, Peplies J, Glöckner FO (2013) The SILVA ribosomal RNA gene database project: improved data processing and web-based tools. Nucl Acids Res 41:590–596

R Development Core Team (2014) R: A language and environment for statistical computing. R Foundation for Statistical Computing, Vienna, Austria. URL http:/\penalty\z@/\penalty\z@www.R-\penalty\z@project.org

Sakurada K, Yamagata S, Oyama N, Itoyama R (2008) The prediction of harmful blooms *Chattonella antiqua* in Yatsushiro Sea. Report of Kumamoto Prefecture Fish Res Center 8:35–45. (in Japanese)

Schloss PD, Gevers D, Westcott SL (2011) Reducing the effects of PCR amplification and sequencing artifacts on 16S rRNA-based studies. PLoS One 6:e27310

Shikata T, Skamoto S, Onitsuka G, Aoki K, Yamaguchi M (2014) Effects of salinity on diel vertical migration behavior in two red-tide algae, *Chattonella antiqua* and *Karenia mikimotoi*. Plank Benth Res 9:42–50

Shikata T, Sakurada K, Jomoto Y, Oyama N, Onji M et al (2011) Growth dynamics of *Chattonella antiqua* in relation to nutrients in the Yatsushiro Sea. Nippon Suisan Gakkaishi 77:40–52. in Japanese with English abstract

Sharples J (2008) Potential impacts of the spring-neap tidal cycle on shelf sea primary production. J Plankton Res 30:183–197

Smayda TJ (1973) The growth of *Skeletonema costatum* during a winter-spring bloom in Narragansett Bay, Rhode Island. Nor J Bot 20:219–247

Smayda TJ (1980) Phytoplankton species succession. In: Morris I (ed) The physiological ecology of phytoplankton. Blackwell Scientific Publication, Oxford, UK, pp 493–570

Smayda TJ (1998) Patterns of variability characterizing marine phytoplankton, with examples from Narragansett Bay. ICES J Mar Sci 55:562–573

Stomp M, van Dijk MA, van Overzee HMJ, Wortel M, Sigon C et al (2008) The timescale of phenotypic plasticity and its impact on competition in fluctuating environments. Am Nat 172:169–185

Takeuchi T (1988) Diel vertical migration of *Protogonyaulax catenella* (dinophyceae). Bull Plank Soc Japan 35:149–157. (in Japanese with English abstract)

Tanabe AS (2012a) Claident v0.1.2013.07.23. software distributed by the author at http:/ \penalty\z@/\penalty\z@www.claident.org/\penalty\z@

Tanabe AS (2012b) Assams v0.1.2013.07.19. software distributed by the author at http:/ \penalty\z@/\penalty\z@www.fifthdimension.jp/\penalty\z@

Watanabe M, Kohata K, Kimura T (1991) Diel vertical migration and nocturnal uptake of nutrients by *Chattonella antiqua* under stable stratification. Limnol Oceanogr 36:593–602

Yamamoto K, Matsuyama Y, Ohmi H, Ariyama H (2010) Diel vertical migration of the toxic dinoflagellate Alexandrium tamarense, temporal changes of associated environmental factors and cell toxin content during the course of a large-scale bloom. Nippon Suisan Gakkaishi 75:877–885. in Japanese with English abstract

Yoshida Y, Numata K (1982) Accumulation and movement of *Chattonella* sp. in Kagoshima Bay and Suho Nada. Bull Japan Soc Sci Fish 48:1401–1405. (in Japanese with English abstract)

Zhang Y, Shi K, Liu X, Zhou Y, Qin B (2014) Lake topography and wind waves determining seasonal-spatial dynamics of Total suspended matter in turbid Lake Taihu, China: assessment using long-term high-resolution MERIS data. PLoS One 9:e98055

Mining of Knowledge Related to Factors Involved in the Aberrant Growth of Plankton

15

Yasuhito Asano, Hiroshi Oikawa, Motoshige Yasuike, Yoji Nakamura, Atushi Fujiwara, Keigo Yamamoto, Satoshi Nagai, Takanori Kobayashi, and Takashi Gojobori

Abstract

We aim to obtain knowledge relating to the causes of aberrant growth of plankton thought to cause problems such as shellfish poisoning, by using data acquired by measuring populations of more than 1000 species of plankton in specific seas areas with a next-generation sequencer. Previous techniques proposed for predicting future time series data from past time series data are difficult to be applied because the number of measurements is small. On the other hand, association rule mining which is one of the classical data mining techniques, is insufficient to obtain knowledge relating to indirect causes, such as "if species B increases, species A increases, and as a result the target species exhibits a characteristic increase." Therefore, we propose a method for finding association rules relating to increase/decrease of species other than the target

Y. Asano (✉)
Faculty of Information for Innovation and Design, Tokyo University, Tokyo, Japan
e-mail: yasuhito.asano@iniad.org

H. Oikawa · M. Yasuike · Y. Nakamura · A. Fujiwara · S. Nagai
National Research Institute of Fisheries Science, Japan Fisheries Research and Education Agency, Yokohama, Kanagawa, Japan
e-mail: yojnakam@affrc.go.jp

K. Yamamoto
Research Institute of Environment, Agriculture and Fisheries, Osaka Prefecture, Sen-Nan, Osaka, Japan
e-mail: snagai@affrc.go.jp

T. Kobayashi
Kitasato University School of Marine Biosciences, Sagamihara, Japan

T. Gojobori
Computational Bioscience Research Center, King Abdullah University of Science and Technology, Thuwal, Saudi Arabia

T. Gojobori et al. (eds.), *Marine Metagenomics*,
https://doi.org/10.1007/978-981-13-8134-8_15

species, and also propose a new model for aggregating those rules, named "time series association graph". We perform knowledge mining using a time series association graph and clustering (community discovery) on the graph to discover knowledge relating to the causes of the aberrant growth of a specified species. We also describe the used codes written in the programming language R.

Keywords
Data mining · Association rules · Time series association graph · Aberrant growth · Programming language R

15.1 Introduction

The purpose of this section is to obtain knowledge relating to the causes of aberrant growth of plankton thought to cause problems such as shellfish poisoning, by using data acquired by measuring populations of more than 1000 species of plankton in specific seas areas with a next-generation sequencer. However, not that much time has passed since the development of the next-generation sequencer, and even if it is used, it is extremely laborious work to measure the populations of more than 1000 species of plankton, and thus the number of measurement sites and the number of measurements inevitably have to be small. In particular, the number of observations of aberrant growth is even smaller.

As related research, numerous techniques have been proposed for predicting future time series data from past time series data (Sims 1980; Matsubara et al. 2015). However, the number of measurements is small, as indicated above, and thus it is extremely difficult to apply these techniques. On the other hand, for the current purpose, it is not necessary to perfectly predict the populations of each species of each organism (in this case plankton), and in the end, it is enough to obtain knowledge relating to causes of aberrant growth, so in this case it was decided to use association rules (Agrawal et al. 1994; Han et al. 2000) the basic method. These rules are a classical data mining technique thought to be suited to the intended purpose.

If association rules are used, then, as explained below, knowledge is obtained along the lines of "if species A exhibits a characteristic increase 2 weeks before, there is a high probability that this will cause a characteristic increase in the target species." However, due to the scale of this data, it was felt that, if association rules were used alone, reliability would be insufficient, and it would be impossible to obtain knowledge relating to indirect causes, such as "if species B increases, species A increases, and as a result the target species exhibits a characteristic increase." To solve these problems, we propose a method for finding association rules relating to increase/decrease of species other than the target species and performing knowledge mining using a *time series association graph* (a model aggregating those rules) and clustering (community discovery) on the graph.

In the following, the specifics of the procedure for processing the provided data, finding association rules, and building a time series association graph are explained by employing the R program which was actually used. The program described here is such that if it is executed in sequence, the results to be explained can all be obtained. However, discussion of the R language itself is omitted because many good sources are already available such as books and web pages.

15.2 Preprocessing

Figure 15.1 shows part of the data collected by the National Research Institute of Fisheries Science, Japan. Actually, this includes measurement data for roughly 15 measurements done about every week (basically from February to March) in each of the 4 years from 2012 to 2015, at 2 specific sites.

Within this data, the parts needed for this research are "each measurement" (st12_2012_Mar06, etc. This indicates a measurement on March 6, 2012, at the site st12), the IDs indicating each species and the numeric values corresponding to the quantity of each species at each measurement. In this study, the technique is applied assuming that the time interval between measurements is basically the same. The actual time intervals between measurements of data are not strictly the same, but there are thought to be no discrepancies large enough to cause problems in practice.

What first comes to mind as IDs are the names of species, but in this data it is permissible for the ID to contain a "set of species names" (e.g., a set of three species names such as *Abra alba*, *Abra nitida* and *Abra prismatica*) as the species name (tophit_name) in a single record. In the relational database (RDB), there is a need to perform a step called "normalization" where these sets are divided into multiple relations, and handling this is somewhat troublesome. Thus here "Representative_seq" will be used as the species ID. In the following, "Species ID" will indicate this "Representative_seq" data. The result of leaving just the necessary

Listing 15.1 Reading data

```
# Read data.
allTable <- read.csv("rarefied_top_mod2.csv",
             header=TRUE, row.names=1)
# Specify the row indices, each of which represents the
   beginning
# of measurements for each year and site.
# The last value is the index of the last row.
sep <- c(1, 14, 30, 45, 62, 75, 91, 106, 123)
# Each table represents one year and one site.
tables <- as.list(NULL)
# length(sep) -1 = 8 tables (4 years, 2 sites)
numTables <- length(sep)-1
for(i in 1:numTables)
{
  tables[[i]] <- allTable[sep[i]:(sep[i+1]-1),]
}
```

	A	I	J	K	L	M	N
1	Represetative_seq	tophit_name	super	phylum	total	st12_2012_Mar06	st12_2012_Mar13
2	JKERKXD02I87L3	Abeoforma whisleri	Opisthokonta	Opisthokonta inc	1	0	0
3	HTSJY7X02GPKLS	Abra alba Abra nitida Abra prismatica	Opisthokonta	Metazoa	95	0	0
4	IB2WHZR01A1DQW	Acanthocorbis unguiculata	Opisthokonta	Choanoflagellida	1	0	0
5	JKERKXD02G6MQZ	Acartia omorii	Opisthokonta	Metazoa	4	0	0
6	JKERKXD02IJ70O	Achlya bisexualis	Stramenopiles	Oomycetes	7	0	0
7	HTSJY7X02GYXPA	Achlya bisexualis Leptolegnia caudata Aplanopsis terrestris	Stramenopiles	Oomycetes	246	0	3
8	I0GO56P01EOX5M	Actinocyclus curvatulus	Stramenopiles	Bacillariophyta	3	0	0
9	IG15ELX02F0VQR	Actinocyclus sp. ECT3672	Stramenopiles	Bacillariophyta	13	1	0
10	IB2WHZR01EXU1G	Akashiwo sanguinea	Alveolata	Dinophyceae	47	0	0
11	HTSJY7X02IKN9F	Alcyonidioides mytili	Opisthokonta	Metazoa	2	0	0
12	JKERKXD02GBEC3	Alexandrium insuetum Alexandrium sp. GD1590bp7	Alveolata	Dinophyceae	9	0	0
13	IB2WHZR01CMO3I	Alexandrium minutum	Alveolata	Dinophyceae	1	0	0
14	IDZDHSF01AJQ01	Alexandrium tamarense	Alveolata	Dinophyceae	6166	0	3
15	JBL7AIO02D1E8Y	Allantion sp. CCAP 1906/1	Rhizaria	Cercozoa	127	1	2

Fig. 15.1 The original data

	A	B	C	D	E	F	G	H
1		JKERKXD02I87L3	HTSJY7X02GPKLS	IB2WHZR01A1DQV	JKERKXD02G6MQ	JKERKXD02IJ70O	HTSJY7X02GYXPA	I0GO56P01EOX5M
2	st12_2012_Mar06	0	0	0	0	0	0	0
3	st12_2012_Mar13	0	0	0	0	0	3	0
4	st12_2012_Mar21	0	0	0	0	0	20	0
5	st12_2012_Mar26	0	2	0	0	0	5	0
6	st12_2012_Apr02	0	0	0	0	0	2	0
7	st12_2012_Apr09	0	0	0	0	0	0	0
8	st12_2012_Apr16	0	0	0	0	0	0	0
9	st12_2012_Apr23	0	0	0	0	0	0	0
10	st12_2012_May01	0	0	0	0	0	0	0
11	st12_2012_May07	0	0	1	0	0	0	0
12	st12_2012_May14	0	0	0	0	0	0	0
13	st12_2012_May22	0	0	0	0	0	0	0
14	st12_2012_May28	0	0	0	0	0	0	0
15	st12_2013_Feb13	0	0	0	0	0	90	0
16	st12_2013_Feb21	0	0	0	0	0	1	0
17	st12_2013_Feb26	0	0	0	0	0	2	0
18	st12_2013_Mar06	0	0	0	0	0	0	0
19	st12_2013_Mar13	0	0	0	0	0	0	0
20	st12_2013_Mar20	0	0	0	0	0	0	0
21	st12_2013_Mar27	0	0	0	0	0	15	0
22	st12_2013_Apr02	0	0	0	0	0	0	0
23	st12_2013_Apr09	0	0	0	0	0	0	0
24	st12_2013_Apr16	0	15	0	0	0	1	0

Fig. 15.2 The "rarefied_top_mod2.csv"

part indicated above is shown in Fig. 15.2. Rows and columns are interchanged from the previous data. In terms of relational database (RDB) terminology, measurements correspond to records (rows), and species IDs correspond to attribute names (columns). This is saved in csv format (here processing was done with Excel, but any software can be used). In the following, this file will be used with the name "rarefied_top_mod2.csv."

15.3 Reading and Breaking Down Data

Here, "rarefied_top_mod2.csv" created above is read in with R, and data is broken down by site and year. Listing 15.1 is the R program for that purpose. The program will be explained by first referring to the listed comments (parts prefixed with #).

The following explains the part not completely explained in the comments. The variable "sep" indicates from which line the separation of sites and years begins. For example, line 1 to line 13 contain data for 2012 at site st12. Line 14 to line 29 contain data for 2013 at site st12. "sep" is made up of the starting lines of each data, so it is: (1, 14, . . .) The reason why separation is included at the same site when the year changes is that data measurement is performed from February to May, and thus even if the year continues, the measurement period does not continue.

15.4 Detection of Increase/Decrease of Each Species

As explained at the beginning, the purpose of this study is to "obtain knowledge relating to the causes of aberrant growth of a certain species (target species)." What is included in this data is the quantity of each species, including the target species, and thus the possibility that those quantities are the cause is investigated, but since the normal or average quantity is different for each species, it was decided, in order to handle them in a uniform fashion, to investigate the increase/decrease of each species and examine whether that is connected with an increase in the target species. This means, for example, we wish to use the association rule discovery technique to obtain knowledge such as "if species A increases 2 weeks before, there is a high probability the target species will increase." Of course, another possibility which can occur in reality is "if the quantity of species A two weeks before was between 15 and 20, there is a high probability the target species will increase," but as explained earlier, there are large differences in the quantity ranges for each species, and thus proper handling is likely to require correct biological knowledge regarding that quantity range or large-scale distribution data. Hence, this paper will present a program focusing only on the increase/decrease of each species, but the basic approach of this program can also be applied if one wishes to employ quantity ranges, and therefore making such changes itself should not be that difficult.

The increase/decrease of each species can be expressed as a difference or ratio of the quantities in two successive measurements. In the case of a difference, there are likely to be large differences between each species, even with the same numeric value, just as in the case of the quantity ranges above, and thus it was decided to use ratios here (naturally, the program can be changed so as to use differences). Measurement results where the quantity is 0 are a problem for calculating ratios, so it is necessary to perform some smoothing. The Code 15.2 performs the simplest additive smoothing for that purpose (a numeric parameter is added to all measurement values). This is applied to all tables obtained in Sect. 15.3. Here, the

Listing 15.2 Smoothing

```
# Apply the smoothing to each table.
smoothedTables <- list()
for(i in 1:numTables)
{
  smoothedTables[[i]] <- tables[[i]] + 0.35
}
```

parameter for additive smoothing is set to 0.35, but from a biological standpoint there is likely to be a more appropriate value, and thus if such a value is found, it should be substituted in. Also, in this case it was decided to use separate code for the data breakdown and smoothing parts for the purpose of explanation, but there is no inherent problem with executing these simultaneously.

Next, the ratio of increase/decrease of each species is calculated. Taking the measurement in the i-th row to be v_i, and the measurement in the $i+1$ st row to be v_{i+1}, the ordinary ratio for this interval is v_{i+1}/v_i. For example, if the measurement in the i-th row is 100 and the measurement in the i+1st row is 120, then the ordinary ratio is 1.2, and if the measurement in the i-th row is 100, and the measurement in the $i+1$-st row is 80, then the ordinary ratio is 0.8. These can be regarded, respectively, as a 20% increase and a 20% decrease. Under ordinary conditions, there is no problem using this ratio, but in this study, it is desirable to provide the same threshold for increase and decrease, and if that is exceeded, to regard it as a "characteristic increase/decrease." If that is done, there is somewhat of a problem with this ordinary ratio. This is the fact that, since symmetry is not satisfied, if a 20% decrease is followed by a 20% increase as in the above example, the original value (100 in the example) is not restored (i.e., it becomes 95 in the example). Therefore, in the case of an increase, $v_{i+1}/v_i - 1$ is used as the increase ratio, and in the case of a decrease, $-v_i/v_{i+1} - 1$ is used as the decrease ratio. If this is done, then the increase ratio in case of an increase from 100 to 120 is 0.2, and the decrease ratio in case of a decrease from 100 to 80 is 0.25. At a glance, this seems odd, but if an increase with an increase ratio of 0.25 occurs after a decrease with a decrease ratio of 0.25, then the original value is restored. This is because, after a decrease from 100 to 80, for example, an increase with an increase ratio of 0.25 means an increase from 80 to 100. What achieves this is Listing 15.3. The function "subRatioMatrix" performs the conversion from an ordinary ratio v_{i+1}/v_i to the increase/decrease ratios (increase ratios and decrease ratios) used here. The function "ratioMatrix" applies the subRatioMatrix function to each row interval (that is, between each two successive measurements) in the given table. For each row interval, a single row is created expressing the increase/decrease ratio, and thus it is necessary to note that the table output by this function has one fewer row than the given table. In the final "for" loop, a table in which increase/decrease ratios are saved is created by applying the ratioMatrix function to each table, and furthermore scaling for each column is performed by applying the built-in scale function of R. Scaling

Listing 15.3 Calculating the increasing ratio

```
# Return the increasing/decreasing ratio from the ordinary ratio
    .
subRatioMatrix <- function(v)
{
  return(ifelse(v>=1.0, v-1.0, -1.0/v - 1.0))
}

# For each row interval, calculate the increasing/decreasing
    ratio
# table: created by the Code 2.
ratioMatrix <- function(table)
{
  m <- nrow(table)
  n <- ncol(table)
  A <- matrix(1, nrow=(m-1), ncol=n)

  # For each row interval, calculate the increasing/decreasing
      ratio
  for(i in 1:(m-1))
  {
    # Require "as.numeric" to prevent to make the result of
    # subRatioMatrix a list.
    A[i,] <- subRatioMatrix(as.numeric(table[i+1,]/table[i,]))
  }
  # Copy the names of columns
  colnames(A) <- colnames(table)
  return(A)
}

ratioTables <- list()
# Apply ratioMatrix to each table in smoothedTables,
# and apply scaling to each table.
for(i in 1:numTables)
{
  temp <- ratioMatrix(smoothedTables[[i]])
  ratioTables[[i]] <- scale(temp, center=rep(0,ncol(temp)))
}
```

is performed assuming that the increase/decrease ratio for each species follows a normal distribution, so that the median is 0 (no increase/decrease) and standard deviation is 1. If the value after scaling is $x > 0$, that means that the median has increased with an increase ratio separated by x times the standard deviation from the median. If the value is negative, it signifies a decrease in the same way.

Points where the increase/decrease of each species is large between two consecutive measurements are found by using these scaled values. The function "identifyChange" in Listing 15.4 looks at each element of the "scaled increase/decrease ratio vector for a certain species" designated by the parameter (this corresponds to one column in each table "ratioTables[[i]]" output from the aforementioned Listing 15.3) and converts an increase with an absolute value at or above the parameter factor (called a *characteristic increase*) to 1, a decrease of similar type (called a *characteristic decrease*) to −1, and other values (*no characteristic increase/decrease*) to 0. For example, suppose that the vector of scaled increase/decrease ratios for species A is (0.3, 0.2, 1.5, −0.4, −1.2). The i-th element of this vector is obtained by scaling, as indicated above, the increase/decrease ratios between the i-th measurement and the $i + 1$-st measurement. Here, if it is assumed tentatively that the factor is 1.0, then the output of this function is a vector $(0, 0, 1, 0, -1)$. This means that the increase between the third and fourth measurements and the decrease between the fifth and sixth measurements were large, and everywhere else there was no characteristic increase/decrease. Whereas in the original vector the elements are real number values, the elements of the output vector are one of three values $\{-1, 0, 1\}$. The reason why this sort of value discretization is performed is because the association rule discovery technique applied below is not good at handling continuous real numbers, and basically it is necessary to separate such values in certain ranges and discretize them into limited types. Naturally, if one wishes to handle increases/decreases with greater refinement, the number of types can be increased a little with a simple correction. In the "for" loop at the end of Listing 15.4, the factor is set to 1.0, and the identifyChange function is applied to each species in each table. This 1.0 value of factor means that increases/decreases equal to or greater than (1.0 times) the standard deviation of the normalized representation are treated as "characteristic increases/decreases." It is likely possible to set more appropriate values using biostatistics, and also to use a distribution other than the normal distribution which is more suitable in the first place for scaling, but here the simplest method is used.

15.5 Conversion to Data Format for Association Rule Discovery

Here, for the purpose of association rule discovery, the characteristic increase/decrease data prepared above is converted to a character string list in a somewhat special format. Therefore, this section will first explain the association rules. In the association rule discovery technique used here, it is assumed that an item matrix like that in Table 15.1 is input. This item matrix represents the example of association rule discovery at a store, and the rows correspond to customers and the columns to products. If the element of the matrix in the i-th row and j-th column is 1, this means customer i purchased product p_j, and if the value is 0, that means i did not purchase p_j. When this sort of item matrix is provided, an association rule corresponding to the meaning "if a person purchases a product p_j, he is also likely to purchase pk,"

Table 15.1 An item matrix

Customer	p_1	p_2	p_3	p_4	p_5
1	1	0	1	1	0
2	0	1	0	1	1
3	0	1	1	1	0
4	1	0	1	0	0

Listing 15.4 Identify characteristic increase/decrease

```
# If the absolute value obtained by the scaling above is
# at least "factor",
# it is regarded as "characteristic" increase or decrease.
# Here, the increase, decrease, and the others are
# represented by 1, -1, and 0, respectively.
# This is the simplest "flag" representation.
# Of course, you can use more levels for increase or decrease.
# scaled: a table containing scaled ratio data.
# factor: a threshold.
identifyChange <- function(scaled, factor)
{
  m <- nrow(scaled)
  n <- ncol(scaled)
  ret <- matrix(0, nrow=m, ncol=n)
  for(i in 1:m)
  {
    x <- scaled[i,]
    ret[i,] <- ifelse(x >= factor, 1, ifelse(x <= -1*factor, -1,
         0))
  }
  colnames(ret) <- colnames(scaled)
  return(ret)
}

# Apply identifyChange by setting the "factor" to 1.
changeTables <- list()
for(i in 1:numTables)
{
  changeTables[[i]] <- identifyChange(ratioTables[[i]], 1)
}

# As a result, each column of changeTables[[i]]
# contains the flags representing all characteristic
# increases/decreases of each species.
```

can be written $\{p_j\} \Rightarrow \{p_k\}$. Here, the set brackets are used so that multiple products can be used as the antecedent and consequent of a rule, e.g., "if a person buys both products p_j and p_h, he is also likely to purchase p_k." That is, in this case it is possible to write $\{p_j, p_h\} \Rightarrow \{p_k\}$.

Good association rules, with a potential of increasing sales for the store are thought to be rules with high *confidence* and *support*, as explained next. To explain these, assume that the symbols X and Y indicate sets of products. First, if described conceptually, the confidence Conf($X \Rightarrow Y$) of rule $X \Rightarrow Y$ is (the number of people who bought both X and Y)/(the number of people who bought X). Expressed in the language probability, if the probability of buying X is assumed to be $P(X)$, then Conf$(X \Rightarrow Y) = P(X, Y)/P(X) = P(Y|X)$, i.e., the likelihood of also purchasing Y given that X will be purchased. Therefore, if this confidence is high, there is a high probability that the person who bought X will buy Y, and thus compared to rules with low confidence, these rules are likely to be more useful for improving sales. If the example in Table 15.1 is used,Conf($\{p_1\} \Rightarrow \{p_3\}$) becomes $2/2 = 1$, and Conf($\{p_1, p_3\} \Rightarrow \{p_4\}$) $= 1/2$. A point which should be noted here is the fact thatConf($X \Rightarrow Y$) is not necessarily equal to Conf($Y \Rightarrow X$). For example, even assuming that people buying bread at a certain food shop have a high probability of also buying milk, that does not mean that people buying milk have a high probability of also buying bread. Actually, in this example of an item matrix, Conf($\{p_3\} \Rightarrow \{p_1\}$) $= 2/3$, and this value is different from Conf($\{p_1\} \Rightarrow \{p_3\}$) $= 1$ found above. Next, support for the association rule $X \Rightarrow Y$, Supp($X \Rightarrow Y$), if described conceptually, is the ratio of people who bought both X and Y. Expressed in the language of probability, Supp($X \Rightarrow Y$) $= P(X, Y)$. Therefore, in the case of support, Supp($X \Rightarrow Y$) $=$ Supp($Y \Rightarrow X$). If this support is high, it means the combination of X and Y sells well, and thus compared to rules with low support, there is potential for helping to increase sales.

In this way, confidence and support are thought to express the goodness of rules, but it is difficult to automatically determine beforehand, for each set of data, how high the confidence and support should be for a rule to be truly good. Therefore, a person who actually attempts to discover association rules must provide, as input beforehand, the *minimum confidence* and *minimum support*. Also, in the association rule discovery technique, all of the rules having confidence at or above the minimum confidence, and support at or above the minimum support, are enumerated. If this enumeration is done using a brute-force approach, it cannot be finished in a realistic amount of time. The reason why is that, if there are n types of products, there are 2^n combinations of product sets used in association rules, and if n is 1000, it will be necessary to examine 2^{1000} combinations. But $2^{1000} > 10^{300}$, so, for example, even if it were possible to check 10^{16} combinations in 1 s, this would take 10^{284} s, easily exceeding the time from the Big Bang to now. When the number of combinations increases exponentially like this as n increases, it is called "combinatorial explosion." Thus a more efficient algorithm is needed, and one well-known option is the Apriori algorithm. With the Apriori algorithm, the product sets to be examined are reduced as far as possible using a property of support called monotonicity, and this makes it possible to enumerate rules having support at or above the minimum support. Using this approach, it is possible to achieve efficient enumeration. Providing an overview of the details of the Apriori algorithm (and techniques making the algorithm more efficient) is not the purpose here and will be omitted, but interested readers should refer to reference or a similar source. There is also a method which uses an indicator called *lift* in addition to confidence and support. With the association rule discovery technique explained below the lift is also calculated, so using lift is easy, but an explanation of that will be omitted here.

The discussion thus far has been a long explanation of association rule discovery, but by applying this to the data in this case, it is possible to obtain knowledge such as "if species A increases 2 weeks before, it is likely that the target species will increase." Therefore, to suit this purpose, it is necessary to create an item matrix as in the above example. To do that, it is necessary to create a list like that in Table 15.2. Each row in this table corresponds to whether or not the target species underwent a characteristic increase between two consecutive measurements at the same site. Different rows indicate different measurement pair intervals. For example, the 1st row is the status between the 1st and 2nd measurements at site 1, and the 2nd row is the status between 2nd and 3rd measurements at site 1. In contrast, the question of whether other species exhibited a characteristic increase/decrease one measurement before or two measurements before is indicated, respectively, in the 2nd column and 3rd column. If one wishes to discover isolation rules by tracing back to even earlier measurements, it is enough to increase the columns.

The structure called a list here takes the elements with value "1" in each row of the item matrix and lines them up for each row. If this is indicated in matrix form, two rows indicating increase and decrease for each species are needed for each measurement, and this results in an extremely large matrix. Here, the size of the data is reduced by listing up only the columns which originally have "1" elements. To be more specific, Table 15.2 is divided into columns for each measurement, but the actual list uses elements such as "species B decreased 1 time before" or "species A increased 2 times before," and the columns for each measurement are not used. To be even more specific, in R, item lists of this sort are processed by replacing them internally with a sparse matrix structure. This data structure enables sparse matrices with an extremely large number of 0 elements, like item matrices, to be handled efficiently.

First, as an example, the following will explain the technique for creating data to discover association rules connected with causes of characteristic increases of the target species (the increase/decrease data prepared above is converted to a character string list with meanings such as "species B decreased 1 time before" or "species A increased 2 times before"). However, it is assumed that characteristic decreases can also be analyzed in the same way simply by changing the program parameters. That program is Listing 15.5.

Table 15.2 Item list for increase and decrease of species

Characteristic increase of target species	Characteristic increase/decrease of other species a week before	Characteristic increase/decrease of other species 2 weeks before	...
Yes	Species A, increase Species B, decrease	Species A, increase Species C, decrease	...
No	Species D, increase Species B, decrease	Species A, increase; Species C, increase	...
No	Species E, decrease	Species A, increase Species C, increase	...
Yes	Species B, increase Species C, decrease	Species A, increase; Species B, increase	...
Yes	Species D, increase	Species A, increase; Species D, decrease	...

Listing 15.5 Conversion into item list

```
# For each row(week), return the names of species which have
# a flag representing characteristic increase/decrease.
# value: -1: decrease, 1: increase, 0: otherwise
scanSingleRow <- function(row, value)
{
  return(names(which(row==value)))
}

# For each species which has characteristic increase/decrease
# "h" weeks before,
# return a string combining the name of the species, "h", and
# a character "I"(for increase) or "D"(for decrease).
# See examples in text below.
# IDs: a vector of IDs of species.
# h: represents how long weeks before.
# IorD: a character "I"(for increase) or "D"(for decrease)
createItemName <- function(IDs, h, IorD)
{
  len <- length(IDs)
  return(paste(IDs, rep(as.character(h), len), rep(IorD, len),
  sep="_"))
}


# For the inputs (
# changeTable: a matrix obtained by Listing 4.
# targetname: the ID of the target species.
# targetWeek: a vector representing target weeks.
# window: represents how many weeks before this function checks.
# ), this function returns a list of strings which record
# the characteristic increases/decreases the other species
# within the window.
# The strings are generated by the function "createItemName".
constructChangeList <- function(changeTable, targetname,
    targetWeek,
                                        window, flag)
{
  m <- nrow(changeTable)
  incRecord <- vector("list", m)
  decRecord <- vector("list", m)
  changeList <- vector("list", m)
  # For each row, create a list of species which
  # have a characteristic increase/decrease
  for(i in 1:m)
  {
    vi <- changeTable[i,]
    # Use columns other than the target species.
    vi2 <- vi[names(vi)!=targetname]
    incRecord[[i]] <- scanSingleRow(changeTable[i,], 1)
    decRecord[[i]] <- scanSingleRow(changeTable[i,], -1)
  }
```

```
  # For each week of the target species, investigate the past
      data.
  for(i in 2:m)
  {
    # If "i" is included in "targetWeek",
    # then add "targetname" to the list
    # (corresponding to a positive instance).
    # Otherwise, the list does not include
    # "targetname" (negative instance).
    if(length(targetWeek[targetWeek == i]) > 0)
    {
      IorD <- "D"
      if(flag > 0) IorD <- "I"
      changeList[[i]] <- paste(targetname, IorD, sep="_")
    }
    # Investigate rows within the window
    for(h in 1:window)
    {
      rowID <- i - h
      if(rowID > 0)
      {
        incs <- createItemName(incRecord[[rowID]], h, "I")
        decs <- createItemName(decRecord[[rowID]], h, "D")
        changeList[[i]] <- c(changeList[[i]], incs, decs)
      }
    }
  }
  return(changeList)
}
```

The function "scanSingleRow" takes one designated type from three categories (characteristic increase, no characteristic increase/decrease, characteristic decrease; indicated respectively by the values 1, 0, −1), finds all IDs of species falling under that type from a single row of a matrix containing characteristic increase/decrease data, and returns the result as a vector. The function "scanSingleRow" creates character strings to be the elements of the list described above. For example, "B_1_D" and "A_2_I" mean, respectively, "species B decreased one time before" and "species A increased two times before." The single alphabetic letter at the end is a symbol indicating whether there was an increase (I) or decrease (D). The numeral before that indicates "how many times before." The inputs when creating this character string are the species ID vector, a vector of numeric values indicating how many times before, and a vector of symbols indicating increase/decrease. For example, if the input is

```
c("AAA", "BBB"), c(1, 2), c("I", "D"),
```

then the output is:

```
c("AAA_1_I", "BBB_2_D").
```

The function "constructChangeList" uses scanSingleRow and scanSingleRow to convert the table of characteristic increase/decrease data created in the previous section into the aforementioned character string list. Here, as indicated in the comment, the parameter window is a numeric value indicating up to how many measurements before to consider as the cause of a characteristic increase in the target species. Also, the parameter targetweeks is a vector indicating the location of consecutive measurement pairs where there was a characteristic increase in the target species. For example, if it assumed there was a characteristic increase between measurements the 1st time and 2nd time, and between measurements the 5th time and 6th time, then targetweeks becomes $(1, 5)$.

To apply the association rule discovery technique, it is necessary to convert all tables of characteristic increase/decrease data to the aforementioned character string lists, and after gathering all of those, they must be converted to a data format called "transaction." Code 15.6 is the program which performs that task, and the function "transactionFromChangeList" is its core. The first "for" statement at the start of this function examines the table for each characteristic increase/decrease data, finds the locations where the target species exhibits a characteristic increase or decrease (only whichever is designated with the parameter "flag"), and obtains a character string list by using the aforementioned constructChangeList, while taking that as "targetweeks." The next "for" statement joins all of the lists. The "append" which performs this joining starts from the 2nd row, and the reason for this is that the 1st line of the table indicates the measurement interval between the 1st and 2nd time, but the past data which is the cause of that is not contained in this table (it is an event even further before that). The joined list is converted to the transaction format by using the as function, and output.

This function is actually executed at the last line of Listing 15.6. Here, we tried considering as a cause measurement pairs up to 4 times before (with this data, this extends back to about 4 weeks before). Note that before using the function "transactionFromChangeList," we have to install and load the "arules" package for Apriori algorithm. See the lines just before the function.

15.6 Discovery and Storage of Association Rules

When transactions have been created, preparations are ready for discovery of association rules. Code 15.7 discovers association rules by using the function apriori contained in the package arules installed at the beginning. The explanation of the parameters themselves of the function apriori is written as a comment, but this section explains the method of choosing the parameters maxlen, supp, and conf. maxlen is the maximum number of elements contained in a rule. For example, if the rule is $\{p_j\} \Rightarrow \{p_k\}$, the number of elements is 2, and if the rule is $\{p_j, p_h\} \Rightarrow \{p_k\}$, then the number of elements is 3 (here, what corresponds to the $\{p_k\}$ on the right side is "there is a characteristic increase in the target species"). In this study, the cases of "maxlen" 2 and 3 are tried out. supp in this case is an extremely small value, but this is because there is a small number of measurement pairs exhibiting a characteristic increase in the target species among the number of all measurement pairs. For example, if the total number of measurement pairs is 100, and the number of measurement pairs where the target species exhibits a characteristic increase is 10, then the upper limit of support becomes 0.1 for an association rule which takes the characteristic increase in the target species to be the right side. On the other hand, if "supp" is too low, only meaningless rules can be obtained. For example, if the total number of measurement pairs is 100 and supp is set to 0.01, then for the rule $\{p_j\} \Rightarrow \{p_k\}$ in the store example, if there is just one "customer who bought both $\{p_j\}$ and $\{p_k\}$," the support becomes 0.01. Taking an event for which there is only one example as a rule can be regarded as almost meaningless, so for support it is necessary to adopt a value between these. The actual circumstances regarding "conf" are that it was selected by trial and error, but even so, rules which go below 0.5 have a probability of less than 50%, and thus there is a need to set a value somewhat above that, and if a value which is too large is adopted, the number of rules will be extremely small. Therefore, it was decided to adopt this value for the current study.

Listing 15.6 Construct transactions

```
# Create transactions.
# changeTables: a list of matrices obtained by Listing 4.
# targetID: the ID of the target species.
# window: represents how many weeks before this function checks.
# flag: 1: increase, -1: decrease
transactionFromChangeList <- function(changeTables, targetID,
                                      window, flag)
{
  CLs <- list()
  # For each changeTable,
  for(i in 1:numTables)
  {
    # Find weeks in which the target species has a
    # characteristic increase/decrease.
    targetweeks <- which(changeTables[[i]][,targetID]==flag)
    CLs[[i]] <- constructChangeList(changeTables[[i]], targetID,
                                     targetweeks, window, flag)
  }
  ACL <- list()
  # Combine the lists
  for(i in 1:numTables)
  {
    len <- length(CLs[[i]])
    if(len > 0)
    {
      ACL <- append(ACL, CLs[[i]][2:len])
    }
  }
  # Convert the list into transactions.
  return(as(ACL, "transactions"))
}

# Install the package "arules".
# (not necessary if this package has been installed.)
install.packages("arules")
# Load the package.
```

```
library(arules)
# Construct transactions by investigating at most four weeks
    before
# the characteristic increase of "targetname".
targetname <- "IDZDHSF01AJQ01"
tran4.ACL <- transactionFromChangeList(changeTables,
                                       targetname, 4, 1)
```

Listing 15.7 Compute rules

```
# The right hand side of the rules (increase of the target
    species).
target <- paste(targetname, "I", sep="_")

# maxlen = 2
rules2_4 <- apriori(tran4.ACL, parameter=list(maxlen=2, supp
    =0.02,
conf=0.65, ext=TRUE), appearance=list(rhs=target, default="lhs")
    )
# Obtained 1 rule

# maxlen = 3
rules3_4 <- apriori(tran4.ACL, parameter=list(maxlen=3, supp
    =0.02,
conf=0.65, ext=TRUE), appearance=list(rhs=target, default="lhs")
    )
# Obtained 23 rules
```

Code 15.8 displays the obtained association rules. Note that the code and output are intermixed. The part other than the prompt in lines beginning with the prompt > is code, and the other parts are output. First, the number of obtained rules can be determined if one tries typing the variables containing the obtained rules as is. In this case, 1 rule is obtained if "maxlen" is set to 2, and 23 rules are obtained if "maxlen" is set to 3. Next the rules are displayed using the inspect function. Some may be omitted depending on the paper width situation. In particular, the "rhs" (the right hand sides of the rules) are all the same {IDZDHSF01AJQ01} (target species ID, tophit_name is Alexandrium tamarense), and thus all members are omitted except the first. The [1 : 5] part in the function "inspect" at the end means to display from 1st to 5th. Looking at these, it can be seen, for example, that if the species with the species ID IG15ELX02IK8A7 (*Anteholosticha petzi*, *Anteholosticha warreni*) exhibits a characteristic increase 3 weeks before, the target species exhibits a characteristic increase with a probability of 75%, and the support is 0.028 (only a little less than 3% of the total number of measurement pairs).

The obtained rules can also be written to a file. Listing 15.8 saves the rule set in csv and xml format. The file name of the part other than the extension is given with the parameter "prefix."

Listing 15.8 Inspect the obtained rules

```
> inspect(sort(rules3_4, decreasing=TRUE, by="support")[1:5,])
    lhs                         rhs                     support
        confidence
[1] {IG15ELX02IK8A7_3_I} => {IDZDHSF01AJQ01_I} 0.02830189
    0.75
[2] {IG15ELX02IK8A7_3_I,
     JKERKXD02IBXMK_2_D} => {IDZDHSF01AJQ01_I} 0.02830189
         0.75
[3] {HTSJY7X02IESV6_3_I,
     IB2WHZR01A0AZH_2_D} => {IDZDHSF01AJQ01_I} 0.02830189
         0.75
[4] {HTSJY7X02GSYLY_1_D,
     HTSJY7X02IESV6_3_I} => {IDZDHSF01AJQ01_I} 0.02830189
         1.00
[5] {HTSJY7X02GSYLY_2_I,
     HTSJY7X02IOORG_3_D} => {IDZDHSF01AJQ01_I} 0.02830189
         1.00

# Note that some columns are omitted due to
# the limitation of the space.
```

Listing 15.9 Writing rules

```
install.packages("pmml")
library(pmml)

# The function for writing rules into a file.
# prefix: filename (excluding the extension)
# rules: set of rules
writeRulesCSVandPMML <- function(prefix, rules)
{
  datarules <- as(rules, "data.frame")
  write.csv(datarules, paste(prefix, ".csv", sep=""))
  write.PMML(rules, paste(prefix, ".xml", sep=""))
}

# Write rules
writeRulesCSVandPMML("rules2_4", rules2_4)
writeRulesCSVandPMML("rules3_4", rules3_4)
```

15.7 Construction of Time Series Association Graph

Using the discovery of association rules thus far, a certain degree of knowledge was obtained regarding the direct causes of characteristic increases in the target species such as "if species A exhibits a characteristic increase 2 weeks before, there is a high probability the target species will exhibit a characteristic increase." However, the first problem is that a characteristic increase in the target species is itself an infrequent event, and thus the support of the obtained rule is low. As the second problem, it can be mentioned that knowledge relating to indirect causes has not been obtained, such as "if species B increases, species A increases, and as a result the target species exhibits a characteristic increase." Thus, it was decided consider whether a broader range of knowledge could be obtained by also including species other than the target species, finding all the association rules relating to direct causes of their characteristic increase/decrease, and then consolidating those.

For that purpose, we propose a *time series association graph*. For a given species set S, and an upper limit $t_{\max}$ on up to how many times before a measurement pair should be regarded as a cause, the vertex set for the time series association graph is composed of the $2(t_{\max} + 1)|S|$ vertices $\{s_{t,I}, s_{t,D} \mid s \in S, t \in \{0, 1, 2, \ldots, t_{\max}\}\}$. Here, $s_{t,I}$ indicates a characteristic increase at the t-th measurement pair of the species s (indicating the $t + 1$-st and $t + 2$-nd measurements), and similarly st, $s_{t,D}$ signifies such as characteristic decrease. The edges of the time series association graph express the discovered association rules. For example, if a rule is obtained that "if species A exhibits a characteristic increase 2 weeks before, then species B exhibits a characteristic increase," then edges are placed from $A_{1,I}$ to $B_{3,I}$, form $A_{2,I}$ to $B_{4,I}$ and so on. More specifically, a directed edge $(A_{t,I}, B_{t+2 \bmod t_{\max},I})$ is created for each $t \in \{0, 1, 2, \ldots, t_{\max}\}$. There is a reason why a remainder is used here. In itself, a time series can continue to infinity, but in reality it is impossible to create a static data structure from an infinite number of elements. Thus, next after the $t_{\max}$ measurement, a torus structure is used so that it returns again to the 0-th measurement pair. Therefore, the measurement pair two times after the t-th measurement pair is taken to be the $t + 2 \bmod t_{\max}$ measurement pair.

Listing 15.10 Construction of a time series association graph

```
install.packages("igraph")
library(igraph)

# The IDs of all species.
allIDs <- colnames(allTable)

# Construct the vetices of a time series association graph.
# IDs: a vector of IDs of species.
# window: represents how many weeks before this function checks.
constructVertices <-function(IDs, window)
{
  nIDs <- length(IDs)
  # Create a undirected graph with length(allIDs)*2*(window+1)
  # vertices and no edges.
  g <- graph.empty(n = length(allIDs)*2*(window+1), directed =
      FALSE)
  #Assign labels to vertices.
  t1 <- 1
  for(i in 1:(window+1))
  {
    t2 <- t1 + nIDs -1
    t3 <- t1 + 2* nIDs -1
```

```
    V(g)[t1:t2]$id <- createItemName(allIDs, i-1, "I")
    V(g)[(t2 + 1):t3]$id <- createItemName(allIDs, i-1, "D")
    t1 <- t3 + 1
  }
  return(g)
}

# Compute the weight of edge
# (the product of confidence and support).
computeWeight <-function(confidence, support)
{
  return(confidence * support * 10000)
}

# Add the edges corresponding to rules.
addAssociationEdges <-function(g, rules, window)
{
  # Vector of lhs (left hand side) of each rule,
  # assuming that the lhs of each rule consists of a single ID.
  # (i.e. maxlen = 2)
  lhss <- as(lhs(rules), "list")
  # The rhs (right hand side) of all rules.
  # Note that all the rules must share the same rhs
  # (which corresponds to the destination of edges).
  # Divide the string of the rhs into elements
  # corresponding to ID and flag.
  rhss <- as(rhs(rules), "list")
  rtuple <- strsplit(rhss[[1]], "_")
  rID <- rtuple[[1]][1]
  rIorD <- rtuple[[1]][2]
  # Confidence, support, and lift of the rules
  values <- quality(rules)
  # For each lhs,
  for(i in 1:length(rules))
  {
    # Divide the string of the lhs into elements.
    # "lDelay" is the time interval between the lhs and rhs.
    ltuple <- strsplit(lhss[[i]], "_")
    lID <- ltuple[[1]][1]
    lDelay <- as.integer(ltuple[[1]][2])
    lIorD <- ltuple[[1]][3]
    conf <- values$confidence[i]
    supp <- values$support[i]
    # Compute the weight for edges from the confidence
    # and support of the rule.
    weight <- computeWeight(conf, supp)
    for(j in 0:window)
    {
      source <- paste(lID, as.character(j), lIorD, sep="_")
      u <- V(g)[id == source]
      # Torus structure is inplemented by "%% (window+1)"
      rTime <- (j + lDelay) %% (window+1)
```

```
      destination <- paste(rID, as.character(rTime), rIorD, sep
          ="_")
      v <- V(g)[id == destination]
      g <- add_edges(g, c(u, v), weight = weight)
    }
  }
  return(g)
}

# Construct all edges.
# g: a graph without edges.
# IDs: a vector of IDs of species.
# changeTables: a list of matrices obtained by Listing 4.
# window: represents how many weeks before this function checks.
# minsupp: the minimum support for association rules.
# minconf: the minimum confidence for association rules.
constructEdges <- function(g, IDs, changeTables, window,
                           minsupp, minconf)
{
  for(ID in IDs)
  {
    # Check both the characteristic increase and decrease.
    # flag = 1: increase, flag = 0: decrease
    for(flag in c(1,-1))
    {
      # Construct transactions (see Section 1.5).
      tran <- transactionFromChangeList(changeTables,
                                        ID, window, flag)
      IorD <- "D"
      if(flag > 0) IorD <- "I"
      target <- paste(ID, IorD, sep="_")
      # If "target" does not appear in "tran"
      # (such increase/decrease does not occur in data),
      # "apriori" returns an error.
      # This means there is no edge entering into "target",
      # so we have only to ignore such errors.
      tryCatch({
        rules <- apriori(tran,
                 parameter=list(maxlen=2, supp=minsupp,
                                conf=minconf, ext=TRUE),
                 appearance=list(rhs=target, default="lhs"),
                 control=list(verbose=FALSE))
        if(length(rules) > 0)
        {
          g <- addAssociationEdges(g, rules, window)
        }
      }, error = function(e){
        # Output error messages (can be commented out).
        message(e)
      })
    }
  }
  return(g)
```

```
}

window <- 4
g <- constructVertices(allIDs, window)
g <- constructEdges(g, allIDs, changeTables, 4, 0.02, 0.65)

# Save the graph data.
write_graph(g, "allrules_graph.pajek", format="pajek")
```

Listing 15.10 actually constructs a time series association graph according to the explanation above. First, the function "constructVertices" constructs a new graph and its vertices. Although a directed graph is natural for a time series association graph (the direction of an edge is going from old time to new time), we adopt an undirected graph because the clustering technique used below assumes an undirected graph as input. Then, "constructEdges" constructs edges between vertices from the rules obtained by setting the minimum support and confidence. As a result, we obtain a graph with 5,230 vertices and 6,680 edges from the used data containing $|S| = 523$ species. Note that $t_{max} = 4$ ("window" in Listing 15.10), the number of vertices $2(t_{max} + 1)|S|$ equals to $2 \times 5 \times 523 = 5230$.

15.8 Detection of Communities on a Time Series Association Graph

Roughly speaking, a *community* is a subgraph whose vertices are densely connected by edges. Because an edge between two vertices represents there is a relation between the vertices, a community can be regarded as a set of vertices which have strong relations to each other. Therefore, detecting communities in a time series association graph might lead to detecting species which are related to the increase or decrease of each other. While several definitions of a community and a number of methods for detecting them have been proposed, we here adopt Louvain method (Blondel et al. 2008), one of the most popular methods.

Listing 15.11 Detection of communities

```
# Detect communities by Louvain method.
> louvainCommunities <- cluster_louvain(g)

# Extract communities that contain at least 3 vertices.
> comIDs <- which(as.numeric(sizes(louvainCommunities)) > 2)

# The number of such communities.
> length(comIDs)
[1] 31

# The size of such communities.
# (i.e. the number of vertices in each community)
> sizes(louvainCommunities)[comIDs]
Community sizes
```

```
 836 1162 1288 1740 1792 1892 1979 1983 2015 2055
 107  113   93   91   71   99  123  140  110   83
2066 2084 2109 2114 2117 2138 2140 2155 2159 2162
  24   98  124  113  196   83   94   77   24  121
2175 2209 2210 2221 2225 2237 2252 2257 2263 2272
  48   26  103  148   86  102   96   33  141   74
2280
  79

# The vertices contained in 2159th community.
> V(g)$id[communities(louvainCommunities)$'2159']
 [1] "HTSJY7X02I5S41_0_I" "IB2WHZR01EQNEE_0_I"
 [3] "IB2WHZR01CIY55_0_D" "IB2WHZR01CL5JX_1_I"
 [5] "IB2WHZR01BPK4V_1_I" "JKERKXD02JPRGZ_1_I"
 [7] "IB2WHZR01EQNEE_1_D" "IB2WHZR01AV3A0_2_I"
 [9] "HTSJY7X02F038F_2_I" "HTSJY7X02G8YF4_2_I"
[11] "HTSJY7X02HD4CW_2_D" "HTSJY7X02I5S41_2_D"
[13] "IB2WHZR01CL5JX_2_D" "JKERKXD02JPRGZ_2_D"
[15] "IG15ELX02JQDD9_2_D" "IB2WHZR01CL2PY_3_I"
[17] "I0GO56P01ELQ39_3_I" "HTSJY7X02GZGW3_3_D"
[19] "IB2WHZR01B23V6_3_D" "IG15ELX02G8U9E_3_D"
[21] "IG15ELX02GTX42_3_D" "HTSJY7X02G8YF4_3_D"
[23] "IB2WHZR01CIY55_4_I" "IB2WHZR01CL2PY_4_D"
```

Listing 15.11 detects communities by Louvain method and shows fundamental information of obtained communities. Because the used time series association graph has many vertices connected to no other vertices, there are a number of isolated vertices. Consequently, the result has many communities whose sizes are too small. We extract communities that contain at least 3 vertices here, and then 31 communities are obtained. For example, we show the vertices contained in 2159th community. The corresponding species might be related to the growth of each other.

As explained above, the current graph has many vertices connected to no other vertices; there are a number of isolated vertices. That is, there are no association rules including such vertices. As a result, for several species, related species might not be obtained. If the amount of data increases in the future, then the number of obtained association rules will increase; consequently, vertices connected to no other vertices will decrease, and therefore better results will be obtained.

We do not introduce a way for visualize the obtained communities because "plot" function in igraph package might not be suitable for a graph with thousands of vertices. Readers who are interested in visualizing communities should try a tool for a massive graph, like Gephi.[1]

[1] https://gephi.org/

15.9 Conclusion

We have explained an approach for applying data mining to analysis of aberrant growth of plankton. We first have utilized a well-known data mining technique called association rules to find causal relationships between a characteristic increase of a target species and characteristic increases/decreases of the other species. The key idea is how we define "characteristic increases/decreases." We introduced a simple definition, while a number of candidates can be considered. We then have proposed a time series association graph in order to conduct cluster analysis (detecting communities) for finding related species. Note that there are several approaches proposed for analyzing time series data using association rules (Agrawal and Srikant 1995; Asano et al. 2014). We have been careful to write a simple program by avoiding complicated error checks and speeding up techniques. For example, R has some libraries for parallel programming, such as "pforeach."[2] We will leave utilization of such libraries to the readers.

References

Agrawal R, Srikant R (1995) Mining sequential patterns. In: Proceedings of the eleventh international conference on data engineering, pp 3–14. IEEE

Agrawal R, Srikant R et al (1994) Fast algorithms for mining association rules. In: Proceedings of the 20th international conference on very large data bases (VLDB), vol 1215, pp 487–499

Asano Y, Oshino T, Yoshikawa M (2014) Time graph pattern mining for network analysis and information retrieval. IEICE Trans 97-D(4):733–742

Blondel VD, Guillaume J-L, Lambiotte R, Lefebvre E (2008) Fast unfolding of communities in large networks. J Stat Mech Theory Exp 2008(10):P10008

Han J, Pei J, Yin Y (2000) Mining frequent patterns without candidate generation. In: ACM SIGMOD record, vol 29, pp 1–12

Matsubara Y, Sakurai Y, Faloutsos C (2015) The web as a jungle: non-linear dynamical systems for co-evolving online activities. In: Proceedings of the 24th international conference on world wide web, pp 721–731

Sims CA (1980) Macroeconomics and reality. Econometrica J Econ Soc 48:1–48

[2]https://github.com/hoxo-m/pforeach

The manufacturer's authorised representative in the EU is Springer Nature Customer Service Centre GmbH, Europaplatz 3, 69115 Heidelberg, Germany. If you have any concerns regarding our products, please contact ProductSafety@springernature.com

Printed and bound by CPI Group (UK) Ltd, Croydon, CR0 4YY
30/07/2026
02181706-0001